CREATURES OF CAIN

Creatures of Cain

THE HUNT FOR HUMAN NATURE
IN COLD WAR AMERICA

ERIKA LORRAINE MILAM

PRINCETON UNIVERSITY PRESS
PRINCETON & OXFORD

Copyright © 2019 by Princeton University Press

Published by Princeton University Press
41 William Street, Princeton, New Jersey 08540
6 Oxford Street, Woodstock, Oxfordshire OX20 1TR

press.princeton.edu

LCCN: 2018942717
ISBN 978-0-691-18188-2

British Library Cataloging-in-Publication Data is available

Editorial: Al Bertrand and Kristin Zodrow
Production Editorial: Nathan Carr
Jacket/Cover Design: Lorraine Doneker
Jacket/Cover Credit: "Scientific Debate," by Nathaniel Gold.
Featured in *The Scientific American* (July 9, 2012)
Production: Jacquie Poirier
Publicity: Alyssa Sanford and Julia Hall
Copyeditor: Gail Schmitt

This book has been composed in Arno

Printed on acid-free paper. ∞

Printed in the United States of America

10 9 8 7 6 5 4 3 2 1

To Michael

Is it, indeed, true that the Poet, or the Philosopher, or the Artist whose genius is the glory of his age, is degraded from his high estate by the undoubted historical probability, not to say certainty, that he is the direct descendant of some naked and bestial savage, whose intelligence was just sufficient to make him a little more cunning than the Fox, and by so much more dangerous than the Tiger?

—THOMAS HENRY HUXLEY

CONTENTS

ACKNOWLEDGMENTS

IN THE WORDS OF LOREN EISELEY, "Ideas do not spring full blown from a single brain. There has to be wandering along bypaths, midnight reading, and sustained effort."[1] To his list, I would add conversation, especially with people who take your thoughts seriously and push back. It would be impossible to list everyone whose ears I have bent as this book has come together, but as someone who works through ideas by talking, these interactions were vital. At the Max Planck Institute for the History of Science in Berlin I began work on the project, at the University of Maryland I planned my research and drafted early chapters, at Princeton University I tore it all apart, and upon returning to Berlin for a year of sabbatical I stitched it back together.

I would like to thank the Science, Technology, and Society group at Maryland, including Lindley Darden, Robert Friedel, David Sicilia, and Tom Zeller. Rick Bell, Janna Bianchini, Mike Ross, and David Sartorius kept me cheerful. My formal and informal mentors—Julie Green, Clare Lyons, and Robyn Munch—provided much needed intellectual guidance at crucial junctures.

At Princeton, I joined a dynamic community of historians who pushed the project in new directions, including Graham Burnett, Angela Creager, Michael Gordin, Katja Guenther, Jenny Rampling, Keith Wailoo, and the slowly changing members of our Monday afternoon Program Seminar. Sarah Milov provided a wonderfully bracing comment on an early chapter. Several years later, Elaine Ayers, Emily Kern, and Ingrid Ockert read the entire manuscript and supplied detailed feedback. Joshua Guild, Kevin Kruse, and Sean Wilentz read various pieces. (Yaacob Dweck agreed to read the entire manuscript but then I ran out of time.) When I was an ocean away, Wangui Muigai scoured the David Hamburg Papers at Columbia. Shortly after I arrived, Keith Wailoo convened a small working group of scholars at the intersection of sociology and history which devoted attention to questions of science and social justice, including Betsy Armstrong, Ruha Benjamin, Bridget Gurtler, and Miranda Waggoner. (I still harbor hopes that we might jumpstart that lovely group again.) Angela Creager and Keith Wailoo provided feedback at not one but two different stages of the project. Wendy Warren never let me take myself too seriously. Katja Guenther, from even before I arrived at Princeton, has listened patiently and questioned keenly each dimension of this book. *Creatures of Cain* has benefitted from these colloquies and friendships.

I owe to Lorraine Daston sincere thanks for sponsoring my time at the MPIWG and for creating a generative environment in which to think and write—being a part of Department II was a delight. I relished lunches with Jenny Bangham, Lino Camprubi, Teri Chettiar, Jamie Cohen-Cole, Sebastian Felten, Jacob Gaboury, Yael Geller, Clare Griffin, Evan Hepler-Smith, Judy Kaplan, Katja Krause, Elaine Leong, Minakshi Menon, Emma Molin, Christine von Oertzen, Lisa Onaga, Katy Park, Felix Rietmann, Anja Sattelmacher, David Sepkoski, Ben Wilson, and whomever else was willing to venture forth to the Mensa in one direction or to multi-colored Turkish salads in the other. Toward the end of the year, Jamie Cohen-Cole and Patrick McCray read as much of the manuscript as I had been able to produce. My memories of Berlin are filled with convivial laughter and luxurious swaths of time to write.

For helpful criticism, discussion, and friendship, I thank Gregg Mitman, Lynn Nyhart, and Robert Nye, all of whom have continued to listen, long after politeness dictated they could have stopped. Karen Darling believed in this project from the beginning.

Archivists, librarians, and other historians made it possible for me to access archived papers and manuscripts. I extend particular thanks to Karen Colvard at the Harry Frank Guggenheim Foundation, Ed Copenhagen at the Monroe C. Gutman Library of Harvard University, Tim Horning at the University of Pennsylvania, University Archives, Carleton Jackson and Non-Print Media at the University of Maryland (especially for the sticky process of converting old U-Matic films into digital files), Kevin Luf at the WGBH Media Library and Archives, Ingrid Lennon at the American Museum of Natural History, Judy Meschel at National Geographic Motion, Daisy Njoku at the National Anthropological Archives, and Marc Rothenberg at the National Science Foundation. Charles B. Greifenstein, Valerie-Anne Lutz, Earle Spamer, and Lydia Vazquez-Rivera at the American Philosophical Society Library always remember me, no matter how many years pass between my visits. My deep thanks, too, to the tireless work of the Princeton Interlibrary Loan Office.

When writing about recent events, archived papers only tell part of the story, and I benefitted tremendously from interviews with people who lived through these events: Patsy Asch, Jerome Bruner, Irven DeVore, Phyllis Jay Dolhinow, Peter Dow, Robin Fox, David Hamburg, Karl Heider, Sarah Blaffer Hrdy, Elaine Morgan, Anne Pusey, Lionel Tiger, Robert Trivers, Richard Wrangham, and Adrienne Zihlman. I am grateful for their willingness to share with me their experiences in the field and on campus.

I received financial support from the History Department and the General Research Board at the University of Maryland, the History Department and

University Committee on Research in the Humanities and Social Sciences at Princeton University, and a Scholar's Award from the Science, Technology, and Society Program at the National Science Foundation (SES-1057586). Any opinions, findings, and conclusions or recommendations expressed in this material are those of the author and do not necessarily reflect the views of the National Science Foundation.

In the later moments of preparing the manuscript, Sarah Levy tracked down image permissions. Helene Nguyen and Ariel Kline heroically quote-checked everything. Gail Schmitt's astute editorial eye improved my prose. Nathan Carr guided the manuscript smoothly through the production process. Al Bertrand offered suggestions with a gentle hand. Sincere thanks to all.

Most of all, the span of conceptualizing and writing *Creatures of Cain* has exactly matched the length of the time I have known Michael. He was there when I began to craft utopian chapter outlines, when I rejoiced at an archival gem, and when I despaired of finding the right words. He listened as I worked out ideas and whisked me away to dinner when I needed a break. It is thus to Michael that I dedicate this book.

CREATURES OF CAIN

Introduction

HUMAN NATURE CONTAINS THE SEEDS of humanity's destruction. Or so it seemed to popular consumers of evolutionary theory in the late 1960s who maintained that the essential quality distinguishing the human animal from its simian kin lay in our capacity for murder. This startlingly pessimistic view enjoyed wide currency in the United States between 1966 and 1975 and became known, by its critics, as the killer ape theory.

Readers at the time associated the concept of humans as mere animals with three men. Robert Ardrey published *The Territorial Imperative* in 1966, which leapt off bookshelves across the country. He styled himself an amateur scientist and believed his experience as a playwright gave him unique insights into the composition of human nature. Konrad Lorenz's white-maned visage loaned him a distinguished appearance despite the black rubber boots he favored when showing people around his farm. Lorenz, the author of *On Aggression*, which appeared in English translation the same year as Ardrey's *Territorial Imperative*, would later share the Nobel Prize in Physiology or Medicine for his perceptive contributions to the scientific study of animal behavior. Desmond Morris unknowingly capitalized on the success of both authors when he published *The Naked Ape* the following year. Well known as the host of Granada TV's popular *Zootime* program, based out of the London Zoo, Morris soon gave up scientific work to concentrate on writing scientific nonfiction and refining his surrealist painting. The reading public, including a wide array of budding and established scientists, treated all three men as authoritative voices who used their knowledge of animal behavior to discern unsavory truths of human nature.[1]

Just ten years earlier, their pessimistic vision of humanity would have struck scientists as odd. After the Second World War, liberal American biologists and anthropologists had struggled to make sense of the recent eugenic horrors predicated on the assumption that some human lives were less valuable, less human, than others. In response, they crafted an account of humanity's past that emphasized a common evolutionary heritage bonded through continued interbreeding into a universal family of man. They sought out and proselytized the best features of humanity: our quick intelligence and problem-solving skills, our capacity to cooperate with and learn from strangers, and the resulting exponential accumulation of knowledge. We had invented agriculture. We had built pyramids. We orbited Earth and, within a decade, landed on the moon. By working together, humans were capable of creating objects of surpassing beauty and technological complexity.

This book tells the story of how definitions of human nature came to grip the American public with such force and why purported scientific insights shifted, so dramatically and in such a short time, from seeing humanity as characterized by our unique capacity for reasoned cooperation to emphasizing, even lauding, our propensity to violence. By 1971, S. Dillon Ripley, then secretary of the Smithsonian Institution, remarked that despite Americans' remarkable achievement in sending men to the moon, future historians would look back on this period and be struck by the "enormous awakening of public and scholarly interest in what one anthropologist has called 'the humanity of animals and the bestiality of men.'"[2] "Curiosity about lunar rocks," Ripley continued, would appear alongside the "organized and capricious human violence" that marred the era—from Memphis to Algiers, Los Angeles to My Lai. The deepening quagmire of Vietnam lighting up television screens in homes across the United States fueled broad discontent and, eventually, anger with military adventurism. The slow pace of change produced by the Civil Rights movement, coupled with economic hardship, precipitated urban unrest and riots in Newark, Detroit, Baltimore, and other major American cities. Newspapers carried accounts of political revolutions in Africa, Asia, and Latin America, not to mention the assassinations of President John F. Kennedy, Malcolm X, Martin Luther King Jr., and Robert Kennedy. Popular accounts of human evolution emphasized that the violence of humanity was too widespread to be an aberration and too common to dismiss as being against our better nature.

Debates over the malleability of human morality had a long tradition. Producers of publicly engaged science in these years—both those who supported and those who abhorred this new vision of humanity as innately aggressive—recognized this. Commentators on books by Ardrey, Lorenz, and Morris speculated that these authors had rewritten the battle between brothers at Eden's gate in evolutionary guise (Figure 1).[3] Just as the biblical story in which Cain slew his brother Abel had introduced murder as a human vice, contemporary evolutionists sought to inscribe in human nature the moral depravity of Cain's descendants. Readers noticed, too, that Ardrey had positioned his contention that aggression was ingrained in human nature against Jean Jacques Rousseau's eighteenth-century conviction that humans were born

FIGURE 1. Illustration accompanying Harry F. Guggenheim's editorial, "The Mark of Cain," *Newsday*, 25 September 1967, 33. Drawing on his reading of Robert Ardrey's *Territorial Imperative* and Konrad Lorenz's *On Aggression*, Guggenheim wrote that "man bears an evolutionary mark of Cain"—instinctual aggression. With this essential nature, each harm against Cain would be magnified against his transgressors until presumably the world fell into chaos. Illustration by Ken Crook. © 1967 Newsday. All rights reserved. Used by permission and protected by the Copyright Laws of the United States. The printing, copying, redistribution, or retransmission of the Content without express written permission is prohibited.

virtuous but corrupted by the society in which they matured.[4] The question of human nature in the 1960s was thus infused with both moral and scientific valence, sketched through evolutionary time. At issue was not merely if men needed to be taught to kill or to compromise, but whether humanity's capacity for interpersonal violence had provided the crucial ingredient that caused our evolutionary lineage to diverge from those of the other great apes, making us truly human.

The question of humanity's biological nature carried implications for the social and political concerns permeating college campuses, from civil rights to the feminist movement. Were race and sex, intelligence and charisma, indelibly etched in our bones, bodies, or genes? The cultural reverberations of such questions contributed to the authority that an evolutionary vision of a universal human nature played in the development of educational programs for American youth, domestic civil legislation, Hollywood movies, and reconfigured research programs across the social and natural sciences. The psychologist Charles Osgood, for example, argued that because the biological tools with which we understand and control our own actions were limited by our Stone Age nature, the pace at which we invented new methods of destruction had long ago outstripped our capacity to deal with these weapons. He echoed the concerns of an entire generation when he wondered, "Perhaps Modern Man, with his head in the sky, still has Neanderthal feet that are stuck in the mire."[5] By understanding our instinctual urges, Osgood and others hoped, perhaps they could alter humanity's self-destructive course.

In the decades after the Second World War, scientific authors became public figures in the United States, trusted as experts on a range of topics from child-rearing to death.[6] At the same time, violence emerged as a site of particular concern at every level of American society. In literature, politics, film, and science, writers rethought and re-presented the role of violence in modern life.[7] *Creatures of Cain* traces conceptions of aggression and the human animal through the "colloquial science" literature in these years, calling attention to a new kind of public intellectual who wrote, backed by the authority of science, in a style intended to engage readers only passingly familiar with his (or her) subject.[8] Hailed as experts in their respective fields, and sometimes beyond, Cold War scientists spoke on *The Tonight Show*, wrote best-selling paperbacks, produced regular columns in magazines, starred in documentary films, and served as advisors to the president.[9]

By emphasizing the colloquial language of these scientific books, essays, and films, I avoid locating them along a charged continuum of popular and professional publications.[10] Postwar scientists learned to communicate their work in at least two different registers: a professional language they used among peers and a colloquial one used when discussing their work with undergraduates, journalists, family members, and crafting so-called popular publications.[11] When contributing to the technical scientific literature of their field, whether reconstructions of humanity's past or descriptions of chimpanzee behavior in the wild, evolutionary scientists identified specific questions with solidly defensible answers, leaving little space for speculating about larger issues like, what does it mean to be human? Even a middlebrow intellectual magazine like *Scientific American*, for example, allowed its authors a scant final paragraph relating their research findings to social questions of the day. Colloquial scientific discussions embraced this larger context, including enough technical detail to establish the plausibility of their claims and directly addressing the social or political implications of their work. Put another way, readers enjoyed colloquial scientific works precisely because of the privilege they granted to the expertise of the author and the everyday, accessible language of the publication itself.[12] Drawing a sharp distinction between specialist and nonspecialist audiences would distort the history of ideas about human nature in these decades. After all, scientists read (and reviewed) colloquial scientific publications, too, especially when exploring new ideas outside their immediate expertise.[13]

This book begins in the years after the Second World War as scientists writing in a colloquial voice from a wide range of disciplines—cultural and biological anthropology, paleontology, primatology, and zoology—crafted a historical trajectory for humanity that was self-consciously anti-eugenic.[14] The best of humanity had not degenerated from living in the artificial constructs of civilization, would not dissolve because of the overbreeding of the lower classes, and could not be corrupted through miscegenation. Instead, these evolutionists (a useful term capturing their shared sense of enterprise) argued that our common past provided evidence of our continued remarkable success as a species. Our diachronic passage from mere ape to fully human rendered humanity the culmination of hardscrabble victories in the unforgiving environment of the open savannah. Behaviorally, we had learned to avoid predators, hunt cooperatively, and share food. We had changed structurally, allowing our ancestors to walk upright and carry weapons. Physiologically, females had developed hidden estrous, and we had become more adept at digesting meat. Linguistically, we spoke to communicate with each other. All of these factors cemented the pair bonds uniting families and, over time, led to a new fully conscious self-articulation of how we differed from other animals. In essence, so these scientists reasoned, our present human nature resulted from the synergy of biology and culture, both in dynamic flux throughout our development as a species.[15] We had become the most recent manifestation of a human lineage destined for even greater things in the future.[16] Through their work, an evolutionary perspective wended its way into each discipline perched at the intersection of the natural and social sciences.

Evolutionists felt a duty to communicate their ideas to people lacking expertise in that discipline and to enter into public dialogue about the social and political implications of various theories of human nature.[17] Their imagined readers included laymen (the contemporary term used to describe readers of both sexes without training in any of the sciences), scientists with expertise in other fields, and their colleagues. To reach all of these readers, they adopted authoritative voices crafted to be accessible to nonspecialists. In the words of one anthropologist, "The world of science is now so diverse that an expert in one branch is hardly more than a well-informed layman in another." He continued, "The credulity gap between scientific disciplines is perhaps in ever greater need of bridging than the rapidly shrinking chasm between science and the citizen."[18] Aspiring scientists also read and seriously debated books written in colloquial language, including those of Ardrey, Lorenz, and Morris. By the end of the 1960s, changing politics tinged earlier diffusionist models of educating the masses with an arrogant elitism. Some professional scientists imagined that members of an elusive public, with grassroots ideals and commonsense truths, would provide important insights into the proper jurisdiction of scientific expertise. In this climate, nonacademic writers vied for sales and prestige alongside professional zoologists, anthropologists, and paleontologists.[19]

By the mid-1970s, however—where this book ends—a new generation of evolutionists who called themselves sociobiologists (reflecting their avid interest in biological analyses of social behavior) defined human nature primarily through comparisons with animal behavior. Sociobiologists devoted their research to understanding the inner workings of the mechanisms by which evolution had brought about the great diversity of living forms. How evolution *worked* became more important than what had *happened*.[20] In making synchronic comparisons of human behavior with the behavior of baboons, chimpanzees, and other animals alive today, sociobiologists portrayed human nature as static: having arisen in our evolutionary past, it had become fixed when our ancestors achieved full humanness. This shift in perspective granted only minor explanatory heft to other scientists who studied humanity's past or present variation, including the cultural anthropologists, paleontologists, and primatologists who had earlier been key participants in the scientific reconstructions of the human animal. By dispensing with the historical development of human nature as irrelevant to understanding the consequences of its final (i.e., current) form, sociobiological theories of what it meant to be human de-emphasized the sense of progress that had characterized postwar scientists' visions of humanity. Stripped of progress, stripped of cultural variation, humans became yet another species of animal.

The halcyon years of colloquial science were also drawing to a close by the mid-1970s. Especially in the evolutionary sciences, books that had been hailed as respectable paperbacks a decade earlier were now denigrated as sensational popularizations. Contemplating the centrality of scientists to public discussions of political and social issues, Rae Goodell in 1977 used the term "visible scientists" to refer to figures who had been willing to venture into the public eye, such as Barry Commoner, Paul Ehrlich, Margaret Mead, Linus Pauling, Carl Sagan, and B. F. Skinner.[21] After the Second World War, scientists had welcomed colloquial discussions of their research in order to gain a wide audience for their ideas and out of a sincerely felt obligation to improve the science literacy of all Americans—not just schoolchildren. Some succeeded in attaining significant public visibility because of four factors, she suggested: they had embraced controversy; they communicated in clear, quotable language; their public reputation had been bolstered by professional recognition; and they had exhibited unabashed charisma in person, on screen, and on the page. However, Goodell worried that the visibility of these scientists no longer depended on the persuasive power of their research but simply on their willingness to engage in "the messy world of politics and controversy."[22] More specifically, evolutionists began to dismiss colloquial scientific publications as "popular" potboilers wrapped in scientific covers. Both sociobiologists and their critics blamed the media for extolling books by authors like Ardrey, Lorenz, and Morris that drove nonscientific enthusiasm for evolutionary theories of humanity, accusing journalists of repeating salacious details to sell copy at the expense of scientific accuracy. When the paleontologist Stephen Jay Gould—young, charismatic, willing to court controversy, and poised to become a scientific celebrity himself—contemplated his career that same year, he self-consciously harkened back to the golden years of science popularization of the late nineteenth century.[23] He would, he told an interviewer, write both for his colleagues and for a general audience.[24] It never occurred to him that these audiences might enjoy identical books or essays.

Between these two endpoints lies the rise and fall of the killer ape theory of humanity, its fate determined by two intertwined transformations: one in evolutionary conceptions of humanity's essential nature and the other in the texture of American intellectual life during the Cold War. Did humanity's capacity for violence explain our exceptional success as a species? Why was that a question worth asking? How did evolutionists become trusted experts on questions of humanity's fundamental essence? What evidence did readers find persuasive? Why did scientists and their readers eventually turn to other conceptions of human nature? These questions occupy the pages of this book. In its broadest scope, *Creatures of Cain* demonstrates that understanding the historical fate of any scientific vision of human nature requires attending to the political and social concerns that endowed that vision with persuasive power (or undermined it). It also illustrates the centrality of scientists and their colloquial engagements to the intellectual fabric of the country during the Cold War.

In the tumultuous atmosphere of the later 1960s and early 1970s, anthropologists, paleontologists, and zoologists did not shy away from public engagement, even though they rarely intervened in policy, manufactured weapons, or received funding from the Department of Defense. (Physicists and politicians had created a structure for discussing science policy—the President's Science Advisory Committee, or PSAC—but no one highlighted in this book ever served on it.[25]) During this decade, both molecular and organismal biologists struggled for authority, defining themselves as the cutting edge, potentially providing key answers to social difficulties besetting the country.[26] Cybernetics captured the attention of social scientists and, especially, molecular biologists looking to ground their discipline in the authority of reductionism.[27] Rather than analogizing life with machines, evolutionists imagined human nature as continuous with animal behavior. With the juggernaut of molecular biology nipping at their heels, they insisted that their fields, too, were modern and politically relevant, even if their research had not solved the structure of DNA or decoded the genetic language comprising the basic building blocks of life. Evolutionary theory could speak to a more fundamental question—what did it mean to be human? In terms of research money, positions at universities, and new departments, the molecular biologists won.[28] In other arenas, however, we cannot say that organismal biologists lost. They maintained a visible presence in the intellectual life of the country through a sustained insistence that because humans were by nature animals, studying animal behavior allowed a fuller understanding of what it meant to be human.[29] During these same years, anthropologists witnessed the cultures they studied in Africa, Asia, and Latin America transformed by decolonization and war. As a result, they increasingly distanced themselves from the goal of defining the universal characteristics shared by all peoples. Of course humans shared a common nature, anthropologists insisted, but variations between cultures and the intricacies of traditional customs better explained how and why humans act the way we do—that was where they concentrated their research, before these cultures vanished in the face of Westernization.[30] Primatologists similarly questioned models of human behavior that relied on comparisons with a single species.[31] These positions, these scientists believed, necessitated educating lay and scientific audiences alike as to the relevance of their research for interpreting the latest iterations of the human dilemma.

In seeking to define the characteristics of a universal human nature, post-war evolutionists wanted to know more than just what extinct "prehuman" individuals, or "hominids," in the convention of the time, had looked like.[32] They also sought to understand how they had behaved and interacted with one another. Doing so required triangulating between several kinds of evidence: the fossilized remains of extinct hominids, contemporary studies of hunter-gatherer cultures, and careful observations of animal behavior in the wild.[33] Fossils offered the most direct access to our ancestors' lives. Skulls and jaws provided insights into what they might have eaten, from the size of the sites where jaw muscles attach to the skull to the shapes and wear patterns of teeth. The orientation of hips, the length of femurs, and even foot shape (although this was rarely preserved) could indicate whether an individual had favored walking on two legs or four. Paleontologists could analyze shoulder joints to see if the creature brachiated, that is, swung from branch to branch, or had already abandoned the trees for the flat savannah. In short, fossilized bones contained an endless series of clues to the ecological environment fossil hominids favored, how they moved, and what they ate.

Although paleontologists in these years had access to a great many more fossils than had scientists fifty years earlier, the rarity of paleoanthropological specimens left enormous gaps in the fossil record. Almost every new find led to a plethora of interpretations, and it could take years, even decades, for paleoanthropologists to reach a consensus about its implications. Fickle preservation meant that when paleontologists unearthed fossilized bone from the surrounding sediment, they recovered only fragments: a partial skull here, a scapula there. This is why Donald Johanson and Maurice Taub's discovery of Lucy in 1974 made such news—they had recovered an astounding 40 percent of her *Australopithecine* skeleton. Additionally, determining if a fossil had been left by a member of a stable species—one that persisted for a long period of time across a wide geographic range—required many fossils of the same type. With only a few specimens, it remained possible that a new find might have preserved a fleeting transitional form. Paleontologists fought hardest, however, over the question of whether any given fossil represented a direct ancestor of modern humans or an extinct offshoot to the human lineage.[34]

Anthropologists believed that additional clues to how early hominids behaved could be inferred from the careful study of contemporary hunter-gatherer societies and perhaps also from the study of primate species in the wild. Certain human cultures, from the !Kung San of the Kalahari Desert to the Mbuti pygmies of the Congo region, lived in environments quite similar to those that evolutionists conjectured were occupied by the earliest humans.[35]

In the absence of direct evidence from paleoanthropological excavations, these living communities seemed to offer one of the only sources of information on how early humans might have organized their social lives. Many studies from the late 1960s and early 1970s emphasized the gentle cooperation of human societies before the intrusion of agriculture and notions of ownership, reinforcing assumptions of humanity's essentially cooperative nature. Other cultural anthropologists, however, had turned their attention to cultures in which interpersonal violence and warfare were far more common.[36] In the resulting morass of conflicting signals, cultural anthropologists rarely drew simple connections between the study of any one group of people and understanding humanity as a whole—they were far more interested in documenting diversity.

Contemporaneously, new details of the behavior of animals in the wild (especially primates) gained considerable traction as a foil against which to define humanity. In the 1960s, *National Geographic* lovingly adorned its pages with colorful images of Jane Goodall and chimpanzees. A decade later, her articles were joined by Dian Fossey's accounts of mountain gorillas, Biruté Galdikas's explorations of the life of orangutans, and Shirley Strum's engaging stories of baboons. This new generation of experts on animal behavior also included university-based scientists, such as baboon expert Irven DeVore and Edward O. Wilson, a zoologist with extensive knowledge of the social behavior of ants. Most scientists who incorporated animals as models of early human behavior emphasized the importance of either a shared environment (both baboons and early hominids lived in an environment that bridged the open savannah and nearby stands of trees) or a shared genetic history (chimpanzees were the closest living relatives to *Homo sapiens*). Only a few voices from inside the academy, or from very near it, gained a recognizable public voice as experts on human nature, but together they called attention to the rise of animal behavior as a discipline of note in the postwar life sciences.[37] By securing intellectual space for expertise in the evolution of different forms of behavior—territoriality, mating habits, foraging patterns, etc.—these evolutionists generalized from their species-specific knowledge to theorizing the role of behavior in all animals, including humans.

Postwar assumptions that humanity was by nature altruistic simultaneously gave way to a darker vision of humans as innately aggressive. In the 1950s and into the early 1960s, American scientists had largely believed that humans were instinctually cooperative. Just think, for example, of Edward Steichen's iconic *Family of Man* exhibit that opened at the Museum of Modern Art in New York in 1955 that depicted the so-called nuclear family as the heart of all human cultures.[38] These scientists struggled to comprehend how human groups could be capable of the incredible prejudice and slaughter evidenced in the Second World War internationally and the violent clashes of the struggle for civil rights at home. They sought to understand, and by understanding, prevent, the aberration of human violence in our otherwise peaceful lives. Given the changed political reality of the following decade, in contrast, a new generation of evolutionists instead conceptualized human nature primarily through comparisons with other animals in order to understand our unusual capacity for cooperation. Sociobiologists sought to understand why humans (or other animals) ever behaved *unselfishly*.[39] Using a new set of tools from mathematical and economic game theory, they emphasized the importance of maximizing individual genetic contributions to the next generation. From this perspective, it struck sociobiologists as deeply puzzling that individuals sometimes sacrificed their own well-being and genetic future to protect others.[40] By asking how such altruism could have evolved, sociobiologists naturalized violence as essential, but not unique, to human nature. The image of humanity bearing the mark of Cain thus enjoyed a brief but influential life, helping set the groundwork for how scientists conceptualize human nature today.

Constructed as a series of chronologically overlapping episodes, *Creatures of Cain* explores the racialized, gendered, and political landscapes in which conversations about human nature took place in the United States between 1955 and 1975. In seeking to reach nonspecialist audiences, publications exploring the nature of humanity often contained illustrations depicting the theories under discussion or scientists hard at work and deep in thought. The visual styles of these striking images reflected the artistic conventions of the era and call attention to the intellectual work required to sustain the plausibility of the scientific theories they depicted.[41] Visual depictions of humanity's evolutionary past required artists, like scientists, to triangulate between different forms of evidence to reconstruct the ecology, behavior, and physical appearance of fossil species as they had lived. For these reasons, the book includes a wide range of artists' illustrations. They convey the highly visual nature of colloquial scientific publications and individually offer glimpses into the changing fate of the human animal in these two decades.

Part 1, "The Ascent of Man," explores how, after the Second World War, an influential group of liberal anthropologists and biologists together articulated a non-teleological and progressive vision of transformations in the organic world, anchored in the ascent of humanity out of a bestial past. In this context, the concept of innate aggression posed a grave difficulty. Evolutionists like Loren Eiseley and Theodosius Dobzhansky invoked an unlimited anti-racist future for humanity and ascribed to evolution the capacity to explain the quantum emergence of human culture. Against the background of the Civil

Rights movement, anthropologists and biologists strove to change the American public's understanding of race by emphasizing the essential unity of humanity. Writing to convince lay audiences, they dismissed humans' capacity to regularly and brutally murder other members of their own species as the result of psychological or cultural deviance. For both Eiseley and Dobzhansky, imparting their scientific knowledge to members of the general public constituted a moral obligation and a form of intellectual activism.

In the developing Cold War, and especially after the Soviet Union's successful launch of the Sputnik satellite in 1957, dismantling the cultural divide between the sciences and the humanities seemed imperative for building the country's social future.[42] Congress set aside new pockets of money to fund innovative science curricula, and high-profile scientists, including the psychologist and pedagogue Jerome Bruner, joined the national effort to improve science education. The program Bruner directed—Man: A Course of Study (or MACOS)—hailed film as an exciting new medium through which to reach new audiences, including schoolchildren. Capturing animal behavior and human rituals on camera meant audiences could virtually experience the excitement of observing baboons on the savannah or watching Louis Leakey uncover fossils from the comfort of their home or classroom. MACOS, its designers believed, mobilized anthropological and biological knowledge in the service of training citizens to think like scientists, even if as adults they never ventured into a laboratory or museum.

This progressive postwar consensus unraveled in the later 1960s, as elucidated in Part 2, "Naturalizing Violence." By following the publication and immediate reception of Ardrey's *Territorial Imperative*, Lorenz's *On Aggression*, and Morris's *Naked Ape*, these chapters track the rise of a new view of human evolution that presented male aggression as not only natural but also as making possible the continued social evolution of humanity. Each book approached the question of the human animal from a different analytical angle, incorporating insights from recent work in ethology, psychobiology, and human sexology. Yet their confluence led readers to identify a shared assertion that studies of animal behavior provided crucial information for understanding human nature. Scientists read and reviewed these books; so too did captains of industry like the philanthropist Harry Frank Guggenheim. Caught by the passion of these authors' prose, Guggenheim planned to provide private support for such research through his foundation devoted to solving the problem of "man's relation to man."

Scientific audiences greeted the books with more skepticism than had Guggenheim. Eiseley, for example, suggested that their insistence on the animalistic nature of humanity failed to take into account the transcendence of human culture. At the same time, other scientists appreciated the popular attention Ardrey, Lorenz, and Morris brought to the field of animal behavior. This recognition came with increased funding (thanks to patrons like Guggenheim) and opportunities to showcase their research in magazines and commercial films. By emphasizing comparative behavior as the key source of reliable evidence of humanity's essential nature, these books began the process of transmuting the progressive postwar model of human evolution into synchronic comparisons between humanity and our living primate relatives.

Part 3—"Unmaking Man"—turns to the expansion of economic and biological agency to females in evolutionary models of the late 1960s. In both the postwar progressivist and the ascendant killer ape models of human evolution, women stayed at home to raise their offspring and gather food while the men hunted. When scientists began to question the idea that women and men possessed different natures, they complicated the role of cooperation and competition within biological notions of family. If males and females united for conflicting evolutionary reasons, then social cooperation could not emerge from a cultural-biological nexus defined by the family unit. At the same time, cultural anthropologists began to distance themselves from the question of a universal human nature, exploding the notion that Western family structures were to be found among all human cultures. Cultural anthropologists now largely agreed that variations between cultures and the intricacies of traditional customs better explained how and why humans act the way we do than did an abstract human nature.

As a function of these discussions, feminist and masculinist interpretations of human evolution co-emerged in the late 1960s. When some anthropologists, primatologists, and paleontologists began to challenge the emphasis that older evolutionary theories had given to hunting in early human groups, others redoubled their arguments that the sexes possessed different biological natures.[43] In writing in a colloquial register about science, men and women faced disparate challenges. Female writers found it difficult to be taken seriously if they also tried to be funny. Feminist readers enjoyed the BBC Radio writer Elaine Morgan's biting treatment of existing narratives of human evolution, for example, but found it difficult to take seriously her alternative hypothesis positing an aquatic phase in hominid evolutionary history. In the fraught sexual politics of the early 1970s, Morgan's critics called her a radical feminist, while she in turn considered scientists who asserted the necessity of a biological perspective on sex difference to be reactionary conservatives. These dynamics never divided cleanly by sex, however; female scientists contributed vociferously to both sides of the nature-nurture debates.

Evolutionary conceptions of humanity came under fire in the early 1970s from two distinct directions, as described in Part 4—"Political Animals." Scientists on the New Left questioned the reduction of human experience to any biological explanation (whether environmental, genetic, or evolutionary), as this appeared to deny individual agency and reified social prejudices in biological language. Seeking to redress racial discrimination, these anthropologists argued that evolutionary theories were determinist and therefore inconsistent with an egalitarian vision of human diversity. Part of their concern came from the widespread attention evolutionary perspectives were receiving in Hollywood. Evolutionary accounts of human nature had spread far beyond university halls, as directors and screenwriters transformed popular scientific visions into images on the silver screen—from Stanley Kubrick's *2001: A Space Odyssey* to Sam Peckinpah's *Straw Dogs* three years later.[44] Directors defended the violence of their films by invoking Ardrey's killer ape hypothesis and suggesting that their recreations of fistfights, shoot-outs, and even rape reflected the truth of human nature. On the New Right, conservative Catholics and Evangelicals agreed that the violence of secular humanism, writ in evolutionary theory, constituted a fundamental threat to moral order. Mobilized by their concerns with the violent content of the educational movies created by MACOS for use in grade-school classrooms, religious conservatives objected to anthropology's association with evolution and the redemptive possibility of science without reference to a Christian God. Caught in the middle, postwar progressive visions of the ascent of humanity unraveled from both ends of the political spectrum.

Part 5—"Death of the Killer Ape"—examines the final collapse of the remaining support for the idea that humanity's capacity for interpersonal violence was linked to our success as a species. The contentious reception of Edward O. Wilson's *Sociobiology*, a rearticulation of human social behavior as the result of evolution, broke the sympathetic alignment of evolutionary perspectives in anthropology, paleontology, and zoology that had characterized visions of the human animal in earlier decades.[45] Sociobiologists in the mid-1970s rebranded and professionalized their discipline so as to dismiss "popular" writers (especially Ardrey, Lorenz, and Morris) as having fundamentally misunderstood the mechanics of evolutionary theory. Unfolding in these same years, Jane Goodall's research team at Gombe Stream National Park observed a series of chimpanzee attacks that resulted in the extermination of one chimpanzee group at the hands and teeth of their neighbors. Humans were not unique; we shared our violent tendencies with at least our closest simian relatives. Whereas earlier writers assumed that the evolutionary process within humans operated at the level of the family group, sociobiologists instead traced the effects of natural selection on individuals. In the process of cleansing human evolution of the last dregs of support for the killer ape account, contemporary evolutionists also severed the interdisciplinary alliance that had bound descriptions of universal human nature to nuanced accounts of its development. Human nature became unmoored from its past.

In the maelstrom of the social, cultural, and political transformations that characterized the American home front in the Cold War, scientific theories of the human animal provided powerful tools for sorting the bewildering violence of the world into sensible order. Long-standing questions about violence and human nature took on an outsized importance, opened colloquial science to new participants, and sustained novel critiques. That readers granted evolutionists the power to settle these questions was neither inevitable nor obvious.

We are still living with this legacy.

PART I

The Ascent of Man

The truth is that if man at heart were not a tender creature toward his kind,
a loving creature in a peculiarly special way, he would long since have left his
bones to the wild dogs that roved the African grasslands where he first essayed
the great adventure of becoming human.

— LOREN EISELEY

LOREN EISELEY penned *The Immense Journey* in a world of silence.[1] He had
lost his hearing thanks to dual middle-ear infections, and although his physician assured him it would eventually return, he felt awkward teaching if he
could not address students' questions. So instead he sat at his kitchen table
and turned his attention to an experiment in writing. Part memoir, part science, Eiseley sought to capture the majesty of human evolution for the lay
reader. In earnest, assured prose he encouraged his readers to imagine the fate
of our ancestors recorded in a "speeded-up motion picture through a million
years of time." His cinematic sweep encompassed the transformation of a flint
ax into a torch grasped in a raised hand. The grassland world in which our
ancestors lived, teeming with bison and mammoths, would vanish to sate their
new appetite for meat. Fire, and therefore cooking, allowed early humans to
more efficiently digest this meat and compensated for stomachs ill-adapted to
an omnivorous diet. Their limbs would grow longer, changing over many generations to stride across vast savannahs. The ice ages and increased hunting
drove many of the large herbivores in Europe and North America to extinction. His imagined scene ended once again with a single man, now holding a
river stone in one hand and contemplating a few seeds of grass in his other. "In
that moment, the golden towers of man, his swarming millions, his turning
wheels, the vast learning of his packed libraries, would glimmer dimly there in
the ancestor of wheat."[2]

After the Second World War, science and technology emerged as transformative forces of both destruction and liberation.[3] As American political elites sought to construct a new national and international order, they relied on the expertise of scientists and engineers to make real the utopian dreams of a gleaming technological future.[4] To help realize everything from bigger atomic bombs to new flavors of Jell-O, Americans needed innovative approaches to traditional grade-school curricula that moved beyond mere memorization and encouraged students to think creatively and precisely about technical problems. Given the rapid pace of scientific change, educators reasoned, citizens in a functioning democracy would also need to stay abreast of scientific and technological innovations throughout their lives. In 1960 *Time Magazine* even dubbed "scientists" their "Men of the Year," noting that they had "transform[ed] the earth and its future. They were surely the adventurers, the explorers, the fortunate ones."[5]

Science education shows appeared on television, geared at children and their parents—from *Watch Mr. Wizard* to the *Wild Kingdom*.[6] Magazines like *National Geographic* and *Scientific American* devoted their content exclusively to science, and paperback science enjoyed a veritable revolution. One publisher noted that "intellectual paperbacks" were "widely available in bookshops, on newsstands, and through mail-order catalogs."[7] He estimated that in 1949 only fifty "intellectual" titles had been available in cheap paperback form. By 1963 that number had grown to fifteen hundred and he expected the number to at least double again by 1968. Commercial and university presses had entered the market with equal vigor, rereleasing classics alongside new colloquial science books that explained science for the "layman." When *National Geographic Specials* debuted with *Miss Goodall and the Wild Chimpanzees* in 1965, they produced booklets that accompanied the television program for intended use in grade schools around the country.[8]

In classrooms, on television, and as authors of "pulp science," scientists saw themselves as engaged in dynamic education of the masses.[9] Americans largely believed that scientists, especially physicists, had won the Second World War through their work on the Manhattan Project.[10] The atomic bomb thus gave scientists both an impetus for wading into the public arena and a reason for doing so—whether or not they supported the mobilization of scientists in creating such weapons.[11] After 1957, when the Soviet Union launched Sputnik, the first man-made satellite to orbit Earth, the importance of cultivating scientifically literate citizens became even more pressing. During his presidency, John F. Kennedy highlighted the importance of scientists in solving the most urgent difficulties besetting the country, from poverty to the nuclear arms race. In this context, the long evolutionary path that led to modern humanity emphasized how far we had already come as a species. No less than physicists, anthropologists and zoologists saw themselves as providing a new set of intellectual resources with which Americans could think about the problems of race and inequality. That they believed they could influence the intellectual and political convictions of the nation from their desks reflects their trust in the nexus of power and knowledge of the era.

Working as a physical anthropologist at the University of Kansas, Eiseley had tried to join military intelligence during the Second World War to put his professional skills to work for the good of the country, but his hearing, which had never been great, kept him out of the military's ranks. (Years later, he speculated that the recruiting officer may also have learned about his mother's lifelong struggle with mental stability.) He remained in Kansas and spent the war teaching anatomy and biology premedical courses to reservists enrolled at the university.[12] By 1957—the year *The Immense Journey* appeared in print and the year Americans listened over radios to Sputnik's slow beep as it passed overhead—Eiseley had moved to Philadelphia and would spend the rest of his career at the University of Pennsylvania. The success of *The Immense Journey* kept Eiseley writing even after his hearing returned. *Darwin's Century*, *The Firmament of Time*, and several other books of prose and poetry followed quickly on its heels.[13] He surmised that the evolutionary past, as revealed in fossil traces scattered in the landscapes he had trudged as a younger man, said as much about failure as it did about success. Many, many organisms exhibited "poor specializations" to the environments in which they lived, he suggested, turning into side shoots "off the main line of progressive evolution."[14] Our human ancestors had not only avoided that fate, they also bridged the past and the future, the evidence of our "immense journey" inscribed forever in our bones. "We are all potential fossils," he wrote, "still carrying within our bodies the crudities of former existences, the marks of a world in which living creatures flow with little more consistency than clouds from age to age."[15]

In less lyrical prose, other postwar accounts of human evolution also spun together a determined optimism and epic historical vistas. These building blocks formed the basis of a universal human nature shared by all living peoples. The anthropologists and biologists who wrote in a colloquial register used the past as a platform to suggest humans were therefore the first species in the history of life on Earth to control its own destiny. Even their book titles sparkled with promise, from Theodosius Dobzhansky's *The Biological Basis of Human Freedom* to Jacob Bronowski's *The Ascent of Man*.[16] Far from naive, their hopes for the future had been forged in horror at the death toll of the Great War and years of economic depression and hardened in response to the atrocities of the Second World War. Family members, friends, and colleagues fought, and some died, to bring an end to the wars. In advancing a theory of humanity's progress, then, postwar scientists emphasized our ancestors' hard-won successes against incredible odds. We humans had clawed our way to the present.

An ardent proponent of this grand evolutionary perspective (for humans, at least), Dobzhansky had moved to the United States in 1927, leaving the Soviet Union to study genetics at Columbia University with the formidable Thomas Hunt Morgan.[17] A reserved man with strangers, his friends remembered him as "conspicuously affectionate and loyal."[18] Dobzhansky's "unceasing energy and enthusiasm for his field work" meant he was equally at home on horseback or at a lab bench. He specialized in population-level variation in fruit flies and loved collecting trips that allowed him to escape the laboratory.[19] Ten years later, he published one of the most important books in evolutionary theory of the century, *Genetics and the Origin of Species*. Recognized as magisterial at the time, biologists today still read it as a worthwhile classic, although it has never attracted an audience outside of academia. Like his colleagues, Dobzhansky worked to separate evolutionary theory from its associations with Nazi eugenic policies and articulate a new biological basis for understanding a universal human nature. He claimed that the principles governing the physical past of humanity differed fundamentally from those that would matter in the coming decades, centuries, or even millennia. In short, he conjectured, when humans became human, a new form of evolutionary process came into being. The very origins of culture changed the way early humans interacted with the world around them and necessitated a new set of conceptual tools for thinking about our evolution. To truly understand what it meant to be human thus required input from an array of scholarly perspectives—cultural anthropologists needed to talk to physical anthropologists, and both with biologists specializing in bird behavior, fruit fly genetics, and primate sociality. Whether he and his colleagues called it cultural, creative, or social evolution (and at various times, they used all these terms), Dobzhansky endowed the ascent of man out of our physical past with hope and self-determination.[20]

Eiseley's vivid image of a man holding in his hands the seeds of the future also conveyed his sense of humanity's biological unity. Dobzhansky would have agreed with Eiseley that all humans shared a common evolutionary history and, perhaps more important, he would have insisted that because of interbreeding between populations, humanity comprised a single shared gene pool. For Dobzhansky, "man" (that is to say, this unitary human population) extended backward in time—to the moment when a split developed within a species of now-extinct apes. One population created by that split would, through adaptation to a unique sequence of environmental circumstances, become modern chimpanzees, and the other, reacting to a different set of environmental conditions, turned into us. This unified human nature also extended forward into the undetermined, unknowable future. Dobzhansky would probably have balked, though, at Eiseley's encapsulation of humanity in the form of a single individual. This image could never capture the cultural and genetic diversity on which he believed the past and future of humanity depended.[21] Underlying these convictions, both men advanced a vision of humanity that depended on a shared essential nature.

What then made us uniquely human? Publications from the 1950s provided a litany of answers spanning the four fields of anthropology, including, but not limited to, our bipedal way of life and large, convoluted brains (physical anthropology); our capacity for true language (linguistics); our ability to use and manufacture tools, skills shared through family and friendship (cultural anthropology); and the wide array of ways humans in the past have shaped and reworked our lived environment (archeology). More disturbing, humans also appeared exceptional in their capacity to kill members of our own species. Genocide, war, and murder were behaviors known only among human cultures. This terrible power fit awkwardly with the altruistic message liberal scientists invested in an evolutionary perspective on humanity and strove to provide to the reading public. An old dilemma, spanning many traditions and cultures, humanity's remarkable capacity for both empathy and evil found new purchase after the Second World War.

By invoking a progressive evolutionary history for humanity, Eiseley and Dobzhansky sought to explain our essential nature, however flawed, through the slow accumulation of these traits. History made us. Only through the interdisciplinary study of our shared past and present, they believed, could scientists come to a full understanding of human nature. However, not all scientists agreed that evolutionary theory offered the most powerful tool for understanding humanity.

According to the cultural anthropologist Margaret Mead, the conditions that caused deadly violence were created and maintained by specific cultural circumstances and were not inherent characteristics of human nature at all.[22] The author of the provocative *Coming of Age in Samoa*, published in 1928, she worked at the American Museum of Natural History in a turret of an office. In this, her first book, Mead had used her research on the sexual behavior of adolescent girls to argue that morals and gender roles varied with a society's traditions. Here was evidence of cultural relativism in a phenomenon as biological as male and female sexual behavior, she argued.[23] Here, too, was the possibility that Americans could change their cultural attitudes toward sex, accepting a wider range of "normal" behavior.

In 1960 she wrenched her ankle and decided to carry a long, forked walking stick rather than use a cane. Conscious of the striking image she had created, she kept the walking stick after her ankle healed. Although her frank discussion of adolescent sex in Samoan and American societies had raised eyebrows among some members of the reading public, for others Mead had not gone far enough. In her 1963 feminist classic *The Feminine Mystique*, Betty Friedan rejected Mead as too mainstream, as having unfairly crystallized universal sex roles for men and women.[24] Yet Mead served, too, as a role model for many girls interested in the sciences—providing evidence that maybe one day they could succeed in an anthropological career. Six years after she died in 1978, the American Museum of Natural History opened the *Margaret Mead Hall of Pacific Peoples*—an exhibit she had a substantial role in planning.[25] When you walk into the exhibit today, you are greeted by a display of Mead's personal effects, including the walking stick and one of her most eye-catching wool capes.

For Mead, a human nature shared by all people did not really exist.[26] In a 1955 essay titled "What Is Human Nature?," she answered that most people used the term to denote the learned characteristics that define our lives.[27] "For we *are* taught to be human," she added. Without social interactions with other people, we might never learn to speak or to walk upright or perhaps even to recognize other humans as kindred creatures. Although basic emotions were universal, behavioral expressions of fear, shame, grief, or anger varied according to the social traditions in which people were raised. These learned differences, in turn, often defined the way cultural groups were stereotyped in common parlance. Even warfare, Mead insisted, was a human invention, a cultural institution in which "rules are learned and modified from one generation to another. All man's most anti-social possibilities—hatred, vengeance, destructiveness—are used in war. But so are his most constructive possibilities—loyalty, cooperativeness, generosity, self-sacrifice for his group and his country, his ideals." If warfare were cultural, that provided a glimmer of hope. "Whatever man has invented, man can change. War can become as obsolete as dueling."[28]

The social and economic transformations wrought by the Second World War led many evolutionists to rearticulate the message a scientific understanding of human nature might hold for the future. Eiseley and Dobzhansky pointed to misguided assumptions that natural selection depended on "the struggle for existence" or "the survival of the fittest"—interpretations, they suggested, that had been popularized in the United States by Herbert Spencer in the late nineteenth century and amplified by racialist thinking in subsequent decades.[29] Even if the physical evolution of humanity had been changed (perhaps even stopped altogether) by the origins of reason and learning, social evolution had gained enormously in strength. Although they quarreled about many other things, anthropologists and biologists writing for a general public, including Eiseley, Dobzhansky, and Mead, agreed that prewar racialist theories supposedly based in science had misunderstood the process of evolution and that in order to counter that outdated picture of humanity, cultural and physical anthropologists would need to work together with paleontologists and zoologists.

The scientific message that all peoples belonged to the same Family of Man carried particular weight given the struggle for civil rights across the country.[30] In 1954, for example, the US Supreme Court declared segregated schools unconstitutional in their landmark *Brown v. Board of Education of Topeka* decision. Chief Justice Earl Warren wrote in the court's unanimous decision that education served as "a principal instrument in awakening the child to cultural values, in preparing him for later professional training, and in helping him to adjust normally to his environment." No child could expect to succeed without "the opportunity of an education." Given the importance of education in society, the court argued that segregated schools were "inherently unequal." This condition held also when tangible factors, like the quality of the buildings, curricula, and teachers, were similar. The act of segregation itself negatively affected the mental development of children excluded from certain schools and therefore violated the Fourteenth Amendment's guarantee of equal protection under the law. As evidence, Warren cited sociological and psychological research on race, prejudice, and the effects of segregation.[31] The Supreme Court's decision thus highlighted both the power of education in individual and national self-realization and the ethical implications of social scientific research on a national stage.

Warren's logic resonated with anthropologists' and biologists' conviction that education provided an important tool for encouraging an antiracist, progressive view of humanity to gain purchase in American society as a whole. Eiseley's personal experiment with writing for that elusive audience of non-scientifically trained readers worked stunningly well. Letters he received from grateful readers expressed their profound indebtedness to the beauty with which he wrote about the natural world. Although he was not a religious man, as he aged he lavished attention on small wonders in a language that approached the spiritual. (He probably received less glowing reviews as well, but those letters never made it into the archive of his papers.) Eiseley even tried his hand at educational television with an NBC-sponsored show for children called *Animal Secrets*. More successful, however, were television programs deemed appropriate for children that also held the attention of their parents, like the National Geographic Specials, which started in 1965 with *Miss Goodall and the Wild Chimpanzees*, and later Jacob Bronowski's thirteen-episode *The Ascent of Man*.[32]

Other scientists sought to teach the hopeful message of a universal human nature by redesigning grade-school education curricula. In 1956 the MIT physicist Jerrold Zacharias helped to found what became known as the Physical Science Study Committee to improve high-school physics courses. When the Soviet Union launched the Sputnik satellite only one year later, the United States Congress increased their budget for science education fivefold. Legislators lamented the Soviet Union's success in engineering the first artificial satellite and worried that without better science education, the United States would eventually fall behind. Over the course of the next two decades, a selection of university-employed scientists turned their attention to updating science education in public schools in hopes of inspiring young Americans to devote their future careers to science and engineering. Zacharias founded an educational nonprofit, Educational Services, Inc., which worked with university professors and local teachers to develop new materials for grade-school science classrooms. ESI's commitment to inquiry-based learning and productive collaborations resulted in an alphabet soup of curricula, including the PSSC (Physical Science Study Committee, released in 1960), SMSG (School Mathematics Study Group, 1961), CHEM Study (Chemical Education Materials Study, 1962), BSCS (Biological Sciences Curriculum Study, 1963), ESCP (Earth Science Curriculum Project, 1965), ESS (Elementary Science Study, 1965), IPS (Introductory Physical Science, 1966), and more.[33] In this mix, a group of Harvard University faculty members interested in the question What is human about human beings? sought to develop a course for social studies students called Man: A Course of Study (or MACOS). MACOS drew on the latest research in cultural anthropology and animal behavior, creating a curriculum that would teach grade-school children how to reason scientifically about the social problems confronting the country, especially the struggle for civil rights.

Taken together, postwar scientists spanning a wide range of disciplinary perspectives constituted an intellectual community deeply committed to revising contemporary assumptions about human nature. They entrusted each other with the responsibility for communicating the virtues of science and a more inclusive vision of humanity, even as they disagreed about the details. The picture of humanity they created reflected contemporary American ideals of a "nuclear family," in which a bread-winning father and nurturing mother lived with a small handful of children, surrounded by suburban bliss.[34] Anthropologists, population geneticists, and paleontologists alike grounded the origins and centrality of this universal family in evolutionary time.

By the late 1960s, scientists not tramping through a field site, spending long hours in the laboratory, or excavating specimens—those without access to new fossils, cultures, and data—looked increasingly old-fashioned. Eiseley wrote his books on the dignity of humanity's evolutionary path years after he had cut his teeth on field research as a young man. Starting in 1959, he served for three years as provost of the University of Pennsylvania, and although he returned to the classroom, he permanently traded the field for the library. By then, the figure of the "armchair anthropologist" already attracted derision, and Eiseley's memoir nostalgically imbued the books and shelves of his research with just enough danger to transform them into a manly past time.[35] "There is a dust one may breathe among old books which can be just as fascinating, the heat as infernally oppressive, as any amount of crawling about in tombs and deserts." "The old Furness Library," he wrote, "was overloaded, lacked air conditioning, and was grimed, particularly in its less-used stacks, by the soot and dust of a century. The lighting in some sections was so poor that the use of a miner's lamp would have been justified. Desiccated red leather bindings had a way, in summer, of ineradicably staining one's suits or collapsing into mummy powder that one breathed."[36] Eiseley's early forays into Darwin's life and ideas had captured his imagination. Over the course of his career, he transitioned from a practicing anthropologist to a writer fascinated by history and science alike.

Despite his scientific training, Eiseley remained skeptical that natural selection, genetics, or any solely natural process could sufficiently explain the complex pattern of the living world. He worried, too, about the invocation of Darwin's name in the same breath as Copernicus and Galileo as three men who had secularized the natural world. This loss of awe at the intricacies of life troubled Eiseley, even though he never thought organized religion provided a viable solution. The answer, he believed, must lie in the friction generated between awe and reductionism.[37] Dobzhansky could not have disagreed more. A member of the Eastern Orthodox church, when Dobzhansky's wife passed away in 1969 his faith became even more important to him. He came to see evolution and theological belief as more than mutually compatible. For Dobzhansky, the origins of culture and our impressive intellectual capacity allowed humans to be the only of God's creations to appreciate the majesty of the natural world. With our self-conscious awareness of our place in the universe, he believed, evolution could act as beacon of hope. (After his death, a former student described him as a "metaphysical optimist."[38]) Even if the cosmic order could never be geocentric, Dobzhansky thought there might yet be a case for its anthropocentric nature.[39] Eiseley, on the other hand, feared that own his equivocation over the power of biology as a sufficient explanation for the complexity of the natural world meant that "I'm not a very good scientist;

I'm not sufficiently proud, nor confident of my powers, nor of any human powers." Neither, he suggested, was Darwin, who had also "wobbled" in his awe at nature—a claim that tells us far more about Eiseley's convictions than Darwin's history.[40]

Eiseley helped popularize a robust interdisciplinary scientific understanding of what it meant to be human. When he was a young anthropologist, researching the evolution of humanity required training in the study of modern and extinct species of humans—their bones, cultures, and traditions. After the Second World War, scientists trained in the study of animal behavior, even specialists in bird systematics and the genetics of fruit flies, joined physical and cultural anthropologists in their endeavor. Hoping to counter claims that races differed constitutionally, this interdisciplinary community of evolutionists provided evidence of our ancestors' collective struggle for survival and remarkable progress through several million years of shared biological history. In so doing, postwar evolutionists emphasized the historical origins of a universal human nature shared by all peoples and cultures.

1

Humanity in Hindsight

IN 1859 CHARLES DARWIN published *On the Origin of Species* at the age of fifty. Centennial celebrations of *Origin* thus doubled as the sesquicentennial of his birth. Historical retrospectives published for the centenary provided a welcome opportunity for postwar anthropologists, biologists, and paleontologists to signal their conviction in the progress of science and trust that evolutionary theory provided key insights into human nature. Contested news of fossil discoveries and archeology were favorite subjects of journalists covering anthropological research—then as now.[1] Postwar evolutionists used discussions of these fossils in the context of the Darwin centennial to craft a vision of a united humanity, a historical past shared by all living humans before our species had been riven by cultural and political identities. Looking back at the previous century also gave paleoanthropologists a boost of confidence. Whereas Darwin had built his theory of human descent on comparative anatomy and logic, they had access to recent studies in blood typing, genetics, and a wide array of hominid fossils.[2]

Publications began to appear in 1958 honoring the occasion on which Charles Lyell and Joseph Hooker presented to the Linnaean Society a "brief sketch" from Darwin and a short essay by Alfred Russel Wallace, each outlining a "natural means of selection" that explained how, over the course of the history of life on Earth, new organic species had emerged and been transformed.[3] *Origin* followed seventeen months later, summarizing the voluminous evidence and keen argumentation Darwin had been accumulating and honing for years. In the 1950s, scientists looked back, both acknowledging *Origin* as Darwin's "great contribution to biology" and pointing to how much more they now knew about genetics, paleontology, and zoology.[4] According to Margaret Mead, the celebrations of Darwin's publication of *On the Origin of Species* had refocused anthropologists' collective attention on evolution as a potential framework for thinking about human nature.[5] In measuring Darwin's successes and failures, anthropologists used their historical reflections to highlight what they identified as the most promising components of evolutionary thought and evidence from recent decades and the evidentiary holes that remained to be filled when describing the emergence of humanity in the geological past.

Darwin's dark jacket and white-bearded face—for he was almost always depicted as an elder sage—stood for a generation of Victorian gentlemen already out of place in the modern scientific world they helped create.[6] Depending on the perspective of recent historians, Darwin has been a warrior in the heroic battle against the backward forces of slavery, a villainous instigator of secular fervor that resulted in dehumanization and genocide, and just about everything in between.[7] In the 1950s, though, Darwin's legacy was far less contested. Retrospective appraisals of (to invoke its full title) *On the Origin of Species by Means of Natural Selection, or the Preservation of Favoured Races in the Struggle for Life* both celebrated Darwin's achievements and noted how much science had advanced in the intervening years.[8]

Loren Eiseley's *Darwin's Century* was one of the first and best-selling forays into the centenary genre, and he fared well in the growing market for "intellectual" paperbacks.[9] His version of Darwin's intellectual development and legacy began the way many classes on the history of modern biology still begin: by emphasizing early-modern scientific empiricism and the desire of Enlightenment natural historians to catalog and classify all living species according to a great scale of nature. Against this background, Eiseley posited, evolutionary thinking—the idea that species have not been static in time, but some have gone extinct and others slowly evolved into new forms—emerged in France, rising from the secularized ashes of the revolutionary republic. Eiseley made clear that Darwin's legacy therefore rested on his innovative mechanism explaining the transformation of species. Like Eiseley, the retrospective essays and books published in the years around the centenary pointed to Darwin's theory of natural selection as his "most important generalization."[10] University presses ensured that a variety of editions were produced for academic consumption, from facsimile reprints of *Origin's* first edition to a variorum comparison of all editions published during Darwin's life.[11]

Natural selection is an elegant, simple idea. If species are composed of many individuals that compete for access to limited resources, then those individuals who are better able to marshal the resources around them will survive and raise a greater number of offspring than their neighbors. As a result, the next generation of that species will look slightly different, because the traits that helped the previous generation succeed will be more commonly represented. Over the eons, as the physical landscapes of Earth inexorably changed, so too did species adapt to the new conditions in which they lived. Most dramatically for the British public, the sandstones of southern England

and around Paris had once accumulated at the bottom of a warm freshwater sea, fossilizing now-extinct shells and the skeletons of ichthyosaurs in layers of sediment.[12] Documenting evidence that natural selection was the best explanation for these changes in the living world required far more work.

Darwin had been developing his "big book" about species transformation since the 1840s and was induced to publish it earlier than planned by a letter from Wallace, who had written from the island of Ternate—one of the Maluku Islands that now comprise northeastern Indonesia. Despite their differences in social class, Darwin and Wallace's friendship was cemented in their correspondence, where they "met on the equal ground of pen and paper." Wallace wrote to Darwin about his idea of explaining how one species could come to be replaced with a superior, "more perfectly adapted and more highly organized form" as a result of environmental pressures differentially killing some individuals and not others. Darwin was surprised to find the essential features of his theory of natural selection in the words of Wallace's letter, perhaps because of his relative isolation from the avant-garde chatter of his friends. In 1842 Darwin had moved from London to Down (then a rural parish in Kent) and convinced himself despite Lyell's warning that other researchers were not wrestling with issues of species, varieties, transformation, and adaptation. After Darwin and Wallace's brief contributions were read to the Linnaean Society in 1858, Darwin compressed his ideas and wrote what he viewed as an "abstract" of his longer manuscript—*On the Origin of Species*.[13]

A century later, the developmental biologist Conrad Hal Waddington wrote that despite Darwin's brilliance in describing speciation, he had never fully understood the problem of variation, which left a large conceptual gap in natural selection. Natural selection only works if the traits that help some individuals survive and reproduce more than other members of the species are passed along to their offspring, who in turn exhibit the same traits—but in the late nineteenth century, the mechanisms of heredity had yet to be worked out. Darwin remained unaware of Gregor Mendel's careful breeding of peas. (Long after Darwin died, Mendel's research would be rediscovered in 1900 by a growing community of researchers exploring the physical basis of heredity, a field that came to be called genetics.) In Waddington's judgment, without a mechanism for heredity, natural selection as a theory failed to measure up to Isaac Newton's far more "complete and self-sufficient" contribution to science: universal gravitation.[14] For Waddington, biologists were able to complete the conceptual work Darwin began in the nineteenth century only when they finally dismissed the possibility that characteristics acquired over the lifespan of an individual could be biologically inherited by their offspring.[15] Also a geneticist by training, Dobzhansky was more generous, noting that first Mendelian genetics and then molecular genetics had "borne out Darwin's conclusions" and "opened to view the causal processes that bring about evolution."[16] Dobzhansky claimed Darwin had instigated "one of the greatest revolutions in the history of human thought" by demonstrating that "man is a part of nature and kin to all life."[17] This idea was not new with Darwin, Dobzhansky quickly added—other natural historians had made similar arguments at least two generations earlier, including Darwin's paternal grandfather, Erasmus, who had died several years before Charles's birth—but he had made the concept of humanity's long evolutionary history convincing.[18]

The physical anthropologist and paleontologist Wilfrid Le Gros Clark praised Darwin's *Descent of Man and Selection in Relation to Sex*, published twelve years after *Origin*, as "an act of great moral courage" for championing the antiquity of man with almost no direct fossil evidence to back his claims.[19] For Darwin, as for his contemporaries, primates had played a key role in scientific reconstructions of humanity's past. Lacking access to paleontological specimens, Darwin had based his own thoughts on primate and human evolution almost entirely on species still alive today. In *Descent of Man*, Darwin had further suggested that humanity's intellectual development made possible all other specializations that make us human, from our capacity to manufacture and use tools, to our conceptual appreciation that the past and future may differ from the present. Le Gros Clark commented that "since Darwin's day . . . the evidence relating to human evolution has grown to vast proportions," placing paleoanthropology on more solid scientific ground.[20] Instead of one so-called missing link, anthropologists had several, allowing tentative reconstructions of humanity's successive adaptations.[21] Paleoanthropologists routinely inferred a fossil ancestor's relative intelligence from the size of its brain case and, therefrom, the size of the brain those bones had protected in life. Based on this accumulated evidence, Le Gros Clark argued that the era of extrapolating the history of humanity's evolution by indirectly comparing ourselves to modern apes had finally come to an end—he contended the analysis of bones and fossils were the only sure way to understand the evolutionary past of humanity. Le Gros Clark noted, too, that paleoanthropologists studying human origins still required a healthy dose of courage because their necessarily speculative conclusions were likely to be disproven tomorrow.[22]

Fearful that his colleagues were lionizing Darwin, the zoologist S. Anthony Barnett warned against lapsing into an absurd "cult of the individual." Historians had long noted Darwin's indebtedness to economic theories of competition, paleontologists' fossil evidence documenting species extinction, and geologists' reconstructions of changes in the physical characteristics of landscapes over the course of millennia past. Therefore, Barnett suggested, Darwin alone could not be credited with sparking all the subsequent transformations in biology, since his work and ideas had materially depended on the intellectual climate in which he worked and wrote.[23]

Loren Eiseley agreed with Barnett, and in his own retelling of the mental evolution of humanity, Wallace emerged as a hero equal to Darwin. Wallace's experience in tropical archipelagos had led him to conclude that the "natives" were mentally only a "little inferior" to the average European.[24] This position had presented Wallace with a dilemma, Eiseley suggested. Natural selection could not explain a mental capacity "far in excess" of that required by the environment, nor could natural selection explain the artistic, mathematical, or musical abilities of humanity. Wallace had inferred that a spiritual source unknown to science must have been involved in the development of the human brain. Upon reading Wallace's views, Darwin had written to him expressing his disappointment. Eiseley paraphrased Darwin's reaction: "If you had not told me you had made these remarks, I should have thought they had been added by someone else. I differ grievously from you and am very sorry for it."[25] (Darwin also filled his letter to Wallace with excitement over their mutual interests and the tone of the entirety was much less sharp than Eiseley's brief excerpt implied.[26])

The significant differences between the brains of all humans and those of animals continued to gnaw at Eiseley. He posited that an Ice Age hunter from twenty thousand years ago who painted mammoths on the cave walls of Lascaux shared "precisely the same brain" as a modern aeronautics engineer.[27] The hunter and the engineer differed merely in the cumulative knowledge and ideas humanity had used to shape the natural world generation after generation. According to Eiseley, it was ultimately culture rather than mental endowment that defined their relative intellectual capacities. Eiseley grieved that Darwin never changed his mind and scientists forgot Wallace's challenge.[28] "It isn't precisely that nature tricks us," Eiseley wrote. "We trick ourselves with our own ingenuity. I don't believe in simplicity."[29]

Trickery, indeed, had played a role in the history of paleoanthropology, a fact of which Eiseley was well aware. Quarrymen in the mid-nineteenth century working in the Neander Valley outside of Düsseldorf had found an unusually shaped skull and associated bones. Hermann Schaaffhausen, the comparative anatomist who first described the bones, hailed them as an anatomically distinct "primitive" human ancestor, but the pathologist Rudolf Virchow and the retired physiologist August Franz Mayer dismissed them as the relics of a diseased Russian horseman who had died as his army traversed Germany on the way to assail France earlier in the century. In the following decades, however, more and more specimens were identified that shared the same peculiar morphologies, providing firm evidence of prehistoric humanity. Neanderthals (*Homo neanderthalensis*), as these individuals became known, generated considerable popular and scientific interest. By the 1950s, archeologists' excavations of numerous ancient sites demonstrated that Neanderthals hunted prey with spears and other weapons, controlled fire, buried their dead, and left traces that many archeologists interpreted as signs of religious practice.[30]

Eiseley focused particular attention on an insidious case, known as Piltdown Man, that he believed had led scientists to misunderstand the role of the brain in the physical development of humanity. Discovered in 1912 in gravel beds near the village of Piltdown, south of London, the Piltdown fossils appeared to confirm that human ancestors much older than Neanderthals had lived in Europe. They also reinforced what Darwin had suggested in *Descent of Man*—that prehuman brains had enlarged early in the divergence of humans from other apes, making possible tool use and manufacture, bipedalism, language, and ultimately culture. A controversial find from the beginning, in the early 1950s scientists finally confirmed the Piltdown fossils were frauds.[31] Whoever perpetrated the deception (theories abounded) had combined the cranium of a small adult human with the lower jaw of an orangutan, replaced the canines in the upper jaw with their orangutan equivalents but filed them down to make them look more like ape-human intermediates, and then stained everything to resemble the quarry rocks. Based on differing levels of fluorine in the teeth, as well as dental wear patterns, paleoanthropologists determined that the canine and mandible were definitely modern and deliberately faked. They concluded the bones represented "a most elaborate and carefully prepared hoax."[32]

The removal of Piltdown Man as a viable "missing link" left a gap of several million years between a putative common ancestor belonging to both human and chimpanzee lineages, and the remains of individuals who closely resembled modern humans, like Neanderthals. Paleontologists suggested that a few skull fragments found in India, dubbed *Ramapithecus*, might have come from an ancient ancestor shared by both humans and chimpanzees. Yet *Ramapithecus* remained a highly controversial designation until postcranial specimens could be found. In the 1950s and 1960s, without linked evidence of a pelvis or scapula and arm bones, scientists could only speculate as to whether the species was bipedal or arboreal. Eiseley therefore concluded that Darwin had been wrong when he suggested that the expansion of the brain in our ancestors preceded the origin of other characteristics scientists identified as uniquely human. Instead, Eiseley argued that brain development in humans must have evolved *after* bipedalism, possibly even after our capacity to manufacture and use tools. This, he suggested, was "the real secret of Piltdown"—a conclusion for which he felt "roundly castigated" by his colleagues.[33] Eiseley took care to reassure his readers that despite their protests he was, deeply, an evolutionist and a scientist. His case was helped by the fact that without Piltdown, the sequence of fossils looked cleaner and more linear (Figure 2).[34]

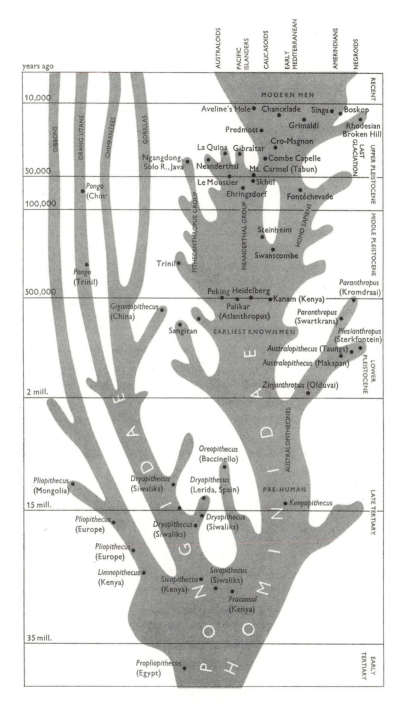

AUSTRALOIDS
PACIFIC ISLANDERS
CAUCASOIDS
EARLY MEDITERRANEAN
AMERINDIANS
NEGROIDS

RECENT

10,000 — MODERN MEN

Aveline's Hole • Chancelade Singa • Boskop •

GIBBONS ORANG UTANS CHIMPANZEES GORILLAS

Grimaldi Rhodesian
Predmost • Broken Hill

Cro-Magnon

La Quina Gibraltar
Ngangdong, • Combe Capelle
Solo R., Java Neanderthal Mt. Carmel (Tabun)

50,000 Le Moustier • Skhul

UPPER PLEISTOCENE LAST GLACIATION

Ehringsdorf Fontéchevade

Pongo
(China)

100,000

Steinheim

HOMO SAPIENS

Swanscombe

Trinil •

MIDDLE PLEISTOCENE

Pongo
(Trinil)

PITHECANTHROPUS GROUP NEANDERTHAL GROUP

Paranthropus
(Kromdraai)

500,000 Peking Heidelberg Kanam (Kenya) •

Gigantopithecus
(China)

Palikar
(Atlanthropus) Paranthropus
(Swartkrans)

Sangiran EARLIEST KNOWN MEN Plesianthropus
(Sterkfontein)

Australopithecus (Taungs) •

Australopithecus (Makapan) •

LOWER PLEISTOCENE

Zinjanthropus (Olduvai) •

2 mill.

AUSTRALOPITHECINES

H O M I N I D A E

Oreopithecus
(Baccinello)

Pliopithecus
(Mongolia) Dryopithecus
(Siwaliks) Dryopithecus
(Lerida, Spain) PRE-HUMAN

15 mill. Kenyapithecus •

Pliopithecus
(Europe) Dryopithecus
(Siwaliks) Dryopithecus
(Siwaliks)

LATE TERTIARY

Pliopithecus
(Europe)

H O M I N O I D E A

Limnopithecus •
(Kenya) Sivapithecus
(Kenya) Sivapithecus
(Siwaliks)

Proconsul
(Kenya)

35 mill.

Propliopithecus
(Egypt)

EARLY TERTIARY

FIGURE 2. Two evolutionary visions of the human family tree. The first (above) hails
from Gavin de Beer's 1964 *Atlas of Evolution* (New York: Thomas Nelson and
Sons, 1964),173, and highlights the diversity of known fossils while maintaining
the visual appeal of a single interbreeding lineage. (*cont. on following page*)

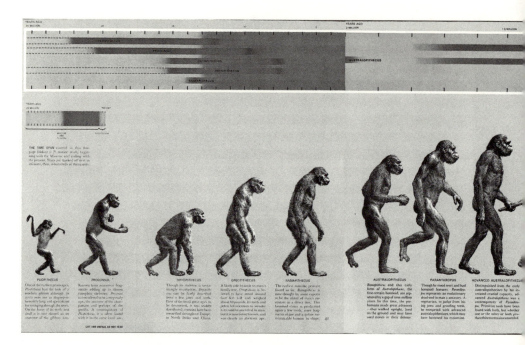

FIGURE 2 (*cont.*) The second (above and following page) appeared the following year in F. Clark Howell's *Early Man* (New York: Time Life Books, 1965), 41–45. Here, the linear progression of human history comes through more strongly. The illustrator, Rudy Zallinger, included a full-body reconstruction of all species even when (in the case of *Ramapithecus*, for example) the only bones yet recovered came from the skull. Note, too, the stone tools grasped in

In the century following Darwin's *Origin,* fossils with substantially preserved skulls, especially those that seemed to have been left behind by previously unknown kinds of "ape men," gained a celebrity of their own, standing out as charismatic individuals in a morass of typically fragmentary bones. Almost all were known by affectionate nicknames, which made it easy for journalists to refer to the fossils while scientists continued to debate their scientific classification. Indeed, paleontologists often took years, even decades, to reach a professional consensus over key factors of interpretation, like the age of the fossil, whether it represented a new species, and if so, whether that species had been a direct ancestor in the human lineage or a so-called offshoot. None of these issues were trivial, and reaching consensus often involved researchers traveling to examine these rare fossils in person. As paleontologists uncovered more and more fossils, the scientific names associated with ancient hominids skyrocketed along with their potential implications for understanding human nature.

In the 1890s, for example, the Dutch paleoanthropologist Eugène Dubois had discovered a partial cranium, tooth, and thighbone on the island of Java,

the hands of *Australopithecus* and the more sophisticated spear carried by Cro-Magnon Man. George V. Kelvin created the accompanying timeline that runs across the top of the pages. Zallinger's encapsulation of human history as a march of progress has been endlessly repeated in visual representations of evolution, both serious and parodic. The Zallinger image is reprinted with kind permission of the Zallinger Family.

then in the Dutch East Indies, and named them *Pithecanthropus erectus*. Dubois argued the specimens came from a missing link between humans and apes that most closely resembled gibbons, a suggestion his colleagues ridiculed. Fixated on the idea that gibbons are capable of walking upright, he assumed humans were more closely related to them than to other great apes, a connotation he inscribed in the name *erectus*. The moniker Java Man stuck, as paleontologists contested for decades Dubois' scientific name for his finds.

Then in 1924, the young Australian paleoanthropologist Raymond Dart identified a fossilized brain case and partial cranium from what he believed was a more ancient "man-ape," calling it *Australopithecus africanus*. The skull became known as the Taung Child. Dart's paleoanthropological colleagues were skeptical. Dart had not found the fossil himself—instead it came from a box of fossils gathered in a South African mine near the University of Witwatersrand, where he worked, which made it difficult to date and impossible to analyze in situ.[35] With Piltdown still a viable human ancestor in the 1920s, it further seemed that the relatively small brain case of *Australopithecus* discounted it as a possible ancestor of humanity. Worse, Taung Child hailed from South Africa, and paleoanthropologists in the 1920s and 1930s remained convinced that humans had originated somewhere in Asia, not Africa—a theory with fossil evidence thanks to several partial skulls and jawbones discovered in central China known as *Sinanthropus pekinensis*, or Peking Man.[36] Additionally, Dart's florid language in describing his conclusions, that *Australopithecus* was "an animal-hunting, flesh-eating, shell-cracking, and bone-breaking ape," only increased the suspicion with which his colleagues treated his claims.[37]

By the 1940s, a new generation of paleoanthropologists unearthed new specimens in both South Africa and on Java. One of Dart's protégés, the stalwart Robert Broom, endeavored to find an adult specimen of *Australopithecus africanus*, and in the 1940s he succeeded. Broom also discovered a set of skulls and skeletons with similar morphologies but more massive jaws and sagittal crests that supported strong jaw muscles, dubbing them *Australopithecus robustus*. Another paleoanthropologist, Ralph von Königswald, followed in Dubois's footsteps and discovered new specimens of *Pithecanthropus erectus* on Java. Because of these finds, it became increasingly apparent that the anatomy of Java Man shared a great many characteristics with the known fossils of Peking Man, and that both sets of specimens came from individuals who had lived more recently than *Australopithecus*. Their discoveries coincided with the demise of Piltdown's respectability, and postwar paleontologists took seriously the notion that human origins might after all lie in Africa, as Darwin had suggested back in 1871.[38]

By the centenary of *Origin*, then, paleontologists agreed that some several millions of years separated *Ramapithecus* and *Australopithecus* (although their opinions continued to diverge over exactly how many). They also believed that it was in this gap that humanity's ancestors had embarked on a journey that led to the present.[39] Taken together, the evidence they had dug out of the earth added up to a view of human history characterized by a single, interbreeding population distributed across vast geographies. Scientists imagined this linear trajectory as punctuated by moments of speciation followed by a brief period of competition between the old and new species. The paleontologist Alfred Romer, for example, used the almost linear sequence of human ancestors to emphasize that as natural selection led to the origins of more successful characteristics, a new species would tend to outcompete its parent type, leading to the original species' extinction. A series of such competitive exclusions would result in an almost linear faunal succession, similar to prehuman fossils. Like many paleoanthropologists, Romer therefore believed it unlikely that multiple species of human ancestors could have coexisted for long.[40] Yet without a greater number of examples of each fossil type, anthropologists remained unsure of whether any one specimen represented a long-lived hominid form—"a relatively stable adaptation"—or a chance fossil preserved during a very rapid period of transitional change.[41]

For evolutionists of the era, humanity's core population extended both backward into the past and forward into the future. Only very recently, Dobzhansky suggested, had *Homo sapiens* subdivided into semi-isolated races. He quickly added that interbreeding among these communities had never stopped. In the rapidly globalizing postwar world, people from different races were meeting, falling in love, and building families together in ever-greater numbers.[42] Any biological characteristics specific to a particular race were thus quickly being dispersed through the rest of the human population. Whether or not the public agreed about the desirability of this process, scientists argued that it would continue unless checked by social factors. In the future, Dobzhansky thus projected, racialized differences would vanish along with cultural inhibitions.[43] This view added a dimension to Dobzhansky's claims that mankind was "in a very real sense, one huge family in which everybody is related, however distantly, to everybody else."[44] Given our common gene pool, mankind evolved as a single genetical system, a single unit of evolutionary change united by a common nature.

For nonanthropological evolutionists like Dobzhansky, the classification rubric for naming fossil humans remained one of the profound difficulties confronting paleoanthropologists.[45] Each new spectacular fossil was given a different name, he lamented, with little appreciation for the variation almost certainly present within species at the time. No one, Dobzhansky wrote, should imagine that fossils from two individuals of the same species would look exactly alike. Such false expectations (combined with self-aggrandizement) had led to a system that ignored whether or not scientific names for extinct hominids represented real biological divisions at the time when a given fossil had been flesh and bone, navigating the savannah with her small troop.

His close colleagues agreed. Ernst Mayr had initially made a name for himself studying the taxonomy and biogeography of living birds, publishing *Systematics and the Origin of Species*, in which he defined species as actually or potentially interbreeding populations that were reproductively isolated from other related groups.[46] Then, at the conference "The Origin and Evolution of Man" at Cold Spring Harbor Laboratory in 1950, Mayr accused the assembled audience of taxonomic "splitting." The numerous scientific names they had created in previous years, he argued, were not warranted—each new fossil was unlikely to represent an entirely new species—as the human lineage appeared to have evolved in a series of iterative transformations with no known instances of two different populations coexisting for long periods of time.[47] As an example, he took a variety of well-known fossils, including Java Man and Peking Man, and united them under a single taxonomic name, *Homo erectus*,

which was characterized by several morphological features as well as the capacity to make fire and manufacture stone tools. *Homo erectus*, he argued, represented the dawn of true man and the onset of the Stone Age. Many paleoanthropologists at the time agreed implicitly with Mayr's contention that all
hominid fossils represented a single lineage (even if they resisted his suggestion that these fossils should be grouped into fewer, more capacious taxonomic categories). In Dobzhansky's colloquial accounts of evolution, this
unified vision of human history traveled far beyond the small group of professional scientists present at the 1950 conference.[48]

Not until the 1970s would new fossil discoveries extend the known varieties and ranges of *Australopithecus* and extinct *Homo* species, demonstrating
that in the past, several prehuman groups had indeed coexisted for extended
periods of time.[49] A generation after that, paleoanthropologists would criti-
cize mid-century scientists for taking Mayr's advice and contributing to a
"saga of a lone hero battling from primitiveness to perfection over the eons,
armed with nothing but natural selection and its own wits."[50] In fact, now our
evolutionary tree more closely resembles a bush than a spruce. These criticisms, however, overlook the persuasive power of this framework in the postwar era because of its resonance with an anti-racist commitment to variation
within a singular, progressive human lineage. For Dobzhansky, as for Mayr,
humanity constituted a coherent interbreeding population—a Family of Man
(Figure 3)—extending back in evolutionary time.[51] In emphasizing humanity's ascent from this original common nature, evolutionists conjured an illimitable potential future for the human species as a whole, as long as we did
not in the meantime ruin Earth or kill ourselves in a nuclear holocaust.

For this postwar cohort of evolutionary thinkers, the fossil evidence assembled since Darwin's publication of *Origin* and *Descent* supported a progressive vision of humanity's past that echoed Eiseley's cinematic retelling of
humanity's transformation from the grassland to packed libraries. Even
though scientists found plenty of examples of evolutionary dead ends and
failures in the fossil record, the overall trend of those species still alive was to
greater complexity, greater beauty, and greater mutual regard.[52] Dobzhansky
somewhat regretted that Darwin had titled his second great book "the 'Descent,' rather than the 'Ascent' of man." He feared it gave readers the wrong
impression. From Dobzhansky's perspective, the "long and toilsome ascent
from animality to humanity" had taken strength and purpose. To ignore that
risked losing sight of humanity's accomplishments.[53]

Meanwhile, museum visitors and scientists alike yearned for captivating
glimpses of reconstructions of fossil faces, social behavior, and the more tangible (if more speculative) factors that define us as human (Figure 4). As the
young primatologist Irven DeVore wrote in 1964, the long prehistory of hu-

FIGURE 3. A family of *Homo neanderthalis*, pictured in front of caves. Maurice Wilson's watercolors reconstructing the lives of prehistoric humans were reproduced in several publications designed to reach popular audiences in the 1950s, including multiple versions of Time-Life books in the United States. To render the images appropriate for young readers, the company darkened in the genital regions of the adults thereby masking the anatomical details. Natural History Museum Images #7728 © The Trustees of the Natural History Museum, London.

manity had to be reassembled from scattered pieces. Many aspects of early human behavior or social capacities were never inscribed into bones. When our ancestors first spoke, for example, could be estimated only through inference. New evidence for such behaviors would therefore come, DeVore expected, "from the archeologist's spade, from the study of contemporary hunter gatherers, and from the natural history of monkeys and apes" that would together allow scientists to "retrace man's passage from his common ancestry with other primates to his uniquely human way of life."[54] Given the relative paucity of fossil evidence during the crucial millennia when humans became human, anthropologists and biologists created vivid living pictures of early human culture by triangulating speculative answers from more abundant sources, especially studies of so-called Stone Age cultures and free-living

Neanderthal   Cro-Magnon

Australopithecus   Pithecanthropus

FIGURE 4. Facial reconstructions of fossil hominids. Originally labeled: "The man-ape of Africa and early men of Eurasia." From E. H. Colbert's *Evolution of the Vertebrates* (New York: Wiley, 1955), 286. Theodosius Dobzhansky reprinted the image in *Evolution, Genetics, and Man* (New York: Wiley, 1955), 329, with a new caption: "Hypothetical restorations of fossil human and pre-human forms." Reprinted with permission of John Wiley & Sons, Inc.

primates. Each human tribe and animal community proffered scientists insights into how our ancient ancestors might have hunted with others, socialized as a community, or traveled in a small family unit. Were our forefathers and foremothers benevolent or brutal? Answers varied according to which culture or species scientists invoked as comparators in their tales of humanity's past.

2

Battle for the Stone Age

THROUGHOUT THE 1950S AND 1960S, respectable colloquial science magazines like *National Geographic, Natural History,* and *Scientific American* published article after article, each sensationally announcing the "Stone Age" status of different human cultures. Following the decolonization of Africa and Asia and faster, cheaper forms of commercial transportation, these magazines filled with tales of the gentle hunter-gatherers of the Kalahari, the warlike tribes of Papua New Guinea, and isolated communities of Amazonians who had never before encountered Western civilization. These articles revealed a common desire to preserve and communicate knowledge of such "primitive" cultures before their unique customs were forever transformed by the inexorable incursion of space-age technological goods. Anthropologists also believed that these apparently timeless cultures continued to live in conditions reminiscent of those that early humans would have encountered. Whereas most human cultures from Asia to Europe had dramatically transformed in the following millennia, they suggested, small pockets of human groups still lived in relative harmony with nature. The more isolated, the more "primitive," the less "civilized" these groups appeared to be, anthropologists thought, the better examples of early human social organization they might provide. Yet exposure to Western technology in the form of motorized vehicles, zippers, and Band-Aids (to name a few of the modern conveniences that often traveled with anthropologists) threatened to shortcut traditional forms of transportation, clothing, and medical care. This situation generated a profound sense of urgency and excitement among anthropologists recording as much about "Stone Age" cultures as they could before they vanished.[1]

Anthropologists and biologists agreed then, as they do now, that human behavior stems both from the culture in which we are raised and from our biological heritage. By 1964, the cultural anthropologist Clifford Geertz derided anthropologists of earlier decades for promoting the idea of a "critical point" in the origin of humanity.[2] Anthropologists had believed, he wrote, that "man's

41

humanity, like the flare of a struck match, leaped into existence."[3] In this version of the birth of human nature, an evolutionary leap had taken place through the tight interaction of several factors—increased brain size, bipedialism, family structure, a new ecology of life on the savannah, hunting and access to meat, and language—all caught in a maelstrom of positive feedback that resulted in the modern human. Geertz noted that, in the meantime, new fossil finds had revealed that *Australopithecus*, with a brain case about one-third the size of modern humans, appeared to use weapons and exhibited something that resembled proto-culture (both assertions were controversial). New primatological evidence, too, demonstrated that baboon and chimpanzee behavior were more complicated than anthropologists had previously thought possible.[4] The former bright line between human and animal seemed more like a hazy stripe.

These professional debates were profoundly shaped by postwar American politics. As college students around the South protested local segregation through nonviolent sit-ins in at department store lunch counters, as interracial freedom rides called attention to the segregation of interstate busses and bus terminals, and as two hundred and fifty thousand Americans gathered in Washington, DC to stand against racial injustice, anthropologists increasingly worried that studying so-called primitive human cultures as stand-ins for our human ancestors smacked of racism.[5] When President Lyndon B. Johnson signed the Civil Rights Act of 1964, legal segregation ended in public accommodations, and activists turned their attention to getting local governments to enforce the new law and securing the right to vote for all African American citizens. Just as violent attacks on civil rights activists made the news, so too did urban riots in Harlem, Philadelphia, Los Angeles, Cleveland, Newark, and more. Again and again, it seemed, violence within the country had been sparked by tensions over race. Adding recent civil rights struggles to their memories of the Second World War, mid-century liberal social scientists worked to substitute naive essentialist views of race (which they feared contributed to the dehumanization of peoples around the world) with detailed ethnographies illustrating the intricacy, even beauty of other cultures.[6] By publishing their studies as easily accessible intellectual paperbacks and by bringing back with them photographs and film of other cultures, anthropologists imagined they could reach a far larger audience than ever before.

The anthropologist Clyde Kluckhohn had hoped to change the way Americans thought about race. In his well-regarded *Mirror for Man*, Kluckhohn had written that "anthropology holds up a great mirror to man and lets him look at himself in his infinite variety."[7] A prolific author for nonscientific audiences, Kluckhohn twice won the Whittlesey House Science Award for popular writing and was hailed by a reviewer in the *New York Times* as "the prophet of the new anthropology."[8] A social anthropologist at Harvard, he also commanded respect among his professional peers. In his writings, Kluckhohn stressed that the appropriateness of customs always had to be measured against existing group habits. Yet he also suggested that "cultural relativity" as an anthropological principle did not give free license to members of one culture to behave in a fashion determined by the action's acceptability in a another. Moral absolutes might exist, but anthropologists' best chance of uncovering any universally shared moral rules rested in the comparative study of social norms and behaviors across a wide range of cultures.[9]

Rather than thinking of modern races as having formed from long evolutionary isolation, Kluckhohn suggested that the populations postwar scientists recognized as races had formed through the fusion of multiple cultures into conglomerate populations. This explained why, from his perspective, anthropologists could not agree on how many modern races existed. Even passably competent students of race, he suggested, varied in categorizing people in as few as two races or as many as two hundred. As a result, the study of race was peppered with frequent and serious misunderstandings among educated people, many of whom failed to recognize that prior to the nineteenth century variations among humans were rarely attributed to biological heredity.[10] Kluckhohn blamed Americans' pernicious appetite for biologism on the erosion of religious faith in the Western world. Physical science brought creature comforts, and biological science seemed on the verge of ending diseases of the flesh. Between them it appeared possible that scientists would eventually find answers to "all the riddles of the universe."[11]

Although cultural and physical anthropologists both sought to break down the defensive suspicion with which they feared Americans of different races viewed each other, they disagreed over the continued utility of race as a concept. For Kluckhohn, world peace depended on individuals' ability "to minimize and to control aggressive impulses."[12] Ashley Montagu agreed, arguing that the problem of race in American society was social, not biological.[13] In his foreword to a reprinting of Kluckhohn's *Mirror for Man*, Montagu called race "a zoological idea" inappropriately applied to people and "converted into social doctrine."[14] Born Israel Ehrenberg, Montagu was ambitious, verbose, and occasionally strident. He hailed from working-class London and had moved to the United States to complete his training in cultural anthropology at Columbia University. Montagu suggested that the word "race" should be removed from scientific discussions altogether, a position he broadcast through the controversial UNESCO Statement on Race he had helped craft in 1950.[15] Trained as a cultural anthropologist, Montagu embraced a biological determinism grounded in human cooperation. People had to be taught to hate, he insisted. Although Montagu obtained a brief position as a professor of anthropology at Rutgers, he lost it at the height of McCarthyism and spent the rest of his career outside formal academia as a public intellectual.[16]

The physical anthropologist Sherwood Washburn, on the other hand, argued that races were real biologically—he called them "an expression of nature"—rather than a manifestation of language, religion, nationality, social habits, or any other cultural trait.[17] If Montagu remained on the friendly fringes of academia, "Sherry," as he was called by his friends and colleagues, carved a comfortable place at its core. Only a few years younger than Loren Eiseley, Washburn had earned his PhD at Harvard under the guidance of the physical anthropologist and ardent eugenicist Earnest Hooton.[18] Washburn worked at Columbia University for a number of years before joining the anthropology faculty at the University of Chicago in 1949 and then moving to University of California, Berkeley, in 1958, where he built a robust and dynamic department. Washburn's reputation grew as he sought to reorient physical anthropology as a discipline that could contribute to discussions about evolution as a process, with insights from human history.[19] The study of physical anthropology, he argued, ought to become part of a broader interdisciplinary conversation about the evolution of humanity that included social scientists, geneticists, anatomists, and paleontologists. Only then could scientists piece together a meaningful sense of what it meant to be human.

Washburn nevertheless insisted that the best way to understand race was through the tools of comparative anatomy and taxonomy to reconstruct historical ancestry. For him, the difficulty lay not in the quest, which could remain noble, but in the proxy measures scientists often used when trying to quantify race. Just as cranial capacity could never substitute for understanding intelligence, he argued, anatomy alone could never solve the complex reality of human ancestry. What physical anthropology revealed, Washburn suggested, was that the modern races were only a few thousand years old. And even if that calculation were wrong in its specificity, he emphasized that the fossils of ancient men recently discovered by paleoanthropologists differed profoundly from any "modern man." Early *Homo sapiens* fossils would be able to pass unremarked in the New York subway whereas Neanderthals and Australopithecines would stick out from the crowd. In response to an imagined question from his readers—"To which of the living races are these ancient fossils particularly closely related?"—Washburn wrote, "To none."[20] He sympathized with the political motivation behind Montagu's desire to get rid of the word "race" but feared that using another word, like "ethnicity," to stand for the same concept would merely obscure the underlying problem, and lay misunderstandings of race and human nature would continue as before. He concluded that one instantiation of race had proved particularly problematic and was based on useless categories: "Race: Pure Nordic; Location: Nowhere; Method: Imagination; Result: Nonsense."[21]

Debates over the utility of race as a useful scientific concept came to a visible head with Carleton Coon's publication of *The Origin of Races* in 1962. Coon used its pages to mobilize the tools of physical anthropology and argue (contra Washburn) that as many as five races of *Homo erectus* existed over a half million years ago and evolved into *Homo sapiens* at different times, independently passing "from a more brutal to a more *sapient* state."[22] The "Caucasian" race, he suggested, made this transition two hundred thousand years before the "Congoid" race, grounding racial difference in the Pleistocene. Members of the segregationist cause immediately pointed to the book as scientific evidence bearing out their policies.[23] Montagu and Dobzhansky, in turn, renounced Coon's book as politically motivated (a claim he vehemently denied throughout his career).[24] For Dobzhansky, if evolution were a contingent process (which he believed it was) and if a single population evolved in concert because of gene flow between individuals in that population, then it made no biological sense for Coon to suggest that the same evolutionary transformation had taken place in five disparate times and places.[25] Montagu in turn directed his frustration at Coon because he thought Coon had deliberately used pejorative language when comparing subspecies of *Homo erectus*—of course his readers then interpreted "brutal" and "sapient" as emotionally laden concepts. Both men believed Coon had denied the fundamental unity of mankind and claimed that all humanity could not have originated in Africa. Worse, they implied, Coon had understood these nuances and wrote in a fashion that allowed his science to be used for intolerant political ends.[26]

In the heat of the moment, Washburn chose to use his presidential address to the American Anthropological Association as an opportunity to discuss the problem of "race" in anthropology.[27] Washburn and Coon had both trained under the same advisor at Harvard, the anthropologist Earnest Hooton, but their subsequent careers had led in divergent intellectual directions. Although Washburn delivered his address mere months after Coon's book appeared on the shelves of university bookstores, he later insisted the timing had nothing to do with *The Origin of Races* and everything to do with an internal debate among the board about whether or not the AAA should endorse a resolution on race and if so what it should be. The board agreed that if he chose to talk about race in his address, they would adopt it as the society's official resolution. Washburn acquiesced.[28] In his speech, he explicitly took issue with Coon's conservative perspective on human evolution and endorsed Dobzhansky's far more progressive *Mankind Evolving*, which had been released earlier that year.[29] Washburn reasserted that neither the fossil record nor evolutionary theory supported the idea that human races were of great antiquity. Racialization, he insisted, was a process that could not be understood without attention to the history of cultural systems. Humanity had created races in the same way we created everything else—through the positive interaction of biology and culture—which was why the races of humanity merged into one another so that it was impossible to tell where one left off and another began.[30] He wrapped up by telling the gathered audience that racism was "a relic supported by no phase of modern science." Human races were a product of the past. In the future, Washburn envisioned a free and open society that embraced "a culturally determined way of life" and allowed for full expression of "the infinite variety of genetic combinations."[31]

The same social and economic forces that guaranteed the interbreeding of human cultures in the future, however, threatened to destroy traditional cultures through modernization. As steel axes replaced stone, modern technology and economic development reshaped cultures and languages. Traditional ways of life were vanishing, and with them potential clues to humanity's history.[32] When the cultural anthropologist Colin Turnbull went to live with the Mbuti people in the northeastern corner of the Congo, for example, he spoke of their forest life as threatened by the outside world.[33] His advisor at Oxford, where Turnbull read anthropology as an undergraduate and then returned for his graduate work in the 1950s, described him as "tall, fair, handsome; inconsequent, impulsive and elusive."[34] As Turnbull aged, his hair darkened but his mercurial temperament remained.

The Ituri Forest first struck Turnbull as "a vast expanse of dense, damp and inhospitable-looking darkness." With time, however, he came to see the forest as "a cool, restful, shady world with light filtering lazily through the tree-tops." The coolness of the forest corresponded, too, with a gentleness of disposition. Cooperation was key to pygmy society, he suggested. The Mbuti negotiated both interpersonal problems and political decisions with the input of the larger community, from late-night spousal disagreements to deciding where and what to hunt. No clear leader took charge. Everyone could voice an opinion.[35] The men hunted different kinds of antelope and chevrotain with bows or nets. The women gathered edible roots, berries, nuts, and herbs from the forest. After spending almost three years with the Mbuti, Turnbull's perspective on the outside world changed, too—it had become unbearably "hot and dusty and dirty."[36] His research led the American Museum of Natural History to hire him in 1959 as a curator of African ethnology, and *The Forest People* appeared a few years later.

Turnbull's romanticization of the forest as a safe haven from strife echoed anthropological theories that described the evolution of humanity as depending on a crucial transition from the protection of the forest to the relative danger of the savannah. With this ecological shift came our ancestors' capacity to walk on two legs, rather than knuckle walking on all fours, in turn enabling them to carry weapons and hunt. His study contrasted well with anthropological examples of more fearsome hunter-gatherer societies outside the protection of the forest—peoples like the warrior tribes of New Guinea.[37]

In the early months of 1961, a Harvard anthropological expedition discovered, according to coverage in the New York Times, a "savage tribe whose culture is focused on war."[38] Buried in the middle of the paper, the brief account contained details of the anthropologist Robert Gardner's first encounter with the Willigiman-Wallalua people. After he and his colleague had pitched their tent, Gardner related, they kept the front open for the evening breeze. Feeling the press of eyes from the gathering darkness, they proceeded with their normal habits. When Gardner removed his shirt, he reported hearing a collective "hissing intake of breath." He attributed their astonishment to the fact that his whole body was white.

The fact that the expedition merited reporting at all was due to one prominent member of the team—Michael Rockefeller, the youngest son of the sitting governor of New York (and much later vice president of the United States), Nelson Rockefeller. No further updates on the expedition appeared until late November, when the New York Times reported, this time at the top of the front page, that the governor's son was missing.[39] Governor Rockefeller packed his bags and left immediately for New Guinea. (He had announced a few days earlier that he and his wife would be divorcing.[40]) He knew his son had last been seen swimming for shore after capsizing off the southwestern coast of Dutch New Guinea. Michael had been visiting the Asmat region after completing his research on the Willigiman-Wallalua with Gardner. Michael planned to collect local art in the Asmat and, according to the newspaper, had a particular fascination with the region's carvings, totem poles, and human heads (although the latter turned out to be mere rumor).[41]

Over the coming weeks, American newspapers kept their readers abreast of the situation. When the governor and his daughter (Michael's twin sister, Mary) arrived in Honolulu the following day, they learned from local authorities that he had been "swimming against 'powerful currents' through shark-infested waters . . . buoyed by two empty gasoline cans." When his companion, a Dutch ethnologist who had stayed on the raft, was rescued safely in the Arafura Sea, everyone held out hope that Michael, too, would be soon found.[42] With each new article, however, his chances of surviving decreased. He had left the raft against the strong advice of his companion. Even if he had gained shore, some of the swamps were infested with crocodiles. Headhunters and cannibals were reported in the region. The dense mangrove thickets of the coast provided little to eat. After five days, the Netherlands government no longer expected to find him alive.[43] After a week, the now "weary-looking" Governor Rockefeller and his daughter returned to their families.[44]

Michael Rockefeller's disappearance continued to make sporadic headlines and ensured that a broad reading public accidently consumed at least a few anthropological details about the Willigiman-Wallalua. *LIFE* magazine featured an issue devoted to color photographs taken by members of the Peabody Museum of the "Stone Age people in a remote valley who live as they always have, with no intention of changing."[45] Until the anthropologists discovered them, the article intoned, the Willigiman-Wallalua people "survived in a forgotten eddy of history, a people still living as men did before history began." The Willigiman-Wallalua spoke Dani, like their neighbors, with whom they were "caught in a continuous cycle of warfare."[46]

The idea that warfare could be a normal way of life fascinated Gardner.[47] The Dani-speaking peoples, as a larger community, had no expectation that their wars would ever end. Rather than conceptualizing war as a "necessary evil" interrupting a normalized peace, they saw war as sacred, as indispensable. What did violence mean in a culture like this? Gardner wondered. He had organized the expedition to study the Dani's "ritual warfare" before local authorities keen to bring an end to the constant violence stopped the wars and forever altered their culture.[48] The Peabody expedition presented a new set of moral concerns for Gardner. He believed that anthropologists should not interfere in the communities they visited and studied.[49] Difficulties arose, of course, when the Dani engaged in behaviors of which the anthropologists disapproved. "Would we merely watch as a man beat his wife senseless for permitting herself to be 'raped' by another man? Would we attempt to discourage a raid on an enemy garden? Would we refuse to provide medical assistance?"[50]

Unlike the Mbuti, a strict gender hierarchy reigned among the Dani, and the male anthropologists on the expedition warmed far more to the local men than to the women. The formal battles and raids set the pace of male life. When the men went to fight, women spent the day harvesting sweet potatoes, looking after their pigs and children, and preparing for the men's return. "The Dani," mentioned Gardner, "are warriors because they wanted to be since boyhood, not because they are persuaded by political arguments or their own sentimental or patriotic feelings."[51] The men fought at the command of their leader. War constituted a major focus of everyone's interest and energy. The anthropologists attributed the women's sullenness, their boredom, and their lack of initiative to being raised in a social system characterized by vast inequality. The women did not fight but they took care of the men and they grieved when relatives failed to return. Without war, Gardner wrote, Dani culture would have no purpose.

Gardner's expedition had caught the Dani "trembling on the edge" of cultural change.[52] The anthropologists had finished their ethnographic study just as Dutch officials entered to "pacify" the region. So, too, had Turnbull. In many ways, the Mbuti and the Dani appeared as each other's antithesis—one pacific, the other bellicose; one based on cooperation and equality, the other on strict social hierarchy; one protected by the forest, the other warring in sparse highlands. Yet they also shared many traits, including a sexual division of labor. Comparisons like these across cultures divided by environment and tradition highlight one of the conundrums confronting cultural anthropologists of the era: cultural relativity. With so many cultures and so many ways of life, how could moral absolutes exist? Surely, experience proved that an act deemed immoral in one culture could be acceptable in a second.

Despite Kluckhohn's early optimism that through comparative cultural studies anthropologists might discover universal morals shared by all peoples, subsequent research seemed to demonstrate that very few actually existed. Some anthropologists worried that as a result cultural relativism would spread beyond anthropology.[53] Others fretted that anthropology tended to draw an unusual number of social nonconformists who took comfort from the idea that the standards of their own society were no more right than the standards of, for example, the Mbuti or the Dani.[54] The closest, perhaps, was an almost universal ban on marrying or having sexual relations with close relatives—an incest taboo.[55] Another was the ubiquity of the hunting lifestyle before the invention of agriculture.[56]

Given the social complexity of even the most "primitive" human cultures, it struck most postwar scientists that the full manifestation of culture in any human community still differed fundamentally from animal nature. In the grand scheme of the history of life on Earth, the origins of human culture had indeed taken place rather quickly and quite recently. Dobzhansky posited that "man . . . is so much unlike any other biological species that his evolution cannot be adequately understood in terms of only those causative factors which are operating in the biological world outside the human kind."[57] He emphasized the singularity of the evolutionary process in humans thanks to our capacity for culture—part and parcel of a strategy keeping eugenic theory at bay after the Second World War.[58] Dobzhansky's friend and colleague, the paleontologist George Gaylord Simpson, preferred the term "social evolution" and argued that it comprised a kind of inter-thinking, which he contrasted with organic evolution's dependence on interbreeding. "The most brilliant of geniuses is an intellectual eunuch if his knowledge is not disseminated as widely as possible," Simpson wrote.[59]

Simpson was passionate about his work and notoriously cranky with those who disagreed with him. Over the course of his career, though, Dobzhansky became one of the colleagues Simpson respected most. For fourteen years after the war, they both served on the faculty at Columbia University—although Simpson's primary position was as curator at the American Museum of Natural History, just two miles south. Then, a few years after Simpson accepted a position at Harvard in 1959, Dobzhansky moved to the Rockefeller Institution. Simpson reasoned it was immoral for any person, industry, or nation to keep the knowledge they generated to themselves—especially information about humanity's place in the universe, which "must guide us if we are to control the future evolution of mankind." When Dobzhansky described humanity as a "radically new kind of biological organization," he used a concept from Simpson's *Tempo and Mode of Evolution* (Figure 5).[60] Human origins lay in a quantum evolutionary transition, he posited—a "pronounced break in the biological continuity"—that ushered in a "third kind of history."[61] Cosmic history described the physical evolution of the universe, and then everything changed when the origins of life created biological evolution. The origins of humanity had changed things again. Our "superorganic culture" provided an enormously powerful means of adapting to the environment, "the most powerful method . . . ever developed by any species."[62] Simpson wrote, and Dobzhansky quoted, that "man has risen, not fallen. . . . Evolution has no purpose; man must supply this for himself."[63]

Unfortunately, from Margaret Mead's perspective, evolutionists like Dobzhansky and Simpson sought to distinguish cultural from biological phenomena—as if these categories were independent from one another.[64] Perhaps, Mead reasoned, evolutionists invoked this separation as a means of counteracting the popularity of "theories of racial difference."[65] Or perhaps the separation originated with experts in animal behavior beginning to pay attention to humans, inappropriately trespassing on the jurisdiction of scientists with expertise in the study of people. (We should read with gentle skepticism Mead's claim of a long-standing jurisdictional separation between scientists who study people and those who study animals. Both experimental psychologists and naturalists had been observing animals and speculating about people since the fields' professionalization at the turn of the century.[66]) Neither possibility made the divide between cultural and biological any more real from her perspective.

Many cultural anthropologists were wary of claims that human nature somehow underlay culture and dismissed the idea that the components of social behavior could be arranged like a layer cake, with biology on the bottom, anthropology in the middle, and psychology on top—all contributing to what it meant to be human but in a discrete, hierarchical order.[67] In this model, internal biological forces pushed from the bottom up and external environmental forces from the top down, catching individuals in the mid-

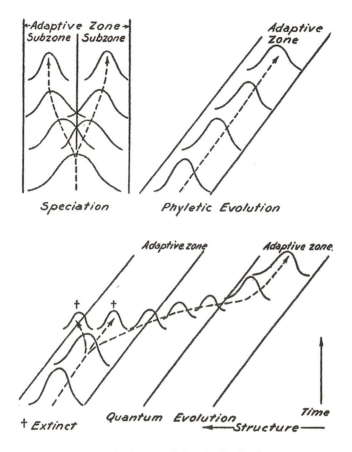

FIGURE 5. George Gaylord Simpson's three kinds of evolution: speciation (associated with specialization to different environmental factors), phyletic evolution (change within an adaptive zone), and quantum evolution (change between adaptive zones). Anthropologists in the 1960s typically associated the quantum origins of humanity with an ecological shift from the trees to the savannah. They posited a second quantum ecological shift with human adaptation to colder climates, leading to the origins of civilization. George Gaylord Simpson, *Tempo and Mode in Evolution* (New York: Columbia University Press, 1984), 198. Reprinted with kind permission of the Simpson estate and Columbia University Press.

dle. Washburn implicitly supported this arrangement, suggesting that "human nature is the foundation upon which cultures must build." He further noted that the fundamental features separating ape and human ways of life were bipedal locomotion, tool use, and our incredibly flexible mental capacity.[68] These physical features, he argued, had appeared sequentially in our evolutionary history: first we learned to walk on two legs and then how to hunt, after which our intelligence increased dramatically, and only then could our ancestors be considered truly human. Frustrated with the division of human evolution into a layer-cake model along these lines, Clifford Geertz later quipped, "Like the cabbage it so much resembles, the *Homo sapiens* brain, having arisen within the framework of human culture, would not be viable outside of it."[69]

In this context, anthropological explorations of cultures ostensibly untouched by Western civilization promised a scientific means of understanding human diversity, but using this research to triangulate answers to how early human cultures might have behaved sat awkwardly with liberal anthropologists' political convictions. Serendipitously, the factors that drove easier access to remote human cultures also increased biologists' access to remote free-living populations of primates. It seemed possible that the study of primates who lived in circumstances similar to those of early humans could also provide useful insights into the relationship between ecological pressures and social organization that would have governed their lives. This could be accomplished without the shadow of racism hanging over evolutionary comparisons of modern cultures with the origins of humanity. Baboons could be observed in Kenya, gorillas in the central African Virunga Mountains, and chimpanzees in Tanzania. Between 1955 and 1965, the number of total months researchers spent studying primates in the field skyrocketed exponentially, growing far faster than other areas of science.[70] These scientists devoted the lion's share of their attention to the *Cercopithecidae*, or Old World monkeys, which include baboons and macaques. Media coverage of this research kept pace, as scientific magazines like *National Geographic* and *Natural History* began devoting as many pages to primate behavior as to exotic human cultures, often describing both as potential keys that could help scientists unlock the secrets of human nature from different angles.

One of the primatological luminaries of his generation, Irven DeVore started his dissertation at the University of Chicago under Washburn and bounced around for a number of years before settling into a tenured position at Harvard in the late 1960s. DeVore hailed from Texas and his personality was about as big. In comparison to Washburn, who had been born and educated in the Northeast, maintaining a dignified composure throughout his career, a journalist once quipped that DeVore had "the soul of a vaudevillian."[71] He loved telling stories, adored teaching, and was never happier than when he had an audience, no matter how small. Washburn encouraged a generation of primatologists, many of whom trained with him at UC Berkeley and worked on an increasing number of different species.[72] In the late 1950s, baboons struck Washburn and DeVore as the most fruitful primate to serve as a foil for early humans. Baboons occupied the same kind of woodland savanna that anthropologists reasoned would have been an "ideal nursery for evolving hominids" since it combined the challenges of the open grassland with the protection of the forest.[73] They are highly social, ground-dwelling primates. Males and females differ in their size and in their behavior, traveling together in sizable troops. DeVore emphasized substantial differences, too. Humans were unique, he argued, not only for our ability to amble on two legs and communicate with language, but also in our sharing, cooperation, and play. He hypothesized that the basis for these differences might reside in the origins of group hunting and food-sharing practices in early human cultures that extended the range of a community's territory.[74]

Together with Richard Lee, a friend and colleague he came to know at Berkeley, DeVore helped to organize a conference on "the hunting way of life" in 1966. Hosted at the University of Chicago and sponsored by Sol Tax (who had also been crucial to Chicago's Darwin centenary celebration only a few years earlier), "Man the Hunter" lasted for four days, and some seventy-five scientists spoke to the recent efflorescence of research on hunter-gatherer and primate societies.[75] Hunting, for the organizers, had dominated humanity's way of life for hundreds of thousands of years—far longer than the recent invention of agriculture. Washburn participated in the conference and cowrote a paper that posited, "In a very real sense our intellect, interests, emotions, and basic social life—all are evolutionary products of the success of the hunting adaptation."[76] For Washburn, accepting the biological unity of mankind meant accepting that all human lineages could be traced back to a previously universal hunting lifestyle. "The biology, psychology, and customs that separate us from the apes," he wrote, "all these we owe to the hunters of time past."[77] Could anthropological studies of modern hunter-gatherers be used to reconstruct Paleolithic cultural characteristics? No clear answer emerged from the conference, except that to understand the evolution of humanity, anthropologists would need to study modern hunter-gatherer societies in aggregate. Any one culture could never provide a singular convincing model for evolutionary reconstructions of our shared past.[78]

Reviewers of the published volume lavished particular attention on Lee's paper, in which he analyzed the diet of the Ju/'hoansi, who live in the Kalahari Desert.[79] Anthropologists identified the Ju/'hoansi as one of the !Kung peoples, which is how Lee referred to them at the time. He argued that most of the calories consumed by hunter-gatherer societies like the !Kung came from women's efforts in gathering edible vegetable matter—as much as 75 percent. Successful hunts for animal protein were unpredictable and men often returned with nothing to show for their efforts. Meat, therefore, could not be relied upon as a regular component of daily meals. (Hunter-gatherer societies living in arctic conditions constituted a significant exception to Lee's generalization.) Equally astonishing, Lee noted that hunter-gatherers spent only twelve to nineteen hours per week obtaining food and taking care of other necessities for living. His work called into question existing assumptions about the harshness of living conditions for our early ancestors and the degree to which they had struggled to survive.[80] Lee also sought to draw attention to the importance of female contributions to the present (and past) survival of human groups, even those for whom hunting remained an important component of their daily life. Hunting would not have been viable until after our ancestors learned to share food.

Humans became human, the *Man the Hunter* volume asserted, because learning to hunt also entailed learning to cooperate, to manufacture tools, to communicate using spoken language, and to divide labor among the group—in short, to be highly social and familial creatures of culture. In summarizing the conference, Claude Lévi-Strauss applauded the work of the cultural anthropologists. He also reserved special appreciation for DeVore's research on baboons. Lévi-Strauss suggested that detailed knowledge of primate behavior revealed fundamental facts that "if they hold true for primates, they also hold true for the whole of mankind," including "early forms of incipient human cultures."[81]

DeVore believed he could identify the roots of human society in baboons as well. Baboon males are significantly larger than females and also sport enormous canines, which grow to around five centimeters in length. In the presence of a threat to the group, like a roving cheetah, the adult males will mobilize against the threat as a unit, while the juvenile, adolescent, and female members of the troop move to safety in nearby trees if possible. The adult males will then rush the cheetah at full gallop, screaming and baring their teeth. Usually the cheetah lopes away. Ecologically, baboons and their smaller relatives, macaques, are one of the only primate groups to have successfully colonized the open savannah—the other was humans. DeVore deduced that the troop worked according to a dominance hierarchy. All members of the troop, from infants to the alpha male, knew their social position. Older males also formed coalitions, DeVore observed, keeping younger males in less dominant positions by working together. The nonhuman great apes and monkeys prefer the relative safety of the trees, where they can easily elude ground-dwelling predators and even those who venture into the treetops themselves. So long as these species remained in the forest, DeVore reasoned, their best defense remained individual flight. In the grasslands, however, where trees are sparse, the cooperative attacks of adult males became the troop's only means of protection.[82] DeVore surmised that the social behavior of early humans, similarly conditioned by their recent move to the open savannah, would have resembled that of baboons.

Whereas baboons functioned as an ecological model for early human ancestors, humans are genetically more closely related to chimpanzees and the other great apes. Humans and chimpanzees are like evolutionary siblings, both descended from a common ancestral parent and both changed over millions of years according to their respective environmental and social circumstances.[83] Chimpanzees' reclusive forest habitat made studying their behavior in the wild a risky choice for anyone seeking an academic career.[84] Then chimpanzees too found their champion.

Depicted as slim, blond, and unguarded, Jane Goodall captured the hearts of an adoring television viewership, earning a far brighter public spotlight than did DeVore (Figure 6).[85] Encouraged by the gregarious Louis S. B. Leakey, a paleoanthropologist seeking to prove that the earliest hominids originated in Africa rather than Asia, Goodall spent decades in Tanzania observing chimpanzees in their natural habitat. In the 1960s and 1970s, the National Geographic Society supported her research and in exchange produced a series of articles and documentaries expounding a universal human nature to which chimpanzees were the key. By stripping away the contaminating influences of modern civilization (culturally, temporally, biologically), it was the remaining bare human nature that most fascinated the public. With chimpanzees, that nature looked rather congenial. Based on Goodall's early years of research, one anthropologist dubbed them "the flower children of the apes, irresponsible and given to bursts of noisy, but apparently happy activity."[86]

Goodall had met Leakey in 1957 at the age of twenty-three. She immediately expressed her interest in studying wild chimpanzees, but it took Leakey two years to raise the funds for such a risky venture. In the meantime, she worked as his secretary, first traveling with him and his family to Olduvai Gorge, and then back to England to take informal anatomy classes and study primate behavior at the London Zoo.[87] Goodall's first six months studying chimpanzees were funded by a $3000 check from the Wilkie Foundation. Under the direction of the tool manufacturer Leighton Wilkie, the foundation sponsored a wide array of anthropological projects.[88] Goodall's observations of chimpanzees stripping twigs of their leaves to fish for termites—not just tool use, but manufacture—combined with Leakey's charismatic fund-raising, soon secured money for longer-term research in the area. Nevertheless, Leakey worried about Goodall's capacity to obtain independent funds once he passed away. Writing to her in Gombe a few years later, he noted that she would need a PhD, probably in ethology.[89] This requirement presented a small conundrum, as Goodall had never earned an undergraduate degree. The

WEDNESDAY, DECEMBER 22

Miss Goodall and the Wild Chimpanzees

FIGURE 6. In the mid-1960s, the National Geographic Society began producing educational films for television. Their first two projects centered on Jane Goodall's research with the chimpanzees of Gombe and Louis Leakey's paleoanthropological research in Kenya. Baron Hugo van Lawick, who would become Goodall's first husband, caught on film many of the observations she had made about chimpanzees during her first five years on site, including the fact that they ate meat and their manufacture of termite-fishing sticks. This advertisement for the documentary accompanied an article on Goodall's research in the magazine: *National Geographic* 128, no. 6 (1965): 831. Photograph by Hugo Van Lawick. © National Geographic Creative. Reprinted with permission of the Jane Goodall Institute.

ethologist Robert Hinde later recalled that one of his colleagues, Bill Thorpe, found a loophole in the University of Cambridge regulations—"If you have done something distinguished, like being the master of a ship for instance, then you would go straight into a Ph.D. programme."[90] Goodall qualified.

Goodall took an entirely different approach to primate behavior than had DeVore. The more she observed the chimpanzees, the more she sought to understand the complex emotional differences that gave individual chimpanzees personalities. Goodall noted that when a female chimpanzee gave birth to a stillborn infant, she would sometimes carry the corpse with her for a few days before leaving it behind. Evolutionary biologists and animal behavior experts, however, lacked theories to explain individual behavior, so she began to draw instead on psychological theories.[91] She also gave each of the chimpanzees a name. This made it easy for the public to relate to her research but also led to accusations from professional colleagues that she anthropomorphized her research subjects.[92] When she enrolled at Cambridge, Goodall worked with Hinde, who taught her to use behavioral check sheets to standardize her observations and to write scientifically about the animals she now knew well. Goodall completed her dissertation, "Behaviour of the Free-Ranging Chimpanzee," in 1965 and eventually secured funding of her own outside the National Geographic Society.

DeVore remembers meeting Leakey for the first time in 1959, just before Goodall left for Tanzania. His first impression of Leakey was of a man sloppily dressed in coveralls, his shock of white hair mussed over a creased face.[93] Around the same time, a journalist and admirer described Leakey as "a slope-shouldered, powerful, restless man with a bold sentient face," and he pictured him sitting hunched on his chair and comfortable in his workman clothes.[94] Both men found Leakey's confidence infectious. With perseverance and stealth, Leakey said, you could sneak up on a gazelle and grab it barehanded. He flinted his own knife blades. He showed how even fleet prey could be brought down with a long strip of leather with rocks tied at either end. He often noted how little technology one needed to hunt.[95] In short, Leakey was a consummate performer. He also believed firmly that the fossils he and his wife Mary were in the process of finding in Olduvai were humanity's oldest known ancestors.

That same year, in fact, Mary discovered a spectacular skull, and Louis named it *Zinjanthropus boisei*, which he characterized as a new kind of ancient human not yet known to science. Leakey and Zinj (as he affectionately referred to the skull) made the news because of Zinj's location in Africa, its relative completeness, the backing of the National Geographic Society, and the fossil's spectacular antiquity.[96] Applying a new technique of potassium-argon dating to the rocks in which Zinj had been found, geologists estimated the age of the fossil at 1.75 million years.[97] Leakey welcomed the media attention that accompanied controversy, and this fossil proved a wonderful tool for advancing several of his pet theories, including the idea that Zinj had lived cheek by jowl with species of *Homo*.[98] This claim proved as controversial as his son's later insistence that Zinj was the first hominid to manufacture stone tools.

Leakey's success in finding ancient human fossils in the East African Rift Valley inspired a new generation of paleoanthropologists to establish their own field sites in the region, whether or not they agreed with the substance of his claims. In the next few years, some of his colleagues contended that Zinj was really a kind of *Australopithecus*, others that it should properly be considered part of *Homo habilis*.[99] By the time Leakey passed away in 1972, experts suggested that the species he had called *Zinjanthropus boisei* lived as long ago as four or five million years.[100] These questions of age, identity, and relationship are fundamental to the practice of paleoanthropology and took time and hundreds more fossils to resolve.

Debates over the essential characters of human nature thus emerged in relation to the study of myriad other qualities. Combining evidence gleaned from behavioral studies of primate societies, the cultural traditions of hunter-gatherers, and detailed analyses of fossil bones, primatologists and anthropologists created a series of near-men and true-men. Although they divided living populations of primates and humans easily, understanding the details of when, where, and how the characteristics that define humanity had arisen required speculation. Even answers to the question How old is Man? varied according to how scientists used the word "man." Some anthropologists, like Leakey, used it to refer to any species identified as living more recently than the final split between the evolutionary lineages that led respectively to humans and chimpanzees. This referent included species of *Australopithecus* and *Zinjanthropus* as well as any older yet-to-be-found discoveries. Other anthropologists reserved the term "man" exclusively for species of *Homo*, including *erectus*, *habilis*, and *sapiens*.[101] Part of the difficulty arose when scientists ascribed attributes like behavior and soft tissues to fossil specimens that had never been etched in bone. When had the human lineage developed the prominent calf muscles that reflect our bipedal habit, or enlarged female breasts, or the comparative hairlessness of our bodies? Which of these species could cooperate on a hunt, conceptualize the future, or make fire? Without direct evidence, theories abounded.

The transition from mere animal to truly human depended on a crucial transition in the prehistory of humanity: the dawn of culture and the origins of the Stone Age. Although our capacity for social cohesion quickly outpaced that of our primate cousins, somewhere in this long history we had also learned to kill each other. Leakey, with his usual bravado, announced in 1961 that he had discovered evidence of the world's first murder 600,000 years in our past.[102] His colleagues treated this claim, like so many others, as a wild stab in the dark and an exceptional episode (if it had even occurred) in humanity's path to increased sociality and cooperation.

No matter how much they battled over specifics, postwar cultural and physical anthropologists together with zoologists constituted a robust interdisciplinary community fascinated by the challenge of unraveling humanity's evolutionary past and present. The progressive vision of human nature they constructed, they firmly believed, required working together and should be communicated as widely as possible—through paperback books and television documentaries aimed at adult audiences, as well as educational materials aimed at the nation's younger citizens.

3

Building Citizens

WHEN THE PSYCHOLOGIST Jerome Bruner started talking to someone, even an elevator man, he gave his whole attention, treating that person for the duration of the conversation like the most important thing in the world. This is how Patsy Asch fondly described his singular warmth decades after they had worked together on Man: A Course of Study, MACOS.[1] Bruner had been born with both eyes blinded by cataracts. An experimental operation to remove the lenses from his eyes left him able to see, albeit with the aid of thick glasses. To meet Bruner, then, was to notice in passing his magnified, swimming eyes and then be disarmed by his congeniality and bright intellect.[2] MACOS began in the halcyon years of American grade-school science reform in the 1960s, and Bruner as its guiding light sought to reevaluate "education in the light of our newly gained knowledge of man as a species."[3] Bruner became involved in education reform in the 1960s for the same general reasons Eiseley and Dobzhansky sought a public audience for their books—he sought to transform society from within. The process of turning schoolboys into men with open minds and technical skills in the early Cold War posed an intellectual challenge worthy of expert guidance from both educators and scientists.[4]

Under Bruner's direction, three questions formed the core of MACOS's pedagogical aims: "What is human about human beings? How did they get that way? How can they be made more so?"[5] Filmic depictions virtually transported MACOS students to the far reaches of the world, where they learned to formulate their own answers to these questions. In the eyes of MACOS designers, if crafted carefully these films offered students the possibility of seemingly unmediated access to another culture—so the filmmakers never appeared on camera, and they never used voice-overs from a bodiless narrator in their ethnographic studies.[6] They hoped this would allow students to formulate their own answers to Bruner's questions.

MACOS also wrestled with many contemporary social issues, especially civil rights. Reflecting the concerns of the academics and teachers who helped to develop the program, violence cropped up occasionally but not as a central theme of the course. Instead, the "gut assumptions" guiding the early development of the program—even before Bruner joined—included the goal that through intelligent study of their own lives, students would learn the tools necessary for the continuance of the human species.[7] MACOS designers hoped students would come to appreciate that all people share the same human nature and to conquer their own racial prejudices. Grade school education in the sciences, if done right, promised a means of improving the technical competence of the future American public in scientific subjects and of inspiring bright children to consider careers in science and engineering.[8]

Exploring the difficulties and joy of the research behind MACOS provides insight into why, for the people involved, the curriculum project was so profoundly personal. Many of the researchers and teachers involved in its creation embraced the liberal aspirations of the decade. Anthropology promised a means by which students could come to appreciate the cultural patterns of communities vastly different from their own while recognizing the fundamental equality of all peoples. If education held the key to securing the nation's place in world, they imagined, it would also play a significant role in teaching students how to overcome any racial apprehensions they may have acquired from parents or peers. Making visible the careful labor that went into MACOS's creation at every level reveals the political and personal stakes inherent to the administrative vision, the ethnographic footage, and the finalized classroom booklets and films, where each sought to transform the way students thought about themselves and the society in which they lived.

Curriculum development for MACOS had begun under the guidance of the cultural anthropologist Douglas Oliver, a Harvard professor known for his work in the Solomon Islands.[9] Anthropologists, he believed, concerned themselves "with the history of 'historyless' peoples, with the economics of communities without price-fixing market institutions, with government and politics in 'stateless' societies, with social relations in places where kinship usually outweighs occupation, with the psychology of non-Westerners, with the anatomy and physiology of the whole range of mankind and its primate relatives."[10] In short, anthropologists utilized a panoply of social scientific skills in the service of understanding the cultures of people whose traditional customs were as yet unaltered by exposure to Western civilization. What unified this diverse enterprise into the discipline of anthropology, for Oliver, was fieldwork. Only in the field could anthropologists learn first-hand a wide array of hard facts about another culture (Figure 7). So when Oliver dreamed of using MACOS to turn "all students into little anthropologists," film became a crucial medium that would allow students to mimic fieldwork without leaving the classroom.[11] The funders of the endeavor, including the Ford Foundation

FIGURE 7. Douglas Oliver argued that research in the field provided the only way to truly understand another culture. This quirky cartoon of a bespectacled anthropologist at work illustrates a variety of common contemporary beliefs about life in the field. The cartoon appears in his *Invitation to Anthropology* (Garden City, NY: Natural History Press, 1964), 5. The first in a series of humorous cartoons illustrating the book, this is the only figure to feature an anthropologist. Illustration prepared by the Graphic Arts Division of the American Museum of Natural History.

and the National Science Foundation, imagined that high school was too late to start teaching students to reason scientifically. To reach the greatest number of promising children, it would be better to instill a scientific mindset starting in elementary school, and they hoped MACOS would do exactly that.

Oliver's vision for a new social science curriculum spanned several years, beginning in first grade with the "simplest and oldest" human cultures, with each subsequent unit tackling more "complex cultural forms" captured on film and accompanied by printed materials.[12] Oliver proposed that the first unit would explore the Netsilik Inuit, the second would turn to Aboriginal Australians and the Ju/'hoansi of the Kalahari Desert, and the third would take a more conceptual look at human evolution and the origins of culture. In their fourth year, children would study the origins of maize agriculture in Mexico, in their fifth, the cities and urbanism of ancient Mesopotamia, and in their final year in the program would engage with a more traditional history of ancient Greece and the Bronze Age.[13] MACOS students' travels through this range of human cultures traced a trajectory reminiscent of Eiseley's evolving man. When Oliver proposed this plan for MACOS, however, he envisioned something different. Students, he reasoned, would come to appreciate that civilization had not been born in one place and that the processes of cultural change varied with space and time.[14] This is what could make anthropology so useful in their lives—by providing "deep and fascinating insights of what we are now and where we have been."[15]

From the perspective of MACOS designers, methodological revisions to social science curricula could not be accomplished without first producing new materials for the classroom. Oliver commissioned the recent PhD Asen Balikci to gather footage of the Netsilik for the first unit. For the second unit, he began negotiations with John Marshall, then a graduate student in anthropology at Harvard, and his mother, Lorna Marshall, to use their diaries and John's films of life with the Ju/'hoansi. The Marshall family had first traveled to Namibia in 1950, meeting the Ju/'hoansi in 1951. At the age of eighteen, John had begun filming for his father using a 16 mm camera that had to be wound by hand and could record only 30 seconds at a time. Over the next several years, both cameras and Marshall's filming technique improved, but in 1958 he was denied reentry to the country. Unable to return, in the early 1960s he enrolled in graduate study at Harvard and served as director of the !Kung unit for MACOS from 1960 to 1963. Oliver contacted Karl Heider, too, another graduate student at Harvard, about filming the agricultural practices of the Dugum Dani people in the central highlands of New Guinea. Heider had participated in the same expedition as Michael Rockefeller but had stayed on when the rest of the team left the area.[16] Heider spent twenty-six months among the Dani from 1961 to 1963, a body of work that formed the basis of his dissertation and a series of short films. These anthropologists and filmmakers envisioned their films as useful both for professional anthropological research and for teaching at all levels of the curriculum.[17]

Balikci joined the project eagerly, taking leave from the Université de Montréal, whose faculty he had recently joined. Although he had already conducted ethnographic fieldwork with the Netsilik, Balikci had never worked with film before and found himself excited at the prospect.[18] He remembered being told to shoot footage of the community's culture in a style simple enough for seven- or eight-year-old children to understand—a daunting task. Additionally Oliver had added that he should record traditional practices only, even though that meant he had to ask the Netsilik to reconstruct practices no longer in widespread use. No rifles, no tea mugs, no snowmobiles. All told, Balikci spent thirteen months in the Arctic, spread over three years, and shot half a million feet of film.[19] Oliver and the planning committee asked Balikci to concentrate on four areas of Netsilik culture: "the severity of Arctic life," emphasizing the harsh ecological conditions to be overcome on a daily basis; "interactions between people," especially intra- and intergenerational customs; "details of technical processes," including close-ups of toolmaking; and "instances in which Eskimos handle a situation differently from the way we would or show attitudes different from ours."[20] This intellectual framework fit with Balikci's belief that the key to understanding Netsilik traditions would be found in cultural ecology.[21] A devotee of the anthropologist Julian Steward, Balikci treated the unique aspects of Netsilik culture as adaptations to the arctic environment in which they lived.[22] When filming, he thus paid close attention to Netsilik technologies and behaviors associated with hunting, fishing, and trapping but largely ignored their religious traditions.

In interviews about his experiences, Balikci later described the field conditions in Pelly Bay as nothing short of brutal. He had "barely survived," he said.[23] In his estimation, most men would have found the conditions equally impossible. Father Guy Mary-Rouselière, a Catholic missionary who had been living in the area for almost fifteen years, brokered Balikci's initial access to the Netsilik community who lived in Pelly Bay. Rouselière also served as the local trading post, selling ammunition, tea, axes, knives, tobacco, and (on the holidays) sugar. To accommodate these goods, he had built a small stone hut—when Balikci arrived, it was the only permanent building for hundreds of miles. Rouselière also possessed a two-way radio, his sole means of communicating with his suppliers. Balikci would radio that he was coming, and Rouselière would made sure someone met him with a dogsled to transport him from the landing area to the safety of the community that had grown up around the hut.

Education Services, Inc. (ESI), MACOS's administrative home in Cambridge, Massachusetts, received a $50,000 grant from the Ford Foundation that paid for the first summer of arctic filming and hired the experienced cameraman Doug Wilkinson to work with Balikci. ESI agreed that if Wilkinson were already in the field, then he could take still photographs on his own time that would not count as part of the material he needed to relinquish at the end of the summer. So Wilkinson traveled north, having been paid $20,000 for the use of his plane and expertise. Balikci and Wilkinson did not get along. Kevin Smith (the executive director of the films and located back in Cambridge) attributed some of the antipathy to their respective dispositions: Balikci was "a little fey . . . he's hard to pin" and Wilkinson was "one of those guys with a long tundra stare. He was always looking at mountains, or way off someplace."[24] Two sets of stories started to come back from the Arctic, Smith recalled—one set from Balikci, claiming that Wilkinson was a crook who refused to work and was probably cheating ESI out of unexposed film; the other from Wilkinson, accusing Balikci of being a terrible ethnographer who had no idea what he was doing. Having no other recourse, Smith waited it out. Wilkinson, he eventually discovered, had indeed been filming and photographing on his own using ESI's unexposed film and sold this footage to a Canadian television station. Smith estimated that Wilkinson took ESI "for about $30,000." Despite Balikci's frustration and initial skepticism of Wilkinson's skills, he said he learned an important lesson from him—to be "hard like nails." Balikci soon appreciated the importance of discipline and the necessity of presenting himself as the person in charge, otherwise his filmic subjects would not show up on time and a more or less regular production schedule would prove impossible to maintain.[25]

In March of 1964, having arduously turned his footage from the previous summer into an ethnographic film for students, Balikci wrote to Margaret Mead with pride, informing her that he had completed the first of several planned color films—*Fishing at the Stone Weir*. The cameras they used could not record sound synchronously, so sound would still need to be added later. He planned to return in May to capture the spring thaw and hoped to obtain feedback from Mead before he left. Balikci wanted to incorporate any suggestions she might have in his next season of filming.[26] Mead's response has not been saved in the archive, but the ESI board loved *Fishing at the Stone Weir*, and the program took off.[27] Balikci planned to shoot and edit seven more films in the next two years, for a total of ten hours on the Netsilik. These films would eventually take their place as part of the MACOS curriculum but in the meantime would also be available for rent or purchase separately.

Just before Balikci traveled north, Oliver left the project for personal reasons and Bruner took his place as director. Bruner had recently published the idiosyncratic and influential *Process of Education* and was widely hailed as a new strong voice in educational theory (Figure 8).[28] He saw his involvement with MACOS as an opportunity to put his ideas into practice. Some advisors worried that the content of the program was becoming too mature for first-grade children, and Bruner decided to pitch the curriculum instead at a fifth-grade level.[29] He even took a semester of leave from Harvard to spend three months personally testing the MACOS materials with fifth graders. He later

FIGURE 8. Jerome Bruner in dynamic conversation. Illustration by Darrel Millsap, for Elizabeth Hall, "Bad Education: A Conversation with Jerome Bruner," *Psychology Today*, December 1970, 50–57.

recalled how exhausted he was each day at 1:15 pm, when he returned to his office. This experience not only endeared him to the other teachers, it also showed him how important intellectual support for teachers unfamiliar with anthropology would be for the program to succeed.

Given the passion of Bruner's vision, it is unsurprising that people with conflicting ideas about the program remember the renegotiations of the MACOS curriculum with more than a little frustration.[30] Although Oliver had originally intended to incorporate paleoanthropological and archeological findings, these ambitions largely vanished by the end of 1964 out of fear they might land MACOS in hot water with parents.[31] Robert Adams, who had been deeply involved in the fifth unit on the origins of cities, lamented the resulting lack of a comparative historical perspective in the curriculum, although he ultimately agreed with the decision to drop the unit.[32] Bruner also reoriented the pedagogical program to exemplify his belief that "knowing is a process, not a product."[33] The ethnographic materials under development thus became a means to an end rather than an end in and of themselves.

When Balikci arrived back in the Arviligjuar area, he found himself in charge of an entirely new crew. Father Rouselière had expected to act as his assistant for the rest of the film series. Right at the end of the first season, however—after Wilkinson had deserted and just before Balikci returned south—Rouselière had fallen into a crevasse and his heavy sledge followed him down, breaking his right shoulder.[34] He survived but lost the function of his right arm for life. During his convalescence, he began working on a master's degree at the Université de Montréal under Balikci's direction.[35] Rouselière returned to Pelly Bay after a year's recuperation and took still photographs. ESI hired two new cameramen the second summer, hoping that at least one would work out. The labor was grueling. Filming "naturally," so that the Netsilik actors did not perform for the camera, required holding the twenty-pound Arriflex 16 camera and filming without a tripod for extended periods of time. Traveling between filming locations, the equipment had to be cleaned, stowed, loaded onto a dogsled, and then transported. Dogsledding itself can require running alongside the sled and helping the dogs pull the weight of the sled free of ice and snow if it becomes mired. After a couple of weeks, only one cameraman remained—Ken Post. Post and Balikci filmed together, through the winter, for nine months.

Back in Cambridge, Marshall developed his own filmic ethnography of the Ju/'hoansi for use as the "Bushman unit" in MACOS. Marshall's father, Laurence, had led a self-financed, Peabody Museum–sponsored expedition to the Kalahari with his whole family. Gardner, equally fascinated by film as a means of providing field experience to those who could not travel themselves, had been in contact with Marshall's father to convince him of the value of footage for "experience learning."[36] Marshall and Gardner subsequently worked together on *The Hunters* from 1954 to 1957, a venture that led Gardner to found the Film Study Center at Harvard. Marshall would later call *The Hunters* "a romantic film by an American kid" that "revealed more about me than about the Ju/'hoansi."[37] Nevertheless, *The Hunters* had caught Oliver's attention. Marshall had 500,000 feet of 16 mm film—surely something useful could be created for young students out of that!

After accompanying Laurence and John Marshall (who had just completed his AB at Harvard) on their 1957–58 expedition, Gardner wrote to Margaret Mead asking if she knew of someone who would be willing to move to Cambridge to work on the Ju/'hoansi footage with them. Mead recommended Timothy Asch. Asch, together with his wife Patsy, moved to Cambridge in 1959, where Tim worked with Marshall editing his footage and Patsy found a position as a kindergarten teacher at a local school. Asch had been a photojournalist during the Korean War and then studied film at Columbia. In a letter to Mead in 1960, he wrote of his enduring interest in "spending a lifetime working with photography and applying such knowledge as I have of film (both stills and movies) to anthropological research."[38] He wrote, too, about how happy he was in his new position creating a series of ethnographic films that could be "as sensitive and as beautiful as they are anthropological." Five years later, he had fully decided to devote his career to exploring "the techniques and use of films in anthropology and in teaching."[39] Having just seen Robert Gardner's footage of tribal warfare among the Dani in New Guinea, Asch believed films of "primitive people" could preserve a record of their customs "before they disappear," although he added quickly that such films needed to be made according to the best modern techniques.[40] Asch bubbled excitedly about using film to explore and record the social structure of ethnographic subjects, not just tell stories in the fashion of "creative" filmmakers. Above all, he dreamt of a comprehensive film library that illustrated the variety of indigenous cultures. For that dream to become a reality, however, a great many anthropology graduate students would need to be trained in ethnographic filmmaking techniques.

Both Oliver and Bruner planned to pair John's films with excerpts from his sister's diaries written during the same family expeditions to the Kalahari. Elizabeth Marshall Thomas had already published a popular book about life with the Ju/'hoansi based on her diaries. Like Turnbull's descriptions of the Mbuti, Thomas's *The Harmless People* drew attention to the community's soft spoken ways and lack of violence.[41] Early in her book, Thomas described the "Bushmen," as they were then known, as a "naked, hungry people" living in sparse desert conditions. She explained, too, that the Ju/'hoansi are more like Asian peoples than like the Bantu communities who bordered their territory, as evidenced by their yellow skin, slight build, and "Mongolian eyefolds." The natural shyness of the Ju/'hoansi, she wrote, led them to run away if they discovered strangers approaching. She described how she had once walked into the middle of an empty village and was initially unable to see their inconspicuous huts made of grass, called scherms, "until I noticed a small skin bag dangling in a shadow, which was a doorway. Then I saw the frame of the scherm around it, then the scherms as well."[42] The Ju/'hoansi, in her estimation, were "one of the most primitive peoples living on earth."[43] In classrooms, they would provide a fascinating comparison with the Netsilik: both were hunters and gatherers but adapted to starkly different environments.

These three men, Asch, Gardner, and Marshall, met because of the Ju/'hoansi footage Marshall had shot in the Nyae Nyae region of Namibia. Together they forged new institutional homes for visual ethnography in American anthropology. Whereas Gardner remained at Harvard throughout his career, Asch and Marshall floated from position to position. They cofounded Documentary Educational Resources in 1968, which served as the filmic archive Asch had dreamt of in his letter to Mead, and they helped to establish the Human Studies Film Archives at the Smithsonian Institution in 1975. Their families were, by then, strongly tied together. In 1961, when Elizabeth Marshall Thomas spent a year in northeastern Uganda among the pastoral Dodoth, she hired Asch to film and take photographs. The material they gathered was published as *Dodoth Morning*, Asch's first film, and his photographs illustrated her book documenting the year, called *Warrior Herdsman*.[44] Both Tim and Patsy Asch worked together on MACOS, devising ways to adapt accounts from Elizabeth's diaries and John's films for the classroom.

Further ethnographic data for the unit came from Richard Lee and Irven DeVore, who ventured to Botswana to study a different community of !Kung-speaking people for the first time in 1963 as part of the newly formed Harvard Kalahari Research Group. Like Balikci, Lee had been influenced as an undergraduate at Toronto by Julian Steward's cultural ecology, although Lee then began his graduate studies with Washburn at Berkeley, picking up a more evolutionary edge to his ethnographic analyses. Recall that his research among the Ju/'hoansi drew attention to the egalitarian contributions of both men and women to a sustainable diet and their surprisingly copious leisure time. Following the anthropological customs of the time, in the 1970s Lee began to refer to the people he studied as San, which he deemed more respectful, and then in the later 1980s adopted the name they used to refer to each other, Ju/'hoansi.[45] Unlike Lee and the rest of the anthropologists hired by MACOS, however, DeVore's main focus of research centered on baboons.

MACOS tackled the "Bushman unit" in fits and starts with varying personnel. The unit seemed promising because of the weight of information available and for the remarkable contrast between the Ju/'hoansi and Netsilik cultures. In both cases, careful ethnographies allowed MACOS designers to build educational materials around a small set of likeable and realistic family members. They reckoned this would allow American students to connect with both Ju/'hoansi and Netsilik children and parents, seeing the ways in which they, too, were motivated by love, envy, fear, and the desire for peaceful solutions to social problems.[46] Yet early on, the developers of the !Kung unit became concerned that in contrast to the Netsilik, Ju/'hoansi families used a simpler and less elegant array of technology, which might limit the kinds of illustrative lessons that could be created for the classroom.

Upon Balikci's return to Cambridge, he discovered that Bruner had asked Asch to recut his hour-long *Fishing at the Stone Weir* into four-minute loops and appeared uninterested in using the longer film in the classroom.[47] When the teachers had tested the materials in the summer school, students had difficulty observing the Netsilik materials with any objective distance. In particular, they found it hard to see the ecological reasons behind the choices people made on film. Why build igloos? Because like all peoples, the Netsilik created shelters in which to live, but in the Arctic they did so with the material at hand, snow, and designed them to trap warmth. By recutting the Netsilik footage into shorter segments, Bruner sought to create more digestible encapsulations of cultural practices or narrative moments that students could discuss and unpack.

Under Bruner, the MACOS team continued to emphasize the widespread use of ethnographic film as a mechanism to replicate the experience of being in the field. They thus refused to overlay directions from an invisible narrator directly telling the students how to interpret what they were witnessing, and they even chose not to include English subtitles translating Natsilingmiutut, the Netsilik language heard in Balikci's films.[48] Skillful editing by Asch and the filmmaker Quentin Brown (who had directed *Miss Goodall and the Wild Chimpanzees* for National Geographic) transformed Balikci's films of the Netsilik into stories of various lengths that the children could figure out for themselves.[49] Ideally, teachers would be able to show each film twice. Initially, the students would see the film uninterrupted and this would be followed by a group discussion. A second viewing would allow students to see things they had missed the first time around. Teachers could stop the projection, noting a close-up or calling attention to the importance of a particular sequence. The MACOS films were designed to startle the students with their realism by avoiding the "hard sell and soft soap" of narration and music that might cushion students from the reality of Ju/'hoansi and Netsilik lives.[50] By allowing teachers to focus on the craft of filmmaking, Asch also hoped that students would adopt a more critical attitude toward other media in their lives that sought to invisibly manipulate their perceptions. Even so, Bruner felt students would need further guidance viewing these materials critically. From the beginning, course-builders had intended the MACOS films to be accompanied by activities, booklets, and a program to train teachers how to use the films in the classroom.[51]

First Oliver and then Bruner began to explore another option, too. In order to introduce students to the process of thinking critically about the ecological pressures on a social community, they thought it would help to include films on the behavior of free-ranging baboons. Doing so might allow students to reason through subsequent human scenarios with greater objectivity. DeVore enjoyed his experiences filming baboon behavior for MACOS. Conditions in Kenya were congenial, and he started the project without a firm sense of where it was headed. Perhaps his professional identity was also less imbricated with the baboon films than was Balikci's with the Netsilik—a testament to their respective personalities and filmic subjects. DeVore had been watching baboon behavior for fifteen months when he started filming, so as his first order of business he sat down and created a list of all the behaviors he wanted to capture on film. As he and the film crew slowly accumulated footage of baboons' behavior in their natural habitat, he checked off the list. When he reached the end, he was done.[52]

DeVore also relished his intended audience of first graders. As he filmed the baboons' natural behavior, he also kept his eye out for amusing scenes that would entertain the children in addition to educating them. For example, when wrestling with how to depict the concepts of food choice and the automatic deference inherent to social relations in a strict hierarchy, he rigged a hatch-cover from the hood of a Land Rover and placed food under it. It was a classic choice test: first he arranged cauliflower, squash, and onions in a neat row under the cover. He'd step back, the camera would roll, and the male baboons would rush up. The most dominant male would take the tastiest food, and the group would work its way down to the least palatable foods. They never touched raw meat. Yet by the end of one afternoon, a male who was sufficiently low in the hierarchy had become frustrated. The other males, in DeVore's recounting, "had absolutely gorged themselves; I mean they had eaten whole cauliflowers, ears of corn, two dozen eggs, and they were sitting there literally burping." The young male ran up, and ate what remained—a raw steak and onion dinner—violating every prediction DeVore had made about baboon eating habits. He added, "It's one of those exceptions that proves the rule. He got nothing except an hour of frustration. He was ready to eat anything." Amid his choice experiments, DeVore also snuck back when the baboons were not looking and coiled a dead snake under the hood to see how the discoverer might react. "It was spectacular," DeVore related. "I mean, this animal turns over the hatch-cover expecting peanuts, and there's a snake! He jumps four feet in the air, straight up, just like a cartoon." Yet when he returned to Cambridge to work with Asch and Brown editing his footage of baboons, they left all of the staged behavior on the cutting-room floor.[53]

Although Balikci had been disappointed in Asch and Brown's adaptation of his films for use in MACOS classrooms, he persisted, returning to the Arctic for a third summer. This time, ESI hired an additional cameraman who concentrated on more intimate footage of people inside of igloos—the commercial filmmaker Bob Young. Young had worked making television specials for National Geographic, and both Balikci and Post noticed his distinctive style as soon as he arrived. Post especially appreciated Young's ability to create delicate, involved scenes with his camera.[54] Years later, Balikci described Young as more of an artist than a cameraman, a painter who moved with facility and elegance.[55] Balikci and Young collaborated, too, on a one-hour film called *The Eskimo: Fight for Life*, which concerned the current difficulties faced by Netsilik culture in an increasingly Westernized world.[56]

Tragedy once again struck, marking the end of Balikci's stay in Pelly Bay. Father Rouselière received a message from a medical doctor who wanted to take x-ray photographs of the community. Balikci was happy to help, offering to let the man set up his equipment in his igloo. Then, a few weeks later, everyone in the community, including Rouselière, fell sick. Balikci, however, was fine. Fourteen people died, out of a total community numbering only one hundred. (Rouselière was among those who survived.) Although no one blamed him officially, Balikci felt the community had become suspicious of him and his research.[57] Balikci was proud of the Netsilik films, but they came at a high personal price. He would not return again for twenty years.

Almost all of the collected materials—with a few exceptions—focused on the kindness and cooperation cementing the Netsilik people into a common culture. After much discussion, MACOS designers chose to include topics they deemed "sensational," like the hunting of animals (in which students discussed "man as a predator"), infanticide, wife-stealing, and polygamy, they but remained conflicted over how to broach these topics within an elementary-school setting.[58] One member of the planning committee worried that by depicting the Netsilik as "cruel predators" or "savages" (as these behaviors might imply), the program ran the risk of "running athwart of all our non-ecological goals, such as showing the children that people in other cultures are human and worthy of respect."[59] In the end, several of these topics were incorporated into the curriculum as legends told by the Netsilik.[60] Many segments of the Netsilik films highlighted aspects of their hunting practices, however, since these sequences aptly illustrated the construction and use of technologies like fishing lines, spears, kayaks, and dogsleds, and these materials beautifully conveyed the arctic environment in which the Netsilik lived.

More recently, anthropologists have described the Netsilik films as mere reenactments because the people in the films were performing a way of life they remembered but no longer led.[61] During the earliest footage shot, which became part of *Fishing at the Stone Weir*, some of the actors wore contemporary underpants, one child sported a yellow Band-Aid, and other small nontraditional items crept into the film. At the first screening of this material back in Cambridge, Massachusetts, viewers noted these details and requested that in the future, all signs of modern technology be removed prior to filming, stating that "there is no creature with so sharp eyes as the elementary school kid. If we are going to reconstruct, we should really reconstruct, so that we have a truly authentic document."[62] Balikci additionally remembers being told to avoid gimmicks, like a long zoom, that could make the sequences look artificial by calling attention to the craft of filmmaking.[63] In recalling his involvement with MACOS, DeVore suggested that the original intention had been to contrast this reconstructed vision of the past with footage illustrating how the Netsilik actually lived in the 1960s—Balikci and Young's *The Eskimo: Fight for Life*. Due to time constraints within the curriculum, that component of the course dropped out by the final version but later aired independently on CBS as a one-hour special.[64] Today the films represent a way of life that is largely unimaginable to younger generations of the Netsilik people, and many of the actors who participated in their creation are glad the films preserved a snapshot of their traditional way of life (even if only a reconstruction) before it passed beyond memory.[65]

As classroom tests of MACOS materials continued, the !Kung unit generated increasing concern. Teachers testing the unit raised all sorts of questions. If the point of MACOS was to understand what was human about human beings, then perhaps contrasting the nomadic Netsilik way of life with a more agrarian society would make more sense? Patsy Asch countered that comparing two hunter-gatherer communities would prevent students from lapsing into a simple ecological determinism in which the arctic cold explained *all* Netsilik traditions and behavior patterns. The most troubling question, she believed, centered instead on the difficulties raised by the color of their skin. Racial issues might obscure the added value of including Ju/'hoansi cultural traditions in MACOS. The fact that the Ju/'hoansi were not "negroes," as Elizabeth Marshall had argued in *The Harmless People*, made no difference in American classrooms. Asch worried that some students (or teachers) might react negatively to the materials because of "direct racial prejudice." Equally problematic, well-meaning teachers might accidentally convey an image of the Ju/'hoansi as "kind primitives of the fairy tale sort that 'we all just love.'"[66] In order to continue with the unit, Asch believed MACOS needed to support additional class trials. They had to finish editing Lorna Marshall's diary as a distributable pamphlet for classroom use (this would match a "field note-book" filled with DeVore's baboon observations already under development for the animal behavior unit).[67] Perhaps most important, the myths and stories of the Ju/'hoansi needed to be "rewritten to match the beauty of the Net-silik songs." The unit also needed illustrations, classroom exercises, and three-dimensional materials. Without strengthening the unit in these ways, Ju/'hoansi culture could not help but pale in comparison to the Netsilik. By then, the MACOS materials emphasized adaptation and evolution as central components of the curriculum, which from Asch's perspective implicitly evoked a hierarchy of development. She became concerned that students were likely to see Ju/'hoansi culture as an evolutionary stepping stone between baboons and American superiority.[68]

DeVore was unsure about the !Kung unit for a different reason. The Ju/'hoansi, according to DeVore, were "utterly unaware of the camera and were often unaware they [were] even being observed." This made it possible to film intimate meetings of friends and shared jokes. Yet, he lamented, Mar-shall relished challenging his viewers, as when he filmed "the snot running out of the men's noses, drooling, that sort of thing, and the camera holds lovingly on it." The implied lesson, DeVore worried, was inherently confrontational: "If you believe we're all human, consider this as your brother." If so, then the fundamental philosophy of MACOS—"that if everyone in the world knew as much as the best informed people that we could get ahold of, they would also be (as) liberal and humanistic"—was naive and likely to backfire.[69]

A second problem presented itself to MACOS designers while they were deciding what to do with the !Kung unit, this time regarding the baboons. When the students watched the films of baboon behavior, they anthropomor-phized the baboons, attributing to them the same desires and motivations people would have in the same situation. At first, teachers tried to counteract this tendency by including a booklet focused on baboon communication and by emphasizing the fundamental gulf between animal communication and human language.[70] Whereas animals cannot lie because they communicate through gestures and "automatic messages," the booklet stated, humans are capable of conscious deception. This both reinforced a vision of man as an animal and suggested that baboons and humans possessed distinct kinds of mind: one capable of mere communication, the other of complex language that could express the conceptual state of an individual, including decep-tion.[71] It did not help. The designers decided to add even more material on animal behavior, starting simple, with a film on the salmon's struggle to sur-vive and reproduce, another exploring parenting behavior in herring gulls, and

FIGURE 9. Many of the MACOS booklets were illustrated with stark woodcuts.
Here, a man yells in frustration and shakes his fist rather than fight. "We can see
this same pattern throughout the animal kingdom," the book on herring gulls
noted. "There is much more threatening than there is fighting." This image
originally appeared in *Herring Gulls* (Washington, DC: Curriculum Development
Associates, 1970), 22. © 1967, 1968 Education Development Center, Inc.

a wide array of booklets on animal behavior, adaptation, and communication.
They even incorporated a shortened version of *Miss Goodall and the Wild
Chimpanzees*, accompanied by a MACOS booklet explaining the familial rela-
tions of the chimpanzees students saw in the film.[72]

The MACOS booklet accompanying *Herring Gull Behavior* developed a
theme that the film only hinted at—aggression. The film briefly mentioned
that although males could be highly aggressive and territorial, rarely did encounters between two birds lead to killing. Sometimes gulls instead opted to
"vigorously peck grass instead of each other" (a form of displacement behavior). The booklet explicitly related this theme to human behavior, with a stark
image of a man screaming and shaking his fist (Figure 9). The accompanying
text elaborated, "Humans act this way, too. When you are angry, you may feel
like fighting and you may fight, but you are more likely to shake your fist, slam
a door or scream." Animals rarely fight, the booklet explained, as each time
they do so they risk injury or death. An animal that can get what it wants
without fighting on average fares better than one that resorts to violence.[73]
Heavy-handed? Perhaps. The designers of MACOS imagined that through

early exposure to nonviolent conflict resolution, children would embrace these ideals when conflicts arose in their own lives.

The herring gull material, based intellectually on the ethologist Nikolaas Tinbergen's much loved *The Herring Gull's World: A Study of the Social Behaviour of Birds*, was a huge hit with the students.[74] The MACOS instruction manual recommended Tinbergen's book for the teachers and Louis Darling's more accessible and vividly illustrated *The Gull's Way* for students.[75] Darling had supplied all of the photographs used in the accompanying filmstrip that teachers could show students in their classes, several of which students could recognize from his book. One teacher, Thalia Kitulkais, described how she reconstructed Tinbergen's blind from which to observe the behavior of others undetected: "I have children running up to me and saying, 'Can I be Tinbergen today? Can I get under the blanket? Can I study the other children? Can I study the animals?'"[76] Tinbergen told his readers that any insights he had into human nature he obtained through watching man, birds, and fish. It was as if, he wrote, "the animals are continuously holding a mirror in front of the observer, and it must be said that the reflection, if properly understood, is often rather embarrassing."[77] The vivid image of children hiding under a desk and taking notes on the behavior of their classmates highlights the importance MACOS designers placed on students acquiring both the content and the methods of scientific research as a way of ultimately helping them reach a greater understanding of their own behavior.[78]

Given the pedagogical goals of the program, the political leanings of the designers, and the increasingly prominent Civil Rights movement, MACOS ultimately decided it was "politically unacceptable to use materials that showed partially naked, dark-skinned 'primitives' in a public school classroom."[79] Of the footage and classroom materials developed on the Netsilik and the Ju/'hoansi for inclusion in MACOS, only the Netsilik material was made available in a distributable format. Only the Netsilik were sufficiently "white" to be uncontroversial yet led lives amply different from those of the students to provide an effective intellectual foil. MACOS filled the resulting gap in the curriculum with the new materials on animal behavior.[80]

In its final year-long form, fifth graders began by learning about the progressive development of social organization in animals. "As the children begin to consider what distinguishes human beings from other animals," the instructional manual specified, "questions are raised about language, tool use, and families, all of which are part of the larger questions that are essential to the course." Perhaps surprisingly, from today's emphasis on standardized testing, the manual continued, "There is no need for final answers at this point, or, perhaps, at any time. Most important, throughout the course, questions should be raised which cause the child to reflect upon his own existence, and see himself and everything around him in a new way."[81] In the second half of the course, students imagined what it would be like to live above the Arctic Circle, following the lives of Itimangnak (known to his friends as Zachary), his wife Kringorn (Martha), and their family. Children tracked their lives through the cycle of the seasons, roaming with the Netsilik in search of fish in the summer months and hunting seals and building igloos to keep warm in winter. Throughout the academic year, MACOS encouraged students to apply the analytical skills they developed to understanding the culture in which they themselves lived.[82] Bruner intended the contrast between American and Netsilik lives to illustrate how all cultures developed explanations for the world around them. Whether those explanations were scientific or mythological, they were equally important for the daily functioning of society. He conceptualized social organization as vital to the work of each species. Whereas salmon fought their way upstream in loose aggregations and baboons gathered into troops for protection, the Netsilik films should prompt students to "reflect back on man's way of organizing himself into a society." These lessons, he imagined, would teach the students something about the unique nature of "man's mind."[83]

Bruner was interviewed in several contemporaneous documentaries that illustrated the use of MACOS in classrooms. The first of these featured Kitulkais's classroom, in which she had reconstructed Tinbergen's blind for studying student behavior, and her predominantly white students at the Newton School, located in a wealthy suburb of Boston. The film captured the supplementary supplies present in their classroom: multiple sets of materials with which to work, games, blankets, markers, and live hamsters. The second followed children from Boston's Dearborn Public School—a poorer, primarily African American elementary school with fewer classroom resources.[84] At the end of the year, the children were asked what effect the course had on their thinking. A flurry of answers followed: "How to survive"; "How to grow up . . . and get a good education"; "You can't make nobody do nothing if they don't wanna. . . . If you try and talk them into it, they might agree. 'Cause if you keep beating 'em and beating 'em, that don't get nowhere." The final words of the film, however, rested with a young girl in ponytails, who slowly lowered her head to her desk as she proclaimed, "We was all wild, and then you came and you tamed us."[85] In these films, we can see some of the less explicit aspirations of the social science curriculum reform project; as a result of their education, the next generation of both white and African American citizens would not protest in the streets but would instead seek social progress through constructive civic democracy.

Initial responses to MACOS were largely positive.[86] At the height of its popularity, the program reached over four hundred thousand students, spread between seventeen hundred schools in forty-seven states.[87] A 1970 article announcing the new curriculum in *Time* magazine declared that "few parents have objected to the course, even though it contains rather fundamental information on mating habits and some of the bloodiest film imaginable on the slaughtering of seals." Both Bruner and Peter Dow, who shouldered the administrative responsibility for pulling everything together, conceded the dramatic nature of the material. Bruner commented, "A generation ago, the problem for kids was sex. For this generation, it's violence." Dow, for his part, suggested that "urban kids are much more attuned to questions of survival and not so frightened by some of the gutsier issues like death and reproduction."[88] Given the social context of the times, they argued, both sex and violence were issues that students needed to work through anyway, and MACOS could provide a set of materials and tools that would make it easier for students to imagine creative alternatives to violence as a mechanism of social change. In 1972, satisfied with the curriculum he had helped to build, Bruner left Harvard to take up a post at Oxford University, traveling across the Atlantic on his beloved sailboat, *Westward Till*.[89]

PART II

Naturalizing Violence

But we were born of risen apes, not fallen angels, and the apes were armed killers besides. . . . The miracle of man is not how far he has sunk but how magnificently he has risen.

—ROBERT ARDREY

THE MULTIMILLIONAIRE Harry Frank Guggenheim read Robert Ardrey's *The Territorial Imperative* in November 1967 and found it "of fascinating interest."[1] He wrote to his "old and valued friend" Henry Allen Moe, brimming with enthusiasm for Ardrey's book.[2] Moe had also recommended Konrad Lorenz's *On Aggression*, which Guggenheim consumed with vigor. Now he suggested that Guggenheim was bound to find Desmond Morris's recently released *The Naked Ape* equally worthwhile. Guggenheim reported it was "extraordinary."[3] For years Guggenheim had been preoccupied with the problem of the domination of some men by others but had trouble deciding how best to proceed. Rather than portraying human violence as an exception to the norm of peace—as had Eiseley, Dobzhansky, and Mead—readers of these books saw in them a biological explanation of why spontaneous interpersonal aggression and coordinated wars between nation-states were so naturally persistent. If Ardrey, Lorenz, and Morris were right, Guggenheim surmised, then solving human violence would require more than a simple appeal to our sense of morality.

In 1959 Guggenheim, together with a small cadre of close friends, had begun a conversation with the University of Michigan professor of psychology Paul Fitts. A few years later, Fitts assembled a group of scientists to tackle the problem at a 1964 symposium entitled "Strategies of Dominance and Social Power," which was held at Henry Ford's former home, Fair Lane, which was by then part of the university's Dearborn campus. For both Guggenheim and Fitts, one of the most promising lines of inquiry lay in analyzing the origins, development, and mechanisms of dominance in order to discover ways of controlling its expression. Fitts hoped Guggenheim would fund a research center at Michigan, but Guggenheim remained skeptical of the institutional stagnation he felt would inevitably characterize any university-based center.[4] In the end, it did not matter. Fitts died less than a year after the symposium at Fair Lane, mere months after Guggenheim had formalized arrangements for his philanthropic foundation "to promote the development of knowledge concerning, and the application of such knowledge to the improvement of, man's relation to man for scientific and charitable purposes."[5]

Guggenheim believed that in order for the Harry Frank Guggenheim Foundation (HFGF) to succeed, it was "imperative" to "enlist the interest not only of top-flight men in the field, but the *right* men" and to find "a first class person to head up the project."[6] That his new venture would come to play a significant role in fostering research on the biological basis of human nature owed a great deal to an unlikely friendship that he began with Robert Ardrey. In the last years of his life, Guggenheim came to trust Ardrey as an expert guide to scientific research into animal behavior and human nature. Born in 1908, Ardrey had studied both anthropology and playwriting at the University of Chicago and, a few years after graduating in the midst of the Great Depression in 1930, moved to New York City to launch his career in drama.[7] He became far more successful as a Hollywood writer than a theater man, perhaps most recognized in the 1950s for his screenplays of *The Three Musketeers* (1948) and *Madame Bovary* (1949). In 1967 he was nominated for an Oscar (Best Writing, Story and Screenplay—Written Directly for the Screen) for his work on *Khartoum* the previous year, but that was as yet in the future.

Together with the ethologists Konrad Lorenz and Desmond Morris, Ardrey was often characterized in the popular press as advancing a vision of man as nothing but an animal. Lorenz had been first introduced to American audiences as the authoritative, witty author of *King Solomon's Ring: A New Light on Animal Ways* and later *On Aggression*—but his authority as an expert on animal behavior was girded by his position as the director of the Max Planck Institute for Behavioral Physiology in Seewiesen, Germany, and numerous professional publications.[8] An academic generation younger, Morris earned his DPhil at Oxford under Nikolaas Tinbergen, Lorenz's long-time scientific interlocutor and friend.[9] Rather than accepting a more traditional teaching appointment, however, Morris moved to London to head the Granada TV and Film Unit at the Zoological Society of London before accepting an appointment as the curator of mammals at the Zoological Society, where he penned *The Naked Ape*, an international best seller.[10]

With books by Lorenz and Morris circulating at the same time, the unlikeliness of Ardrey's cachet as an expert on animal behavior and human nature reflects the overlapping systems of authority in colloquial scientific publications of the 1960s and the growing popularity of the "human animal" as a subject. What actually made someone an expert in human nature: the number of books they sold, the passion of their writing, their personal charisma, or their intellectual standing among their peers? The answer to each of these options was yes. Among the resulting forms of public authority, academically trained scientists competed with professional writers and journalists for readers and viewers.

When Harry Guggenheim picked up *The Territorial Imperative*, he was a man who moved through elite New York circles, surrounded by socially and economically powerful men whose mettle, he believed, had been tested by combat and hardened by business. He intended to spend his money on practical solutions to one of the persistent dilemmas confronting all humanity. In 1848 his grandparents had moved to Philadelphia, where they started a successful mining company. Harry's father, Daniel Guggenheim, eventually took over the growing family business, and Daniel and his nine siblings became fixtures in East Coast philanthropic networks, creating the Solomon R. Guggenheim Museum and Foundation in New York City, the John Simon Guggenheim Memorial Foundation, and the Daniel and Florence Guggenheim Aviation Safety Center at Cornell University. Harry served in both world wars as a member of the Naval Aviation Forces, where he met James "Jimmy" Doolittle and Charles "Slim" Lindbergh, as he referred to them in his correspondence. He was appointed US ambassador to Cuba from 1929 to 1933 and cofounded *Newsday* with Alicia Patterson (his third wife) in 1940.[11] Doolittle and Lindbergh were family friends and were awarded Daniel Guggenheim Medals in 1942 and 1953, respectively, honoring their "notable achievements in the advancement of aeronautics." Doolittle retired from active military service in 1959 but remained interested in aviation safety throughout his life (a concern shared by both Guggenheim and Lindbergh). Lindbergh had been an associate of the Guggenheim family at least as early as the 1920s, when he toured the country promoting aviation under the sponsorship of Daniel Guggenheim.

Harry Guggenheim was also strong-willed, perhaps obstinate. He admitted as much in a letter to his friend Henry Moe.[12] Moe was then president of the American Philosophical Society and would soon become interim chairman of the National Endowment for the Humanities (until the first official chairman appointed by President Lyndon Johnson could begin his duties). Moe believed firmly in the importance of universities in cultivating an intellectually healthy nation.[13] In his letter, Guggenheim apologetically noted, "I'm afraid I am perhaps a difficult donor, and perhaps unable to accept the role of a philanthropist who calls in experts to dispense his funds. I have been the head of three Foundations, two of which I still head, not as a philanthropist but as the directing spirit, with some expertise, dispensing funds of others." He wondered whether, due to this experience, he found it "hard to turn over these funds to professionals and say, 'I want to improve man's relation to man; here are X dollars; now get to work.'" Expressing a sentiment he came to repeat often in his correspondence with friends, Guggenheim added, "In the six years that I have been attempting to make some progress on this project I have found that the only suggestions . . . in what I consider a practicable manner were not suggested by professionals, but were the intuitive suggestions of laymen."[14] Practicality, or common sense, was a quality Guggenheim deemed especially lacking in university social scientists.

Guggenheim and his associates expected the HFGF would award about six fellowships a year of between $5,000 and $9,000 each (in 2016 terms, between $38,000 and $68,000). Awardees were to be granted a great deal of leeway with their research projects under the assumption that as vetted men of quality they would produce top-notch results. Moe (chairman), Fitts, and G. Edward Pendray formed the initial fellowship committee, but they hoped additionally to find a part-time director for the fellowship program.[15] Pendray was another long-term associate of the Guggenheim family, having helped develop the Guggenheim Jet Propulsion Laboratory at the California Institute of Technology, among other such ventures. In the wake of Fitts's death, the nascent foundation floundered, waiting for someone to assume responsibility for the entire project or (at least) the fellowship program—part-time initially, but full-time later if he and the project were to "take-fire."[16]

When Guggenheim read Ardrey's *Territorial Imperative*, Lorenz's *On Aggression*, and Morris's *Naked Ape*, he was thus looking for a fresh perspective as well as a new man to spearhead research on "man's relation to man" with the full financial backing of his foundation. For Guggenheim and his board, these three books all made the same point—that to understand why men killed other men, one must consider humanity's long past as animals who had only recently acquired culture and civilization—even as they debated the utility of an evolutionary perspective as a practical means of remediating violence in the international world through which they moved. (For readers more familiar with the interdisciplinary landscape of academic zoology and anthropology, these books actually looked quite distinct.) Professional demarcations between "amateur" and "scientist" mattered less for Guggenheim than whether or not he appreciated the man and the perspective he offered.

Despite his more conservative politics, Guggenheim's approach to scientific ideas, through books and through charisma rather than accreditation, resonated with that of many consumers of science in the age of the counterculture.[17] In their own way Ardrey, Lorenz, and Morris each represented a substantial break with postwar norms of familial masculinity.[18] Such rhetoric, with its core American values depicted in the nuclear families of June and Ward Cleaver (*Leave It to Beaver*) or Ozzie and Harriet Nelson (*The Adventures of Ozzie and Harriet*), had emphasized the importance of reproduction to the country's democratic future.[19] Ardrey instead stressed the importance of male-male interactions in the origins of social stability to the exclusion of families bound together by marriage. At the same time, his version of masculinity stood at odds with the burgeoning counterculture, represented in popular media in the form of long-haired men and women dressed alike in flowing, brightly colored clothes. Lorenz argued that human violence emerged, ironically, from our legacy as a species without natural weapons like claws, horns, or piercing incisors—thus he explained the unique capacity for humans to kill members of their own species. When engineers and scientists had developed the technological means of killing at a distance, Lorenz insisted, this compounded the situation. Morris combined his zoological perspective with that of the sexologists, especially as exemplified in the controversial publication of William H. Masters and Virginia E. Johnson's *Human Sexual Response*.[20] Morris's framing of humans as sexy, naked apes meant that his book resonated with the mores of the sexual revolution. Together these books cultivated animal measures of social success that became essential to public wrangling over human nature in the coming decade, including within Guggenheim's circle of friends.

A remarkable exchange of letters between Guggenheim and the members of his board—Henry Moe, James Doolittle, Charles Lindbergh, and G. Edward Pendray—reveals the merits and difficulties posed by these evolutionary perspectives on human nature to colloquial audiences.[21] It began in the spring of 1966 (before the books by Ardrey, Lorenz, and Morris were available). Guggenheim wrote to Lindbergh that throughout history men had abused their political power. "In pursuit of that primary urge to dominate their fellow man," he suggested, "they have decimated him and caused incalculable destruction to the accumulated works of beauty and utility that man has created." Guggenheim further observed that in the 1960s the world still contained several of these men, who needed to be controlled lest they "continue to cause holocausts of destruction."[22] Lindbergh believed that the quickest, most effective, and reasonable strategy for improving man's relationship to man would be to ameliorate the conditions of human life—especially through the conservation of natural resources.[23] To this Guggenheim replied that the fundamental issue he wished to address was located not in the environment but in the "qualities in man." He asked Lindbergh, "How can we determine the cause of this destructive rather than constructive competitive quality in man? How can we educate him so that we may divert these energies to competition that is good rather than evil?"[24] In the face of Guggenheim's queries, Lindbergh remained firm: "It seems to me there is good domination, and bad domination (possibly 'leadership' would be a better term to work with), and all kinds of forms in between. Again, 'good' and 'bad' vary with frameworks of reference." He staunchly continued, "I think that men who gain great ability to dominate and exert power intertwine with their environment, in addition to being affected by hereditary characteristics. I don't believe you can separate them from institutions of their times any more than you can separate heredity and environment."[25] Lindbergh's objections echoed concerns expressed by anthropologists and zoologists, albeit in a different register. Guggenheim responded once more without addressing Lindbergh's point. Their dialogue, he wrote, helped him "think through the answer to a most difficult question: 'Is there a basic quality in man that can be isolated which is the cause of strife with his fellow man?' "[26]

Guggenheim found Ardrey's *Territorial Imperative*, Lorenz's *On Aggression*, and Morris's *Naked Ape*, so pertinent to his question about humanity's nature (and to his incipient foundation) that he sent copies of each to his entire board.[27] In 1967 Guggenheim also penned an editorial—"Mark of Cain"—in *Newsday* to call readers' attention to the grave need for understanding "the nature of the beast within man" and lavished praise on both Ardrey and Lorenz for their efforts to uncover man's instinctive aggression.[28] Lorenz's conception of aggression as key to human nature, and fundamental to our virtuous qualities (leadership and kindness) as well as our violent tendencies (dictatorship and murder), fit neatly within Guggenheim's vision of human social relations. Ardrey, impressed by the "Mark of Cain" editorial in *Newsday*, wrote to introduce himself to Guggenheim, and they began a lively correspondence. Rather than pessimistically predicting man's inevitable doom, Ardrey closed on a positive note: "I believe that when one regards oneself as a risen ape, the future becomes illimitable. When one regards oneself as a fallen angel, one has no future at all. What man needs in our time, above all else is an elation. I think that's what my ethologist friends are finding."[29] Here, it seemed, was the expert guidance for which Guggenheim had been looking. He replied to Ardrey, informing him of his foundation and asking him to contribute an article to *Newsday* as part of the series, the Condition of the American Spirit.[30] Ardrey responded immediately, taking the opportunity to cultivate a potential patron, and cleverly made his intellectual project about Guggenheim's: "With admirable intuition you as long ago as 1963 grasped the problem of dominance as central to the human predicament, and I have outlined the program of my work to demonstrate that I too regard it as central."[31] Yet by dating Guggenheim's interest to 1963, Ardrey also established his own priority, as *African Genesis* had been published two years earlier. Ardrey followed his compliment with a request—would Guggenheim be so kind as to send him a copy of the bibliography on dominance that Fitts had prepared after the conference?

Guggenheim found Ardrey's ideas enthralling. Still, before responding, he decided he should vet the man with his trusted friends and so telephoned Moe. As Guggenheim later recounted, he was delighted to discover that Ardrey was a former John Simon Guggenheim Fellow and so counted as an "old friend" of Moe's.[32] In his reply to Ardrey, Guggenheim promised to dig out a copy of Fitts's bibliography but worried, "The bibliography is going to be a disappointment to you as it was to me. It consists mainly of references to power in industry." He also forwarded copies of his complete correspondence with Ardrey to Moe and Lindbergh—writing that he anticipated "this contact with Mr. Ardrey opens up a new vista in our Man's Relation to Man Project" and asking for their thoughts and reactions.[33]

Pendray weighed in, too, concerned that if Ardrey were correct (which Pendray doubted), then his arguments seemed to call into question the utility of proceeding with the man's relation to man project. "If human behavior is basically instinct, what can ever be done (short of a long period of evolution) to modify or improve it?" Pendray continued, "I still believe profoundly in the modifiability of human behavior, based on knowledge, understanding, social pressures and education. How else can we account for all the varieties of cultures already to be found in the world?"[34] Guggenheim defended Ardrey, insisting that his studies "confirm my thesis that dominance is a basic instinct of man. Man's actions are governed by the sum of his inheritance and environmental characteristics." By changing the crucial environment for powerful men, he imagined, "we can influence man by directing his instinct to dominate for the progress of rather than for the depravity of mankind. In the former case we had Christ, in the latter a Hitler."[35] This assertion, of the malleability of man in the face of his inherited instincts, would prove to be a sticking point among several of the inner group (as it was for many social scientists).[36]

While waiting for his other friends to respond, Guggenheim and Ardrey, now Harry and Bob, grew closer. The next time he wrote, Ardrey addressed Pendray's criticisms (which Guggenheim had forwarded to him along with his own response). Ardrey reported enthusiastically, "You couldn't have given a better answer.... What we now know about dominance is that in the males of all species it's an instinct. The drive is there, born in, and cannot be obliterated as it cannot be ignored. But the goals are adjustable." Ardrey argued that, therefore, by "denying" the innate drive to dominance in all men, "we lose all control over the goals." Headway on the problem of man's relation to man could be made only by accepting man's base nature. Ever the careful correspondent, however, Ardrey had no wish to ostracize Pendray, so he added that he "retain[ed] a considerable sympathy for Mr. Pendray's question," even if it "rests on the false concept of instinct that we've been taught and is forced on us by every Ashley Montagu in the hope that we'll deny it exists."[37] Ardrey said that given the novelty of his ideas, he understood the resistance to his books. He then discounted the work of European ethologists (including Lorenz and Tinbergen) on whose research he had based much of *The Territorial Imperative*, cryptically noting that "their attitudes towards instinct were formed at the cellular level . . . a level at which nothing so far has proved demonstrable." Ardrey hewed closely to Guggenheim's stated position: "What you wrote is utterly correct. A genetically determined behavior pattern is a cup of determined shape. What rain falls into that cup God and man must decide. But something will fall, and something will be retained. So one d[e]termines the difference between a Christ and a Hitler."[38]

Guggenheim also arranged a "stag dinner (business clothes)" at his five-story New York City town house.[39] For posterity and the HFGF records, Pendray kept minutes of the occasion.[40] Moe attended, as did other New York notables, including Roger Straus (Guggenheim's cousin and later cofounder of publishing house Farrar, Straus and Giroux), Peter Lawson-Johnston (heir to the Guggenheim Brothers business), and Dr. Malachi Fitzmaurice-Martin (a Catholic priest and recent recipient of a grant from the Harry Frank Guggenheim Foundation). Doolittle and Lindbergh regretfully declined. According to the minutes, Ardrey took the opportunity to advance his own cause by arguing that his volumes, together with those of Lorenz and Morris, were the only "worth-while books" about human aggression and dominance. When asked who might be appropriate to direct the man's relation to man project, Ardrey requested more time to think but also suggested "the most important possibility to be Dr. Robin Fox, Chairman of the Department of Anthropology at Rutgers University."[41] All told, the dinner seemed to go quite well.

Lindbergh's letter responding to Guggenheim's query about Ardrey arrived next. Although he began by stating that he largely concurred with Guggenheim's positive assessment, Lindbergh's final opinion turned out to be more complicated. "I could not be in more agreement with Ardrey's statement that 'To me, the nightmare is the denial of instinct, the denigration of competition—' (I think the essential value of competition is deplorably lacking in the marvelous philosophy of Jesus, at least as it has been handed to us.)" Yet he shared Pendray's doubts about applying this philosophy to the foundation's purpose. "When we relate dominance to competition, it seems to me we are relating a fragment to a whole, and I cannot believe this is the best way to approach improving man's relation to man." As he had become both older and more experienced, Lindbergh continued, he also became "more aware of the limitations of man's sciences." The difficulties facing the task ahead of them were tremendous and long-standing. "How do we clarify the issues of war in Viet Nam, of riots in our cities, of our cold-war with Russia?" In sum, he closed, "man's relationship to man expands into the miracle, and here the tools of science are inadequate."[42] The sheer magnitude of these problems belied any easy answer.

When Doolittle chimed in, he pragmatically noted the different timescales required to address the problem of man's relation to man with various methods. The most permanent solution operated on the longest time frame. "There is little question but that improving mankind through evolution will take a very long time indeed; measured in many millennia, or perhaps even millions of years," he wrote. "The ability to successfully change the genes—cut the meanness out and put kindness in—may come in tens or hundreds of years." In the interim, "social pressures cannot change man's instincts but may well serve to suppress the baser of them. With proper planning and implementation this effect might well occur in the relatively short time left to you and men." Like Pendray, Doolittle advocated strategies that would allow a rapid change in human actions, and that meant concentrating the foundation's resources on immediately controllable elements of the human social environment.[43]

Although members of Guggenheim's inner circle seemed inclined to trust the fate of humanity to evolution in the long run, and genetics in the slightly less distant future, research into the social and cultural causes of violence still struck them as entirely more practicable in the short term. Against this background of skepticism, Ardrey continued corresponding with Guggenheim and Moe, and two men recurred in his letters as suggestions for the post of research director. "This pair with the improbable names," as Ardrey dubbed Fox and his colleague Lionel Tiger, planned to employ an evolutionary perspective to "establish an anthropological redoubt against orthodoxy."[44] From Guggenheim's perspective on the general inutility of most social scientists, that made them only more attractive as candidates.

Guggenheim's plans for his foundation stalled in the face of his own battle with cancer.[45] After Ardrey learned of Guggenheim's illness, he wrote to Morris, asking for his "sympathetic attention" in case anyone from the HFGF would contact him. "Old Harry is sick," he typed, "wearying for that immortality which money cannot alone dispose. He is intelligent beyond the carrying point. I love him. Should you become involved, give his efforts every possible boost."[46] When Guggenheim passed away two years later, in January 1971, no decisions had been made as to the fate of his foundation. Guggenheim's legacy lived on because of the board's desire to honor his intentions, even if they disagreed with them intellectually. Survival in the philanthropic savannah required an intuitive sense of the masculine dominance hierarchies governing the social interactions of these men. Ardrey, although a relative newcomer to the group, had quickly figured out an appropriate rubric for maintaining a friendship with a powerful man like Harry Frank Guggenheim—a well-balanced combination of maverick bravado and studied deference.

When read in sequence, the coincidence of the three books by Ardrey, Lorenz, and Morris, published within two years of each other, gave weight to the idea that human nature was indelibly marked by the worst aspects of our animalistic evolutionary past.[47] They served as fodder for discussions among undergraduates and graduate students training in both anthropology and zoology.[48] They also invigorated conversations among nonexpert readers, like Guggenheim and his board of directors, by supplying new facts and an evolutionary vocabulary with which to discuss the fraught relationship between violence and human nature.[49]

4

Cain's Children

ROBERT ARDREY claimed that his youthful experience with Sunday school in Chicago prepared him for learning about anthropology in Africa. While adults gathered peacefully and reverently in the church above, the kids would congregate unsupervised below in the basement. Mostly they discussed sports programs and initiated new members when necessary—after the close of all official business, they "would turn out all the lights and in total darkness hit each other with chairs."[1] As an adult, Ardrey traveled, and although he worked for Hollywood, he eventually settled outside the United States. For many years he preferentially lived in Rome, mere blocks from the east bank of the Tiber River in the Trastevere.[2] Photographs of Ardrey from the 1960s depict a man with rumpled black hair and a bit of a slouch. From this unassuming source stemmed a decade-long controversy about the violent origins of humanity. The killer ape theory, as it became known, was not Ardrey's idea, but he was perhaps its most mellifluous proponent (Figure 10).

To understand Ardrey and his process of translating the work of paleontologists and experts in animal behavior for a general audience, we need to start with *African Genesis*, published in 1961, in which he described his experiences navigating savannahs and deserts in search of experts who could teach him about the nature of early man. "Not in innocence, and not in Asia, was mankind born. The home of our fathers was that African highland reaching north from the Cape to the lakes of the Nile. Here we came about — slowly, ever so slowly — on a sky-swept savannah glowing with menace."[3] With these words, Ardrey began.

Let me call attention to two conventions of his writing. First, Ardrey did not confine his use of "mankind" to *Homo sapiens* or to men. Preferring to recognize the long evolutionary lineage resulting in modern humans, he used "man" to include all of our hominid ancestors, from the moment our evolutionary lineage diverged from the lineages of other apes.[4] The paleoanthropologist and gregarious showman Louis S. B. Leakey followed the same convention, as did much anthropological literature written at the time for a

FIGURE 10. "Ardrey writes again! With verbal pistols flaming, he straddles his hobbyhorse of territorial aggression and charges up and down evolutionary trails, shooting off behavioral anecdotes in all directions." Illustration by Julio C. Fernandez, for Ronald Singer, "Ardrey in Wonderland," review of *The Social Contract* in *Natural History*, 79, no. 9, November 1970, 80–82.

general audience. Second, throughout his writings, but especially in *African Genesis*, Ardrey evoked stereotypes of Africa as a timeless, wild, and primitive continent in which our ancient past had been preserved for the few Westerners (like himself) who were brave enough to confront (in another man's words) "the human past, the dark abattoir of human time and the deep abyss of human experience."[5]

In doing so, Ardrey promoted images of Africans that cultural anthropologists, civil rights leaders, and the designers of Man: A Course of Study were desperately trying to combat but that a reading white public eagerly consumed.[6] Take, for example, a similar description by the budding journalist Richard Rhodes for *Playboy* magazine: "Man began in Africa black, footloose and free. It seemed remarkable that the continent had remained as primitive as the land below looked when its history of near-human and human habitation ran back 20,000,000 years. It should have been worn smooth as an old coin. Instead it was still largely untracked, still wild, had resisted civilizing through millennia."[7] Ardrey's writing similarly echoed the well-worn trope of civilized men finding inner strength while on safari in Africa. Compare his books with contemporary writings by Robert Ruark, for example, who evocatively described his experiences hunting wildlife. Ruark found a masculine resolve in the landscape much like Theodore Roosevelt had decades earlier.[8] "There is something of the safari . . . that you will not be able to find on TV or even in church," he wrote shortly before he died of cirrhosis of the liver, a casualty of a lifetime of heavy drinking. "If you are lucky, you'll be able to find it in yourself."[9] This powerful formulation of masculinity through conquest over self and nature, emerging from our animalistic evolutionary roots, percolated into Hollywood and New York in fairly short order (as we will see in future chapters). I uncovered no reviews that questioned Ardrey's florid descriptions of the continent or the people who lived there.

Ardrey had first ventured to Africa as a freelance author in the 1950s when he accepted a commission from the *Reporter* magazine to write a series of articles on African politics.[10] (Based out of New York, the *Reporter* earned a reputation as fiercely liberal, pro–civil rights, and anti-McCarthyist.) On this trip he met the scientist whose theories he would make fashionable, Raymond Dart.[11] In his published work, Ardrey described Dart as "a small, compact man of far-reaching interests, far-gripping personal magnetism, and appalling durability."[12] His private opinion was perhaps more measured. In a letter to his later critic, the anthropologist Ashley Montagu, Ardrey noted that he found Dart to be "a truly top-drawer, imperishable, from-here-to-eternity artist." A little later he added, "I know you won't quote me about his being a second-rate scientist, but I think you'll know what I mean. He's a wonderful scientist, in his way, but far more important, he's a great man."[13] When Ardrey wrote this, he knew Montagu found some of Dart's theories repellant and was perhaps trying to distance himself from Dart in Montagu's eyes. The truth of his private opinions about the man likely fell somewhere in the middle. That they were friends is beyond question—Ardrey married his second wife, the South African actress Berdine Grunewald, in Dart's living room.[14]

An Australian by birth, Dart had moved to South Africa in 1922 to accept a post as a professor of anthropology at the University of Witwatersrand. In subsequent decades, Dart witnessed the beginning of apartheid legally enforcing segregation based on a strict racial typology. Although the engineers of apartheid relied on the work of Afrikaans-speaking ethnologists, a similar classification undergirded research in physical anthropology at the University of Witwatersrand and the University of Cape Town. The typological interpretation of race and culture made it difficult to challenge apartheid on scientific grounds within South Africa, and Dart thought of himself and his science as apolitical. Even so, during the 1950s he testified in court for several apartheid racial "reclassification" trials. In one particular case, an eighty-year-old woman was on trial for purchasing a bottle of sherry—a legal activity if she were "white," but illegal if she were "colored." Dart carefully measured her skull, looked at her eye color and facial features, and confidently declared her "colored."[15]

Dart believed that human civilization evolved only once and described the KhoeSan peoples living in southern Africa as cultural and physical "living fossils."[16] In his publications regarding archeological sites in the Solwezi area of northwest Namibia, Dart's racial assumptions are clearer than in his discussions of the origins of humanity generally. At these sites, Dart claimed to have found engravings of long-haired or hooded figures belonging to an ancient mining people who had traveled to the area over four thousand years ago, bringing with them the knowledge of how to carve rocks in this fashion. In other words, his argument implied, the neolithic engravings and metallurgical skill of the sites could not have been produced by ancient southern African peoples without outside help.[17] Later researchers who revisited the site disagreed with his interpretation and could not discern the figures Dart described in the engravings nor find any evidence of foreign travelers.[18] These racialized implications of Dart's publications for South African readers were recast for Americans through Ardrey's writing.[19]

In "The Predatory Transition from Ape to Man," a paper from 1953 in which Ardrey took particular interest, Dart had argued that the "blood-bespattered, slaughter-gutted archives of human history" were consistent with man's predatory origins, the "bloodlust" of humans separating us from the rest of our animal relatives.[20] He claimed that Australopithecines, by dint of sharing this

FIGURE 11. Berdine Ardrey, Robert's second wife, accompanied him on his travels and illustrated all of his evolutionary books. Here she captures the "Predatory Transition to Man," associated with humanity's new capacity to use tools for the hunt. Illustration by Berdine Ardrey, for Robert Ardrey, *African Genesis: A Personal Investigation into the Animal Origins and Nature of Man* (New York: Atheneum, 1961), 175.

carnivorous appetite, also shared our essential nature. They were proto-men whose primary "cultural tools" were weapons: clubs made of broken antelope leg bones and sharp pig teeth they could use to cut, chop, and saw—found objects, lethally transformed.[21] The combination of a new appetite for meat with an enlarging brain made it possible for modern humans to evolve.[22] For Ardrey, Dart's theory turned typical assumptions about the origin of humanity on their head: "Far from the truth lay the antique assumption that man had fathered the weapon. The weapon, instead, had fathered man" (Figure 11).[23] Thus, Ardrey concluded, "we are Cain's children, all of us."[24] This was the main argument of *African Genesis*—that humankind was born in Africa and in sin.

Dart, Ardrey implied, was on the side of justice and patient reason. His critics were not. Ardrey described how the editor of the journal in which Dart's "Predatory Transition" paper had been published added a critical note, lest unwary readers assume the journal endorsed Dart's conclusions.[25] "The foreword ended with a pitiful sigh: 'of course, they were only the ancestors of modern Bushman and Negro, and of *nobody else*.'"[26] By calling attention to the editor's polygenist intent, Ardrey cast doubt on the politics of Dart's detractors. In his foreword, the editor had made his intentions more explicit than Ardrey implied, stating that European races "go back to different ape-ancestors."[27] While discussing the predatory origin theory in Dart's office, Ardrey asked what implications his vision of Australopithecines as killer apes had for modern societies, now that humans possessed weapons (like the atom and hydrogen bombs) capable of obliterating life on Earth. As related by Ardrey, "Dart turned from his window and sat down at his desk; and somewhere a tunnel collapsed, a mile down, and skulls jiggled. And he said that since we had tried everything else, we might in last resort try the truth."[28] Ardrey implicitly contrasted Dart's reasoned reflection with the racism of his opposition, eliciting readers' sympathies for his friend.

In its organization, *African Genesis* was a story of adventurous travel in which Ardrey uncovered "secret" knowledge, a tale of scientists as modern-day heroes and their adversaries.[29] Helping his readers through the morass of scientific detail, Ardrey depicted himself as a trustworthy outsider, an every-man's man with whom his readers could identify—not as smart or well-trained, nor as athletic, as any of the men he described. "African anthropology has been the work of wild men," he wrote, "and one must simply hold one's breath for had there been no wild men, there would be no African science of man, and the world would be the loser."[30] If he could understand this material, Ardrey implied, so could any of his readers. The power of these tactics was encapsulated in his accounts of trying to pronounce convoluted scientific names. "It was the first time, to my recollection, that I had ever heard of Australopithecus. I struggled to pronounce it to the rhythm of those-who-do-not-seek-us. I struggled with the preposterous adjective, australopithecine, for which one must master a different rhythm, pass-me-the-pickle-brine."[31] As Ardrey, the character in his book, struggled along, he reinvented himself as representative of a universal human—and as he succeeded in his quest, so did anyone reading the book to its conclusion. Ardrey was his own ultimate hero, having overcome great tribulation to bring his audience this exciting story of human origins.

Within this novelistic structure, Ardrey sought to make two additional points about the state of contemporary understanding of human nature. First, he suggested that in order to draw appropriate lessons from animals, they must be observed in the wild. "Only in a state of nature," he claimed, "can we be sure that we are observing true animal behavior."[32] In asserting this, he echoed one of the major tenets of (predominantly European) ethologists in their battle against the (largely American) laboratory-based behaviorists. Ardrey proclaimed that "only when a new generation of scientists went out into the field could we begin to apprehend, for example, the subtlety of organization in those natural animal societies which cannot exist in the zoo." As a result, laboratory experiments would still be useful, "but only as measured against the new observations in the field could indoor conclusions be accepted as meaningful."[33]

Ardrey's second point emerged from the first; the explanations of human behavior by the noted psychologist and founder of psychoanalysis Sigmund Freud had been handicapped by his lack of access to reliable accounts of animal behavior.[34] In fact, all of the social sciences were similarly stunted, especially psychology and cultural anthropology.[35] It is worth quoting Ardrey at length, to provide a sense of both his rhetorical style and damning intent:

> To conclude that human obsession with the acquisition of social status and material possessions is unrelated to the animal instincts for dominance and territory would be to press notions of special creation to the breaking point. To conclude that the loyalties or animosity of tribes and nations are other than the human expression of the profound territorial instinct would be to push reason over the cliff. To conclude that feminine attraction for wealth and rank, and masculine preoccupation with fortune and power and fame are human aberrations arising from sexual insecurity, hidden physical defects, childhood guilts, environmental deficiencies, the class struggle, or the cumulative moral erosions of advancing civilization, would in the light of our new knowledge of animal behavior be to return man's gift of reason to its Pleistocene sources, unopened.[36]

Ardrey laid the blame for our faltering knowledge of human nature at the feet of the social sciences, which had become "institutionalized, universalized, and sanctified."[37] Only by acknowledging the biological nature of humanity would scientists be able to sort themselves out, yet this Ardrey believed they seemed reluctant to do. He deemed it the "superb paradox of our time that in a single century we have proceeded from the first iron-clad warship to the first hydrogen bomb . . . yet in the understanding of our own natures we have proceeded almost nowhere."[38]

Louis Leakey objected to the substance of Ardrey's argument about toolmaking in early man but did not deign to review the book. In particular, Leakey would not accept "the point about the fashioning of tools and weapons from bones before stones." The crux of Ardrey's argument was flawed, Leakey insisted, "because it was the very process of making and using such tools and weapons that brought about the big human brain."[39] Bones were too soft, he insisted, to make either good weapons or cutting tools. Australopithecines would not have been able to pierce the skin of a dead antelope—he knew this because he had tried himself.[40] Much better were stone tools. Leakey never abandoned his objections to Ardrey's support of Dart's

Australopithecus as the obvious progenitor of humanity, preferring to grant that honor to his own remarkable fossil find—*Zinjanthropus* (later reclassified as *Paranthropus*).[41] Ardrey chose to recall their falling out as a function of Leakey's pettiness. "Leakey and I have not been on speaking terms since 1961, when I correctly diagnosed <u>Zinjanthropus</u> as A. Robustus," he typed in a letter to David Pilbeam in 1969. "I regret the loss," he continued, "but in a way I don't blame him. It shouldn't happen to a dog, let alone a scientist."[42] Only in the final years of Leakey's life did they reach a rapprochement, marked by a convivial exchange of lectures sponsored by the Leakey Foundation in November of 1972.[43]

Although *African Genesis* was a success in terms of sales, it failed to generate the enthusiasm among scientists Ardrey craved. Ardrey's editor sent page proofs of *African Genesis* to a number of professional biologists and anthropologists, seeking blurbs and favorable reviews. Many responded, far more frankly than if they had realized the editor would immediately forward their letters to Ardrey. The well-known paleontologist George Gaylord Simpson refused to lend his name to promoting the book, writing that given Ardrey's personal remarks, "sometimes bordering (at least) on libel," if he did endorse the book he would then have to apologize to "[Raymond] Dart, [Louis] Leakey, [Wilfrid] Le Gros Clark, [Solly] Zuckerman, [Sherwood] Washburn, and other good friends."[44] The zoologist Clarence Ray Carpenter, about whose work Ardrey would write extensively in his next book, thought he had "over-emphasized the importance of my primate studies" and a "biased opposition to Zuckerman."[45] Carleton Coon, author of the controversial *Origin of Races*, noted that "Mr. Ardrey writes . . . very emotionally and dramatically" and believed he had been overly swayed by Dart's version of history. In stark contrast to Ardrey's description of Dart's "appalling durability," Coon claimed that Dart "talks all the time and is wont when overexcited to burst into tears."[46] Ashley Montagu wrote to Ardrey directly, complimenting him on the book "in spite of the fact that I thoroughly disagree with your views on the innate nature of man" and promised to review it and recommend it "enthusiastically."[47] Ardrey interpreted his olive branch less magnanimously, writing to the Dutch primatologist Adriaan Kortlandt, "I know Montagu. And he hasn't raised one peep against AFRICAN GENESIS, not in public. He's frankly afraid of me on a literary level."[48] The physical anthropologist Sherwood Washburn complained that Ardrey's version of the controversy over Dart's evidence was long out of date, writing that recent findings proved Australopithecines could make tools, some even of "complicated form." He added, too, (thanks to Jane Goodall's research on chimpanzees in Tanzania) that as "animals are known to be tool makers, the question of bone tools is of far less importance than before." "I think we would all be better off," Washburn concluded, "to stress our definite knowledge of the stone tools and minimize the highly debatable fragments of bone."[49]

Despite scientists' general lack of regard for the book, it was reviewed in a small handful of venues—perhaps because Ardrey's argument touched on so many contemporary issues in society, like juvenile delinquency and international communism.[50] In general, these reviews emphasized the environmental factors Ardrey largely ignored in his analysis of our aggressive human nature.[51] (He had been careful to mention, albeit only once, that "not all specialists yet agree" on "our total animal legacy."[52]) *Time* magazine, for example, complained that nonviolent behaviors like game traps, digging sticks, and seed mills were equally important in the development of human culture, just as fire and speech had been crucial in the origins of cooperation and the enculturation of humanity. Ardrey had hardly mentioned these things, instead expressing his fascination (perhaps even admiration, the anonymous reviewer intimated) for juvenile delinquents. A quote from Ardrey made his point clear: "There is always the weapon, the gleaming switchblade which the nondelinquent must hide in a closet, or the hissing, flesh-ripping bicycle chain which the family boy can associate only with pedaling to school." "Hardly a scientific observation," the review concluded.[53] Most prominently, in *American Anthropology*, the political scientist Jean E. Havel remained optimistic. He noted that one implication of accepting the concept of territoriality as central to human development "is a condemnation, as against nature, of Communist society."[54] Even in Hungary and the Soviet Union, he noted, differences in social status remained and were actually reinforced. Havel, alone among Ardrey's reviewers, promoted the relevance of recent research on animal behavior in shedding light on human nature.

When Ardrey began his next project—*The Territorial Imperative*—he returned to Africa to produce a new kind of book. In it, he took his criticism of social scientists even further.[55] He eliminated the novelistic structure, along with his self-deprecating "everyman" narration, replacing them with a more traditional approach to popular science. Ardrey referred to this new style as "pretend[ing] to a much stricter authority than AFRICAN GENESIS."[56] He changed his mind on a couple of intellectual points, too. Ardrey rued using the word "instinct" so blithely in *African Genesis*, "chopping off behavioral heads with it in a manner to have delighted Alice's Red Queen."[57] His concern, he wrote, stemmed from the lack of a professional consensus over the term's meaning. On the one hand, he surmised, instinct was a useful concept and not one to be avoided ("in the manner of many psychologists")—the term neatly combined "genetic design and relevant experience."[58] On the other, he warned himself and his readers against using it "like a two-handed axe in the hands of an absent-minded dentist."[59] There he let the matter rest.

Perhaps the most profound difference, however, was the near-total replacement of evidence from paleoanthropology with detailed discussions of animal behavior. Based on recent and decades-old research of ethologists, Ardrey derived lessons from mockingbirds and the kob antelope about territoriality, patterns of mating, and social organization in humans. Ardrey declared that territoriality was a consequence of humans' "evolutionary past, not our cultural present." "Man," he suggested, "is as much a territorial animal as is a mockingbird singing in the clear California night." "The dog barking at you from behind his master's fence acts for a motive indistinguishable from that of his master when the fence was built."[60] Parallels between human and animal sexual desires were as strong as those between human and animal instincts for private space.[61]

Most counterintuitively, Ardrey defined territoriality as an inward pressure, manifested in a drive to achieve dominance over other (male) members of the group. Therein lay the key to understanding differential social status and the organizational stability of hierarchy. He saw each of these instincts as ultimately selfish—through holding a territory, defending their social status, or challenging neighbors, men strove to better themselves.[62] In this way, Ardrey saw social cohesion as emerging from ritualized, inherently aggressive, interactions among individual males. He posited that amity and enmity were two sides of the same coin.[63] The aggressive instinct, when turned against a common enemy, would unite members of a community seeking to defend a shared border. On the other hand, inward-looking territorial fights also united a group of people who constantly renegotiated ownership of contested resources. Following the French primatologist Jean-Jacques Petter, Ardrey called such a community united through mutual antagonism a *noyau*.[64] In French, the word means "nucleus" or "core," like the pit of a fruit, and Ardrey adopted it to distance himself from the positive connotations of English words like "community" or even "society."[65] He applied *noyau* to animal and human communities equally, singling out the bowerbird as "the cultural champion of the nonhuman animate world" and "citizen of a cleanly defined *noyau*," thanks to his thieving, vandalism, and bullying behavior.[66] Ardrey did not imagine that all human or animal societies operated this way nor that *noyaus* were genetic or instinctual. His cultural predilections revealed themselves when he described Italian society as a functional *noyau* (Ardrey was living in Rome at the time) in contrast to the far more cooperative British society. The greater the forces of enmity arrayed against a group, the stronger the bonds of amity binding those individuals together into a community.[67] Readers (and reviewers) rather quickly picked up on the *noyau* as his most, perhaps only, original contribution to theories of social organization.[68]

Ardrey acknowledged that the "territorial imperative" was only one among many evolutionary forces shaping our lives, yet he insisted it was likely the most important, because it provided "the biological law on which we have founded our edifices of human morality." To clarify, he continued, "Our capacities ... for social amity and mutual interdependence have evolved just as surely as the flatness of our feet, the muscularity of our buttocks, and the enlargement of our brains, out of the encounter on African savannahs between the primate potential and the hominid circumstance. Whether morality without territory is possible in man must remain as our final, unanswerable question."[69] Ardrey did not sound a warning of our necessary and impending doom, however. He reasoned that our capacity for morality arose as a result of our most selfish tendencies (not despite them). Thus, he could see no reason why we would retain psychological "subconscious remembrance of the monster" past. Indeed, he insisted, "we clung to and perfected further that most effective of defensive weapons to be found in our primate legacy; for the monster was within us."[70] Ardrey designed his book to help readers understand their deeply engrained territorial instincts and thereby learn self-control.

Ardrey found particular inspiration for his vision of social behavior in the work of the ecologist Vero Copner Wynne-Edwards.[71] Wynne-Edwards argued that birds and other animals rarely caused each other lasting physical harm during their contests for superiority—one of the key lessons herring gull behavior had exemplified for MACOS students. "Instead," Wynne-Edwards suggested, "they merely threaten with aggressive postures, vigorous singing or displays of plumage. The forms of intimidation of rivals by birds range all the way from the naked display of weapons to the triumph of splendor revealed in the peacock's train."[72] Humans, on the other hand, had lost the knack for ritualized combat and too often succumbed to real killing. Yet based on his firm belief that much social behavior was indeed ritualized, even in humans, Wynne-Edwards defined societies as "brotherhood[s] tempered by rivalry," expressed through ritualized combat so as to preserve the longevity of the population.[73] Ardrey made use of Wynne-Edwards's twinned concepts of brotherhood and rivalry in his explanation of *noyaus*.

Ardrey fancied his evolutionary tale of territorial competition to be an inoculation against Western civilization's dual obsessions with sex and Freud. Neither females nor males in the animal kingdom fixated on the sexual attentions of the other, he insisted—females mated with whichever male happened to be occupying the best territory, while males fought each other for territory and therefore for status. Ostensibly describing the mating behavior of the kob antelope, he suggested "the female wants her affection, but she wants it at a good address. Whether or not our human sensibilities are offended or intrigued, it is a harsh truth that the doe is attracted and excited by the qualities of the property, not the qualities of the proprietor."[74] Ardrey dismissed, too, the idea that male courtship had evolved primarily to attract females, contending instead that courtship rituals functioned as competitions for territory with other males, only secondarily garnering females in the process.[75] He submitted that therefore the true objects of a male's thoughts during the mating season were other males—his performance mattered "in the eyes of his fellows"—and discounted Charles Darwin's theory that male courtship displays had evolved to attract females.[76] He followed this explanation by defensively disavowing potential objections that such male-male attentions represented a potentially homosexual tendency, not healthy adult male relationships.[77] He need not have worried. The idea that males were preoccupied with gaining the respect of other males became one of the oft-cited conclusions of his work.

The Territorial Imperative spent three weeks on the *New York Times* bestseller list. For readers of colloquial science, Ardrey's book spoke to the centrality of current research in animal behavior to understanding the dilemmas confronting world politics. Non-ethologically-trained scientists were particularly enamored of its arguments, giving it a stamp of academic legitimacy. One zoologist, for example, suggested that Ardrey was such a good scientific detective, discovering and integrating "important matters that have hitherto been overlooked," that a century earlier, "he would probably have rescued Mendel's work from obscurity well in advance of its actual discovery."[78] Even if reviewers still doubted some of Ardrey's more speculative conjectures, several granted him considerable leeway, given the perceived import and persuasiveness of his larger message. Not everyone agreed.

Rather than greet *The Territorial Imperative* with silent disdain (as they had *African Genesis*), the scientists who disagreed with Ardrey's conclusions chose to negatively review the book—in part because of the attention it was already receiving, but perhaps more importantly its publication coincided with the English translation of Konrad Lorenz's *On Aggression* (and shortly thereafter Desmond Morris's *The Naked Ape*). The anthropologist and historian of science Loren Eiseley, a decade after trading the dust of the desert for the grime of Furness Library, objected to Ardrey's thesis (in Eiseley's recounting) that "man . . . will tenaciously act out the drama forewritten in his bones by evolution."[79] Eiseley preferred to think of *Homo sapiens* hopefully, as "an emerging species" with "an unknown future." The anthropologist and Episcopal minister Earl Count grew equally frustrated with the praise heaped on Ardrey by the popular media and several years later suggested that *The Territorial Imperative* had "appeared just in time to permit a lumping of two very different authors . . . into a target for demolition simultaneously with a faulting of ethology itself—quite a roily agglomeration."[80] In bludgeoning Ardrey, then, some scientists sought to distance the fate of professional ethology, and thereby protect Lorenz, from what they saw as the playwright's amateur mistakes.[81]

The Territorial Imperative would remain Ardrey's most well regarded book exploring the nature of humanity. Ardrey planned to produce three more volumes along the lines of his existing anthropological forays—"a play in five acts," he called it. The next of these would be *The Social Contract*, exploring "the inner mechanisms adjusting the needs of the individual to the necessary demands of the group of which he is a part" (published in 1970). The fourth volume he imagined as *The Loneliness of Man*, introducing "the final animal pattern of such critical importance to the human way—exploratory behavior—and for the first time bring to focus in the light of all our assembled knowledge the legitimate uniqueness of man." Finally, and tentatively, *The Ordeal of Angels* would tease out "the philosophical implications of this new portrait of man, and the dilemmas facing such a remarkable species which in truth is a risen ape."[82] Neither of these final volumes ever saw the light of day. Instead, Ardrey would develop a fourth book based on the most controversial and therefore most salable aspect of his earlier volumes, *The Hunting Hypothesis*.[83]

5

The Human Animal

UNLIKE ROBERT ARDREY, Konrad Lorenz was a credentialed scientist through and through, although he too wrote with an eye to a wide readership. Primarily interested in the scientific study of animal behavior, Lorenz believed that understanding how and why animals behave the way they do would shed light on the predicament of human behavior and the problem of nuclear escalation. *On Aggression* sold well to the general public—first in German and then in English translation—even as it also generated considerable consternation among his ethological peers.[1] In interviews about the book, Lorenz never failed to impress. When he met the American freelance journalist Edward Sheehan writing for *Harper's Magazine* at the Max Planck Institute for Behavioral Physiology (in Bavaria, just 33 km southwest of the center of Munich), he dressed casually for life on the farm and wore large rubber boots, his "rich white mane and bearded chin" nonetheless lending him the distinguished appearance of a "wise man."[2] Another journalist described Lorenz's appearance as that of "a knickerbockered ski instructor grown venerable, hurly, and white-bearded"[3] (Figure 12).

As a young man, Lorenz had earned his medical degree and then enrolled in zoological and psychological seminars at the University of Vienna, where he began researching "animal psychology" in earnest. His contacts with psychologists and psychiatrists in Austria ultimately proved far more rewarding than his anatomical studies, and he hoped to eventually find a position as director of a field station. Until that proved possible, he conducted research on his growing menagerie at home in Altenberg. When Lorenz met Nikolaas Tinbergen, only a few years younger than him and equally excited about animal behavior, they quickly struck up a friendship. Remarkably, they managed to stay friends despite fighting on opposite sides in the Second World War— Lorenz had been drafted as a physician by the Germans (the first time in his life he had practiced medicine), and Tinbergen resisted the nazification of his native Holland. That both men served time as prisoners of war—Lorenz was

FIGURE 12. The *New Yorker*'s 1969 profile of Konrad Lorenz, in which *On Aggression* is
mentioned in the same breath as *The Territorial Imperative* and *The Naked Ape*. The
magazine carried stories about all three men, although Ardrey first appeared in their
pages in the 1930s for his work on Broadway. Illustration by William Hamilton, for "A
Condition of Enormous Improbability," *New Yorker*, 8 March 1969, 39.

captured by the Russians, Tinbergen was arrested by the Germans and sent to a hostage camp—proved important to their eventual reconciliation. After the war, the Max Planck Society established the new Institute for Behavioral Physiology and installed Lorenz as its director. Tinbergen moved to Oxford University, where he founded a program in ethology. By lore, Lorenz's "bold, intuitive" thinking and Tinbergen's "experimental and analytical talents" complemented each other well, which became a polite way of noting that their research programs diverged intellectually in the decades after the war.[4] Between their considerable influence, ethology began to flourish as an academic field devoted to the scientific study of animal behavior.[5]

On Aggression, destined to become the most sensational of Konrad Lorenz's books, opened peacefully, with Lorenz floating in the ocean off the Florida Keys, realizing his childhood dream of flying, moving through the water as most vertebrates move on land, "back upward and head forward."[6] He transported himself and his readers back in time, back to a peaceful, Edenic state in which he was one animal among many, "a stranger in a wonderland far removed from earthly cares."[7] Having awakened in his readers a childlike sense of awe in the natural world, he slowly pulled back, casually noting the behavior of the fishes beneath him. At the end of the chapter, Lorenz emerged, shedding the cool watery world to enter once again the sunny warmth of land and dispassionate analysis. "Never have I seen fish of two different species attacking each other, even if both are highly aggressive by nature," he wrote (he meant something more agonistic than predators eating prey).[8] This opening served as a gentle promise to readers that he would show them both the beauty of the natural world and the violence it contained, guiding them through unseen complexities with the sure hand of science.

Whereas Ardrey had lumped together hunting, cannibalism, and murderous rage into a single entity that defined humanity, Lorenz carefully distinguished the hunger associated with the killing of other species for food (an interspecific behavior) from (intraspecific) aggression inherent to killing a member of one's own species.[9] Hunters and warriors were not the same thing—and between them, Lorenz was interested in only the latter.[10] "My teacher, Oskar Heinroth," Lorenz joked, used to say "next to the wings of the Argus pheasant, the hectic life of Western civilized man is the most stupid product of intra-specific selection!"[11] One of the deepest intellectual splits between Ardrey and Lorenz concerned the timing and causality of man's relationship with tools of war: whereas Ardrey insisted that the accidental discovery of weapons drove our intellectual and social development as humans (the weapon made the man), Lorenz flipped these, asserting the far more commonly held belief that early humans self-consciously developed weapons as tools for hunting.

The remaining thirteen chapters of *On Aggression* followed the kind of reasoning and observations that Lorenz suggested had led him to an appreciation of the importance of aggression in organizing animal and human behavior. He began the book with simple observations of behavior, in both nature and the laboratory, from there moving to discussions about the physiological basis and function of individual behaviors. Ritualization—the process by which behaviors become fixed, stereotyped, and characteristic of a species—followed, mirroring his shift from descriptions of the behavior of individuals to the behaviors of groups and species. Lorenz argued that important animal behaviors, those associated with feeding, mating, and offspring care, for example, existed in order to help individuals survive—they had a function and an evolutionary history. With this point firmly established, he turned to the fundamental basis of the book: four chapters on the social organization of animal species. Presented in order of increasing complexity, Lorenz wended his way through the anonymous crowd without the perception of individuality (most fishes), family-based territories (night herons and other colonially nesting birds), tribal or pack-based societies (rats), and finally, societies with fluid membership in which mutual love and friendship act to maintain a coherent social order among individuals (Lorenz's favorite example, graylag geese[12]).

Toward the end of the book, Lorenz applied to human society the same kind of analysis he had hitherto devoted to animals. In doing so, he readily admitted, he was venturing from "natural science" and verified "recorded facts" to "turn to the question of whether they can teach us something applicable to man and useful in circumventing the dangers arising from his aggressive drives."[13] Lorenz described humans as by nature quite similar in social organization to graylag geese but also noted that much had gone wrong with the causes of aggression in our species. By approaching our nature with a sense of humility, he suggested, we might yet be able to avoid the violent self-destruction offered by the rapid proliferation of atomic weapons. Our ability to do so, however, required acknowledging that humans had a problem, working to understand humanity's aggressive natures so as to more reasonably interact with those around us, and whenever possible sublimating or redirecting aggressive drives through alternative channels.

When Lorenz's English translation of *On Aggression* appeared in 1966, it initially resonated with antiwar activists because he argued that humans are by nature dovelike—we lack natural weapons like fangs or claws—and it was the mechanized weapons of our own manufacture that drove our tendency to engage in world wars and outstripped our instinctive capacity to defuse conflict.[14] For members of the counterculture, Lorenz's most important message was that man lacked the "killer instinct" characterizing Ardrey's conception of humanity. Instead, Lorenz argued, men were born without inhibitions to violence. According to one editor, "the rebellious young . . . loved it when he told them they were storming in the wrong direction."[15] Give up the weapons, his book seemed to imply, and the violence between nations could be stopped. When asked about Ardrey, Lorenz remarked that he had "started with an interesting theory and then went out and found convincing evidence for it, rather than the other way around, which would have been more scientific." "You can't explain everything in terms of territory," Lorenz added.[16]

The novel *Ecotopia*, written by the committed environmentalist Ernest Callenbach and published in 1975, took another of Lorenz's suggestions seriously—that the Olympics and other sports competitions could replace the battlefield.[17] It may therefore have come as a surprise to some of its readers (as it was for the narrator of the book) to discover the quiet tree-lined avenues of Ecotopia occasionally marred by aggressive screaming fights between its citizens and sporadic "war games" between towns. The narrator suggests that by releasing all their anger and aggression at the moment it occurred, people were far less likely to engage in acts of unprovoked violence, and crime had vanished completely. (Unlike the dominant rhetoric of the era, though, the narrator becomes a real Ecotopian man not by harming anyone else in the games but by being run through with a spear and surviving. As he recovered in the hospital, he discovered too that places of heading could be joyous) Lorenz agreed that rationality could trump our animal instincts, but only if people began building international and inter-ideological friendships and abandoned military enthusiasm.[18]

Although fictional Ecotopians had readily adopted Lorenz's advice, many actual reviewers of Lorenz's work questioned his sense of optimism by either berating him for a lack of "detailed arguments" justifying his trust in science or lamenting the inevitable persistence of the problem of aggression in modern society.[19] Additionally, critiques of Ardrey in American magazines, newspapers, and journals soon embroiled Lorenz, because their books shared a common vocabulary, if not sympathy. Like Ardrey, Lorenz dismissed a Freudian "death wish" as a potential cause of aggression, insisting that most violent acts emerged from the natural functioning of an aggressive drive (not individual psychoses)—itself a Freudian concept.[20] Both argued that aggression could be a useful behavioral instinct, acting to space individuals out over the range of the species and producing "impressive" male fighters and "courageous defenders of the family and herd."[21] Both dated the time at which the aggressive drive became important in human history to the evolutionary moment when humans diverged from our animal heritage. For Lorenz, when killing became more than just finding food to survive, it became a factor in competitions between males of the same species. He argued, sounding much like Ardrey, "When man had reached the stage of having weapons, clothing, and social organization, so overcoming the dangers of starving, freezing, and being eaten by wild animals, and these dangers ceased to be the essential factors influencing selection, an evil intra-specific selection must have set in. The factor influencing selection was now the wars waged between hostile neighboring tribes."[22] Some reviewers deemed these likenesses cosmetic, celebrating Lorenz as a modern-day scientific Saint Francis of Assisi and dismissing Ardrey as a mere popularizer.[23] Others saw both authors as perpetuating a new mythology of man as innately and irreparably violent.[24] Looking back in 1972, one anthropologist noted, "The author known theretofore to English-readers from his gentle and wise *King Solomon's Ring* and *Man Meets Dog*, suddenly appeared in monstrous guise for his readings of human nature in the light of the non-human spotted by ethology."[25] Perhaps most important, both Lorenz and Ardrey oriented their moral tales to readers outside the confines of academic discussion.

For American scientists who studied animal behavior, Lorenz's book raised a host of issues. Experimental psychologists criticized the observations filling the pages of *On Aggression* as nonrepeatable and "old-fashioned," reinforcing divides among scientists invested in animal behavior.[26] For at least a few observers, these reactions ironically reflected the deeply territorial nature of academic science. B. F. Skinner, from the epicenter of postwar American behaviorism, noted that aggressive behavior could be experimentally induced in animals through the administration of electric shocks. Thus, he concluded, aggression as a term combined behaviors that could stem from a variety of causes, not all of which were instinctual.[27] He worried that the effect of *On Aggression* might be to seriously mislead Americans if it "encourage[d] a nothing-can-be-done-about-it attitude."[28] Theodore Schneirla and Ethel Tobach worked as comparative psychologists at the American Museum of Natural History in New York City and both wrote negative reviews, taking a position against Lorenz's description of animal "instincts" and the anecdotalism of the book.[29] The zoologist and author of colloquial science Marston Bates remarked that "scientists are not nearly as impersonal and impartial as they like to make out—after all, they are people, and they can be quite aggressive when their particular intellectual territories are invaded."[30] But objections to the book were both professional and personal, as at least several of his peers believed Lorenz's ideas carried more than a whiff of his involvement with National Socialism during the Second World War.

Lorenz had joined the Nazi Party in 1938, and historians have attributed his decision to a combination of careerism and genuine enthusiasm.[31] In 1941 he was drafted into military service and served two years as a military psychiatrist in the German army, was then transferred to the Eastern front as a physician, and less than three months later was captured by the Soviet army. Lorenz spent nearly four years as a Soviet prisoner of war. In publications between 1938 and 1943 (he wrote the final manuscript while serving as a psychiatrist in Poznan), Lorenz used his observations of domesticated and wild animals to emphasize the importance of racial purity and family harmony in producing healthy, natural behavior.[32] Domestication of animals (and humans), Lorenz had suggested, would lead to physical degeneration and sexual deviance. He never wrote derogatory statements about Jews, nor did he explicitly endorse Nordic racial supremacy.[33] Yet after the war he also seemed hesitant to acknowledge that by actively contributing to the scientific community during the war, his publications legitimized more incendiary claims.

Written in German, Lorenz's publications received little notice from Americans until after the war. In 1953 Danny Lehrman—a graduate student at NYU working at the American Museum of Natural History under Schneirla—critiqued Lorenz's concept of "innate" behaviors in animals. Lorenz and Tinbergen both classified behaviors as either "innate" (due to the biological inner workings of the animal) or "learned" (due to experience cultivated over its life). Yet Lehrman, following Schneirla, believed this was a logical fallacy; all behaviors were a combination of both.[34] Supposedly innate behaviors are only expressed under particular environmental or social conditions. Similarly, all supposedly learned behaviors depended on the biological capacity of the animal to learn, he reasoned. Once scientists took seriously the idea that behaviors develop in individuals, the distinction between the two categories quickly broke down. Lehrman could read German fluently thanks to his wartime service as a cryptologist, so in preparing the manuscript for publication, he read all of Lorenz's articles that had not yet been translated into English. In the full draft that he showed to Schneirla and submitted to the journal, Lehrman made a second critique too. He argued that ethologists had been too quick to leap from their discussion of animals to people, and he used quotes from Lorenz's publications in the 1930s and 1940s to prove his point, focusing especially on those moments when Lorenz discussed racial purity. The editors at the *Quarterly Review of Biology* persuaded him to tone it down.[35] They worried that too much attention to Lorenz's politics would obscure the scientific points Lehrman hoped to make. Although Lehrman's translations of Lorenz's wartime publications never made it to print, at least a handful of American behavioral scientists read the manuscript in draft. Still others heard about its contents. As it was, the published version of Lehrman's paper infuriated Lorenz, although Tinbergen read it (at least the second time) far more receptively.[36] In the coming years, Tinbergen took Lehrman's critique to heart and dramatically changed the way he thought about the origins of behavior in animals and people.[37] Lorenz dug in his heels.

When *On Aggression* had appeared in translation, ethologists abroad and comparative psychologists in the United States thus read it as evidence that Lorenz was continuing to frame his analyses of animal and human behavior in much the same way as he had before Lehrman's attack. At Cambridge University, Robert Hinde suggested that the central point on which Lorenz's argument failed was his "insistence on the *spontaneity* of aggression."[38] Recall that Hinde served as Jane Goodall's advisor when she worked on her dissertation at Cambridge. Together with Shaw and Lehrman, he also cofounded *Advances in the Study of Behavior* as an explicitly international journal and maintained a close working relationship with Lehrman throughout the latter's career. In *On Aggression*, Lorenz had suggested that lacking a proper outlet, the aggressive drive welled up within men until it unpredictably caused them to act. For Hinde, psychoanalysts might be amenable to such explanations, but "they are not accepted by the majority of students of behaviour." "Let us not" pin "all our hopes on redirection in situations which may in fact be arousing aggressiveness, or on sublimation," argued Hinde. "Let us not merely plan a society to cope with man at his worst, but remember that society can influence the nature of its ingredients."[39]

Anthony Barnett, of Glasgow University, reviewed *On Aggression* for *Scientific American* and complained about the many positive reviews the book had already received: "A book prominently reviewed by celebrated writers becomes important regardless of its intrinsic merits. The reviews mentioned (all by nonzoologists) agree on one thing: that at least Lorenz's accounts of animal behavior are authoritative and reliable." But, Barnett contended, "*On Aggression* does not in fact represent the methods or opinions current in ethology (the science of animal behavior)."[40] Upon reading Barnett's review, Ardrey took it upon himself to write Gerard Piel, the editor of *Scientific American*, in Lorenz's defense and called the review "an unforgettable disgrace to American science." In describing his missive to Desmond Morris, Ardrey called it an "old style glove slap" and complained the review was "worse than what Piel did to me (via Sahlins) for African Genesis."[41] (Marshall Sahlins had called Ardrey's first book "naturalistic fiction" and wrote that "the plausibility of Ardrey's thesis . . . depends as much on ignorance of mankind as it does on knowledge of animal-kind."[42] It degenerated from there.)

Solly Zuckerman, a prominent figure of British postwar science—thoroughly agreed with Barnett's assessment, declaring that "judged as a piece of writing, as a work of rich and compelling description, the book deserves all praise. But it is hardly a serious work of science, which one assumes is what Lorenz intended."[43] Like Lorenz, Zuckerman had trained in medicine but never intended to practice, devoting his professional career before the war to comparative anatomy and physiology (especially baboons). After the war, he returned to academic life. Zuckerman critiqued *On Aggression* by positing that "the complexities of human social and political behavior cannot be explained on the basis of oversimplified homologies and analogies with highly selective aspects of animal behavior."[44] This was indeed the most common criticism of the book: Lorenz analogized too readily between animal and human behavior and between the actions of individuals and the agendas of nations. At the core of scientists' frustration with Lorenz's book were real methodological disagreements over how to understand the origins of behavior in animals and whether the results of such studies could be usefully applied to modern human societies (and if so, how).

The anthropologist Ashley Montagu relentlessly attacked Lorenz for contributing to the sensationalism generated by Ardrey's overly simplistic biologism. As Lorenz was a scientist, Montagu held him more responsible for the tone of his publications.[45] Montagu argued that the scientific basis of either author's assertion was far from incontrovertible, insisting instead that "man, the animal, as a consequence of his unique evolution within the adaptive zone of the man-made part of the environment, culture, is born instinctless with a highly generalized capacity for learning. This constitutes his innate nature as an animal, as a hominid." For Montagu, human nature was not the result of man's biological legacy but "what *Homo sapiens* learns from his culture."[46] Throughout his career, Montagu had argued that education was the best method by which to change America's attitudes toward race and other social problems. Consistent with these long-standing political convictions, Montagu attempted to counter the rising popularity of humans as innately violent through articles in newspapers and magazines (with a wide readership).[47] In such essays, he posited that the new conception of men as aggressive apes was nothing more than a return to older, religious discourses of original sin (almost certainly this analogy stemmed from Ardrey's declaiming of modern humans as the children of Cain). In both cases, Montagu suggested, humans were supposed to overcome their base nature in order to become cooperative, contributing members of society. Yet "the fault" of violent Western culture lay not in human nature, he contended, but in nurture, thus it was "in the latter that the remedy also lies."[48] Many scholars picked up on Montagu's arguments, citing them favorably as evidence in their own attacks on the concept of man's aggressive nature. Sandra McPherson, for example, suggested that cooperation was more innate than antagonism as it was required for the procreation of the species. As a child psychologist, she agreed that there was no place for religious ideas like "original sin" in science.[49]

Despite the criticism heaped on Lorenz by some of his peers, however, other scientists defended both the man and their positive impressions of the book. The ornithologist Peter Driver, for example, portrayed Barnett's critique as suffering from fatal "deficiencies" and "hoped that people interested in behavior will not be discouraged from reading *On Aggression* by this particular warning display."[50] Additionally, in his *New York Times* review of Lorenz's new book, the British social anthropologist, and expert rhododendron cultivator, Geoffroy Gorer suggested that it was difficult to maintain a "belief in the essential gentleness of 'human nature'" when surrounded by the compelling evidence from history, social anthropology, and the daily news, of "man's willingness to hurt and kill his fellows and to take pride and pleasure in doing so."[51] He found persuasive Lorenz's argument that modern circumstances had forced people into closer proximity; because humans lacked the natural inhibitions against killing members of their own species, when crowded their aggressive drives caused greater strife and increased the incidence of violent crime. As a scientist speaking in favor of the book, Gorer's review was picked up and cited by writers like Driver who sought evidence of Lorenz's legitimacy.[52]

Lorenz, of course, was quick to defend himself: "Assertions that these are false analogies or anthropomorphizations betray a lack of understanding of functional conceptions. To call the animal jealous is just as legitimate as to call an octopus' eye an eye or a lobster's leg a leg."[53] He firmly believed that careful observations of animal behavior could help scientists reason through human actions. In fact, he deemed it necessary "if we are to avert destruction."[54] It was time for the moralists to give scientists their chance.

Even so, Lorenz would come to regret his use of the word "aggression" because he felt it obscured distinctions between his own theories and those advanced by Ardrey.[55] In an interview he conducted with the French magazine *L'Express*, later translated into English and published in the *New York Times*, he was asked whether *On Aggression* "justified" human aggression. He became quite upset: "Excuse aggression? Defend violence? I was trying to do just the opposite! I filled 499 pages in an attempt to explain that violence and war are a derailment of the normal instinct. I tried to show the existence of internal forces that man must know in order to master. I said that reason could conquer aggression."[56]

In the United States more so than in Germany, criticisms of Lorenz's work from fellow scientists hinted at the specter of his past as they sought to discredit his ideas on scientific grounds. Critics of Lorenz's position during the Second World War made easy connections between his insistence on innate behavioral mechanisms in animals and the idea that human behavior was crafted by biology, not culture. Readers unfamiliar with professional ethology found very different messages in his books, including a dire warning against nuclear proliferation. The positive synergy many readers had found between

Ardrey and Lorenz received a boost with the publication, fifteen months later, of a third book describing humans as animals—Desmond Morris's *The Naked Ape*. Neither *On Aggression* nor *The Naked Ape* ever earned the professional recognition equivalent to those of the intellectual paperbacks penned by Eiseley or Dobzhansky, but both reached large audiences thanks to the expanding market in colloquial science and the timeliness of the issues they raised.

6

Man and Beast

IN JANUARY 1968, before a single copy of Desmond Morris's *The Naked Ape: A Zoologist's View of the Human Animal* had been sold in the United States, it made national news.[1] A review in *Book World*, the shared book supplement of the *Chicago Tribune* and the *Washington Post*, used the word "penis" when discussing the anatomy of chimpanzees. The editors of these papers, who had not seen the text of the review until it was already in print, judged this language "in bad taste," and they recalled and reprinted the supplement—a costly decision that ultimately drew more attention to the book than the review itself would probably have merited.[2] Later that month, on Morris's American book tour, which was timed to coincide with the week of *The Naked Ape*'s release by McGraw-Hill, he mentioned the incident to Johnny Carson while being interviewed on *The Tonight Show*. Carson matter-of-factly replied, "You discuss the fact that man is one of the primates. You talked about his penis. What other word could you use for that?"[3] Despite Carson's calm demeanor, he had just spoken the word "penis" for the first time on (live) American television and (again) raised quite a few eyebrows.

When Morris stopped in Chicago a few days later, a representative of *Playboy* enterprises, the appropriately named Mr. Ravage, called and offered—at the behest of Hugh Hefner himself—to ensure that Morris's stay was "a pleasurable one" and offered him "anything you want (pause) *anything at all*."[4] By April, the film producer Zev Bufman and the director Donald Driver signed a contract for three films with Universal Pictures, including the movie rights to Morris's now best-selling book.[5] Morris began to receive annual gifts imprinted with *Playboy*'s iconic bunny logo. Two years later, Hefner, whose interest in Morris had not waned in the meantime, agreed to back the project to the tune of $1.1 million.[6]

When Morris was young, he wanted to be an artist. His interest in wild animals (sparked by a schoolteacher) combined with his fascination with surrealist art (engendered by friends) to produce fantastical images on his canvases.

He later described his art as creating "a private world in which my own invented organisms evolved and developed."[7] Morris ended up seeking a position as a graduate student at Oxford in the zoology department both because he was excited to work with Tinbergen, Lorenz's career-long interlocutor, and because it allowed him to follow his new found love, Ramona, who was reading history as an undergraduate student at St. Hilda's College (founded in 1893 as an Oxford Hall for women).[8] He later recalled his first lecture by the well-known ethologist as akin to a "religious conversion." Tinbergen showed him "in sixty minutes that seemed like five" how to engage in "rigorous research without having to turn one's back on the natural world of the living animal."[9] In Morris's recounting, Tinbergen was the careful thinker Lorenz could never be, a point he emphasized with a humorous tale of Lorenz's accidental partial circumcision by a crow and an uncomfortable public airing of Lorenz's affair with Helen Spurway.[10] Morris never lost his fascination with surrealism. With each animal he studied, he wrote in his autobiographical *Watching*, he "*became that animal.*" "Instead of viewing the animal from a human standpoint—and making serious anthropomorphic errors in the process—I attempted . . . to put myself in the animal's place, so that *its* problems became my problems, and I read nothing into its lifestyle that was alien to its particular species."[11] When contemplating the behavior of people, he continued this perspective, attempting, as it were, to study humans zoomorphically.

Morris took a more light-hearted approach to human nature than had either Ardrey or Lorenz, choosing to emphasize the pleasure-seeking aspect of human nature and the resulting sexual dilemmas of modern man—in effect, placing recent research by Alfred Kinsey, William Masters, and Virginia Johnson in evolutionary perspective.[12] According to Morris, "the naked ape is the sexiest primate alive."[13] More than in any other species, he suggested, human social bonding resulted from sexual attraction and interactions. Morris earned his DPhil under Tinbergen at Oxford, but rather than follow a traditional academic career, for years he hosted the popular British television show *Zootime*, broadcast at teatime and intended for younger viewers, and he published an astounding number of colloquial science books. In *The Naked Ape*, released in the UK in 1967 and still his most famous book, Morris provocatively suggested that humans lost the fur covering the bodies of most other mammals because it facilitated sexual caresses and made possible the development of other, now accessible, sexual signals: the rounded breasts and buttocks of women, the larger size of the male penis in comparison to other ape species, female orgasm, and the increased sensitivity of human nipples and genitalia. All told, these features made sex a lot more fun for humans and rewarded pair bonding with increased sexual pleasure.[14] Morris's *Naked Ape* concentrated on those aspects of the human physique that caused individual sexual attrac-

FIGURE 13. "The Expectant Valley," 1972. After his first exhibition, Desmond Morris decided he would never be able to make a living as an artist. He then devoted his career to science and painted in the free time he could carve out from his schedule. Following the surprise success of *The Naked Ape* and Morris's departure for Malta, his painting style changed for a few years, before he once again returned to his familiar "biomorphs," seen here. Richard Dawkins purchased this painting and used the image for the cover of *The Selfish Gene* (New York: Oxford University Press, 1976). Oil on canvas, 24 × 30 in. Collection of Dr. Richard Dawkins, reprinted with kind permission.

tion and desire, noting that human sexual partnerships formed the basis of our more general sociality.[15] A sexually active species, like humans, would be naturally more altruistic than a less sexualized species.[16]

The Naked Ape brought Morris such unexpected financial success that he fled to Malta, a tax haven from the British Isles, where (according to Ardrey in 1969) "he lies in the sun, gets phenomenally rich and appallingly bored." When they met, Ardrey had been impressed with Morris and described him as "an elegant scientist . . . and an elegant man, with an elegant miniskirted wife, Ramona"—an enviable life (Figure 13).[17] Morris's life in Malta was also quite productive. In his five years there, he published *The Human Zoo, Intimate Behavior,* and a collection of his scientific papers, *Patterns of Reproductive*

Behaviour.[18] Morris moved his family back to Oxford so that his son, born soon after the move to Malta, could attend British schools.

The Human Zoo represented a change in tone for Morris, who chose not to focus on the sexual habits of individuals, but instead explore the difficulties besetting urban populations around the globe. Were cities like urban jungles? No, Morris insisted; cities were like cages. When concentrated in large, over-populated urban centers, the normal dominance relations that governed human social interactions broke down. "The leaders of the packs, prides, colonies, or tribes come under severe strain. . . . So much time has to be spent sorting out the unnaturally complex status relationships that other aspects of social life, such as parental care, become seriously and damagingly neglected."[19] Asked (thanks to Hefner's interest in *The Naked Ape*) to contribute a précis to *Playboy* of his most recent book, Morris spent the bulk of his space offering lessons for managing human patterns of power, a vital skill in the unnaturally crowded environments of city life.[20] He provided his own answer to the same problem of dominance that Ardrey and Lorenz were eager to sort out—a zoological manual for winning the human rat-race. Unlike those authors (and unlike his previous work), Morris concentrated his entire effort on primates—baboons, chimpanzees, and humans. "The point is that baboons, like our early human forebears, have moved out of the lush forest environment into the tougher world of the open savannah, where tighter group control is necessary."[21] In so arguing, he followed the anthropologist Irven DeVore's contention that baboons made excellent models for understanding early man because they occupied similar ecological habitats.

According to "Man and Beast," an article published in *Playboy* in 1970, it was during the mid-1960s—thanks largely to the popularity of books by Ardrey, Lorenz, and Morris—that ethology "emerged from relative obscurity into the glare of world-wide attention, and its practitioners, once viewed as harmless bird watchers, [we]re now regarded as scientist prophets at whose feet modern man sits all atremble, waiting for the word."[22] Morton Hunt, the author, continued, "If those who study man—psychologists and sociologists—have not been able to tell us what we need to know, perhaps we can find it out from those who study animals."[23] A prolific nonfiction writer, Hunt was especially fascinated by the nature of American men and women. Eight years earlier, he had argued that all humans were a "hopeless tangle of heredity and environment," a position he continued to advocate through the 1970s.[24] Despite the widespread acclaim with which these recent books on human nature were received, Hunt refused to regard Ardrey, Lorenz, and Morris as legitimate prophets and remained deeply skeptical of their conclusions. Lumping their research together, common by 1970 despite rather dramatic differences among them, Hunt claimed their books epitomized man as genetically destined to be "the most brutal and uninhibitedly aggressive of all animals."[25]

In his critique, Hunt fixated on the aggressive drive he believed was common to all theories of human nature derived from observations of animal behavior, and he expressed dismay at readers' fascination and sympathy with "this depressing news."[26] He implied that the appeal of books by authors like Ardrey, Lorenz, and Morris lay in their promise of a new basis for understanding human nature where social scientists had failed. Hunt identified these very same individuals, including Montagu, Margaret Mead, and the philosopher Susan Langer, as opponents of the aggressive view of human nature who agreed that humans had no instinctive behaviors at all. Yet between these two extremes, Hunt suggested, lay most researchers interested in animal behavior, who insisted that "the entire nature-nurture, innate-learned, instinct-experience issue is outmoded, if not meaningless. . . . By far the largest part of it [behavior] results from interactions between genotypic tendencies and environmental influences."[27]

Hunt's lesson for *Playboy* readers? "For better or for worse," animal-human comparisons were here to stay. "The study of animal behavior will never again be a quiet backwater of zoology. Men now fervently hope, and almost demand, that animal-behavior researchers help them understand themselves and one another; and, given the present human condition, who can blame them?" So what was needed, in Hunt's view, was a more careful study of the functions of animal behavior without resorting to overly simplistic anthropomorphism.[28] "Man does have an aggressive instinct, but it is not naturally or inevitably directed to killing his own kind. He is a beast and perhaps at times the cruelest beast of all—but sometimes he is also the kindest beast of all. He is not all good and perfectible, but he is not all bad and not wholly unchangeable or unimprovable. That is the only basis on which one can have hope for him; but it is enough."[29]

Perhaps surprisingly to us now, Hunt's article attracted a series of Dear Playboy letters to the editor penned by professional scientists. For example, Evelyn Shaw, the biological psychologist and curator of Animal Behavior at the American Museum of Natural History, hailed the piece as a "brilliant rebuttal" to the simplistic determinism of Ardrey, Lorenz, and Morris. Julian Huxley wrote in, describing Hunt's "Man and Beast" as "interesting and rather provocative," while Irven DeVore sent compliments on the "provocative and judicious treatment" of animal behavior. Even Sally Carrighar, a prolific and well-known wildlife writer, added her praise.[30] After reading the piece, Montagu wrote to Hunt asking to reprint the article in a forthcoming volume—he readily agreed.[31] Support for Hunt's point—that books geared at the general public generalized too easily from animals to humans—came from across the scholarly community. All the while, these books continued to sell and people continued to read them.

Professional concern with the books tended to emphasize that Ardrey, Lorenz, and Morris were popularizing out-of-date ideas. Eiseley, for example, divided readers into two opposing camps. On one side, he associated excitement for the books with a conservative paradigm invested in naturalizing man; on the other, readers hesitant to accept the authors' negative essentialized view of human nature were likely committed to a liberal belief in humanity's perfectability. Eiseley placed himself in the second group and intimated that "the conservative school of thought has a lingering affection for the determinist machine of nineteenth-century law," into which we can read inklings of Eiseley's concern with the persistence of racialized thinking. In direct opposition to Ardrey, he declared that in the process of becoming self-conscious and self-measuring, humans had "in some strange sense" become "transcendent."[32] Eiseley penned these words in a 1972 centennial volume commemorating Darwin's 1871 publication of *The Descent of Man and Selection in Relation to Sex*. Other contributions also mentioned the recent books by Ardrey, Lorenz, and Morris, and none of the scientific authors were entirely pleased. The notoriously grouchy paleontologist George Gaylord Simpson chose Morris as his main target, arguing that "man is *not* a naked ape; he is a different species (and genus, and family) altogether."[33] Such "bad zoology" had led to the "naked ape fallacy," he insisted. Simpson argued that "aggression and territorialism are not ineradicable parts of human biological nature. They are maladaptive and we must use an evolutionary adaptability to counteract or readjust them where they occur harmfully among us."[34] The British ethologist J. H. Crook, meanwhile, equivocated in his evaluation of *The Naked Ape*—he clearly admired Morris's engaging writing but also worried that his skill as a writer obscured fundamental difficulties with his reasoning.[35] George Schaller, who had published intellectual paperbacks on the behavior of gorillas and lions, espoused a similar middle path: "Man is inextricably linked to his dual past, carrying within his frame both the terrible power of the predator and the frailty of the ape."[36] Regardless of their perspective, everyone suggested that yet-to-be-discovered fossils and more thoroughgoing analysis of behavioral data already at hand would resolve any lasting dilemma in their own favor.

The reaction of professional zoologists and anthropologists to *The Territorial Imperative, On Aggression,* and *The Naked Ape* was complicated by the fact that they appreciated the attention their disciplines received as a result of their popularity. When the Smithsonian Institution organized a conference on human and animal behavior in 1969, they titled it "Man and Beast." In the inevitable edited collection that followed, S. Dillon Ripley, then secretary to the

Smithsonian, asked, "What is man? Is the study of animals, including insects, fish and birds, also the study of man? Should the study of man be based on man alone?"[37] He did not answer these questions—that was the job of the contributors—but his preface highlighted the dramatic place animal behavior had taken in scientific deliberations over human nature. The book, he stated, stemmed from public demand for evaluating the contentious claims made in the new literature on animal behavior. He listed in this orbit a wide range of books from ethologists (even Tinbergen's *The Herring Gull's World* that had inspired the MACOS curriculum), psychologists, anthropologists, paleontologists, and primatologists. Understanding human nature was deeply interdisciplinary work.

The organizers of "Man and Beast" invited the anthropological doyenne Margaret Mead to comment, and predictably, she objected to the idea that the biological mechanisms at work in humanity could not be easily modified. Mead had been exchanging letters with the anthropologist Geoffrey Gorer for years, and when his review of Lorenz's *On Aggression* had appeared in the pages of the *New York Times*, she had written expressing her congratulations and offered to send him a copy of Ardrey's *The Territorial Imperative* (he gratefully accepted since the book appeared in the United States months before it was available in England).[38] "Because of scanty fossil evidence," she reasoned, scientific speculations about early human behavior relied on inferences from "recent studies of baboons, chimpanzees, gorillas, and others whose ancestors and ours sprang from the same stock."[39] Yet she urged the participants, and her later readers, to attend to the "rewards and limitations of how we extrapolate across species" and to draw equally on anthropological investigations of human cultures. One of her primary objections to the Ardrey-Lorenz logic rested on its use of animal models to explain, and even justify, negative human emotions like despair, helplessness, and pessimism. The entomologist Edward O. Wilson and various geneticists had tried to paint a sunnier picture of humanity's future, she noted, by arguing that it was possible to change animal behavior through selective mating in a very short number of generations— perhaps as few as ten. But even claims like these, Mead contended, focused public attention on genetic traits rather than environmental factors like nutrition and education that could alleviate social problems far more quickly. She feared this attention to biology, rather than the environment, as the primary source of (and solution to) human ills would cause readers to lapse "into a kind of passive despair."[40] Mead understandably objected to Ardrey's conclusions, but she found equally frustrating his insistence that he had single-handedly brought ethology into the public limelight.[41]

Despite the distinct identities and disparate training of Ardrey, Lorenz, and Morris, their fates became intricately linked in the eyes of the nonscientific public, such as Harry Guggenheim, who never appears to have appreciated the significant differences between the three authors. That Lorenz and Morris were trained scientists neither excluded nor protected them from public scrutiny. When Lorenz was asked about Morris's *Naked Ape* and Ardrey's *Territorial Imperative*, he responded that both authors were good friends. However, he added, if he "were forced to call the human species apes," he would choose instead "the culture apes, the cumulating tradition apes, [or] the ideal conception of all apes."[42] Ardrey, according to Lorenz, had "become a very good ethologist in his own right," but he still found *The Territorial Imperative* to be "a bit too daring."[43] All three men participated in a single conversation about human nature, both amateur and professional, popular and scientific, but above all public.

The utility of an ethological perspective in understanding human nature received a glowing recommendation when in 1973 Lorenz, Tinbergen, and Karl von Frisch (who discovered the honey bee dance language) won the Nobel Prize for Physiology or Medicine. By then Lorenz had become more careful to emphasize the malleability of humanity's future and argued that we all had a role to play in enabling that future. "For man is certainly an animal, but man, although identifiably a primate, is also a primate of a unique—and uniquely dangerous—species."[44] He was no longer as optimistic as he had sounded at the end of *On Aggression*. Building on another religious metaphor, in "Civilized Man's Eight Mortal Sins" he enumerated the most important transgressions perpetrated by modern civilization against nature and humanity: "overpopulation," "pollution," "uncontrolled growth," "the numbing of consciousness," "genetic decay," "the breakdown of tradition," "susceptibility to indoctrination," and "nuclear weapons."[45] Without attention to these issues, Lorenz insisted, humans were heading to premature extinction. After 1973, articles discussing research on animal behavior never failed to mention the Nobel Prize and reveled in Lorenz's slight-outsider status among other scientists: "Like fellow Nobel laureate Einstein, Konrad Lorenz, the father of ethology, has never been afraid to speak his mind—even when it caused his enemies to rise up and attack him like rabid beasts."[46] Reporters suggested that the prize served to mark the widespread acceptance of humanity's place within nature, rather than outside it, "subject to the principles that mold the biology, adaptability, and survival of other organisms."[47]

1973 also marked the release of *The Naked Ape* film (Figure 14). The movie combined live action and cartoon montages and took care to repeat much of Morris in his own words. For example, at one point in the film, a professor asks each student in his course to bring in an example of erotic classic literature. After embarking on a series of extended daydreams about dating fellow student Cathy (played by Victoria Principal), Lee (Johnny Crawford) apolo-

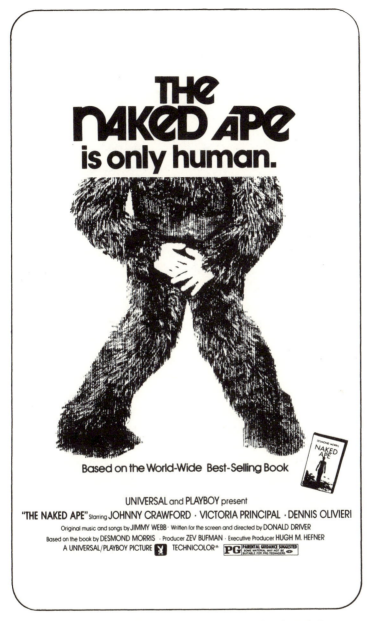

FIGURE 14. Advertising image from the Pressbook for *The Naked Ape* (Universal Pictures and Playboy Enterprises, 1973), directed by Donald Driver and based on the book by Desmond Morris. The Pressbook described the film as "a story about man's evolution, [that] traces his development from 10 million years ago to tomorrow with the integration of live action and animation."

© 1973 Universal Pictures and Playboy Productions, Inc.

gizes for not having completed the assignment and instead reads a passage from Morris's *Naked Ape* about biologically natural attitudes toward sexual experimentation. His performance is such a success that the whole class breaks into spontaneous applause, and the professor can do little but nod in approval and remark, "Right."

To convey even more of Morris's message, the comic relief of the movie, Arnie (Dennis Olivieri), composes aloud two letters he wants to write to the US president from the front lines of Vietnam, where he had been sent along with Lee. In the second of these letters, he laments the lack of a true pair-bond and wonders about the fate of the girls GIs had left behind. As Arnie begins to speak, he and the steamy jungle disappear, replaced by an idyllic Henri Rousseauesque cartoon. Meanwhile, in the cartoon jungle, an early human male is hiding in the bushes, watching a naked woman pick berries and frolic with butterflies, alone because her mate had just embarked on a hunting trip. Arnie's speculation that "true love" allowed men for millions of years to leave home to hunt or wage war, confident that their wives would remain faithful, was overlaid with images of an early human male sneaking out of the bushes to chase the beautiful, solitary woman. After a giggling romp through the lush foliage, the couple disappears from view, but the sounds of their lovemaking can be heard, as creatures of the forest turn out to watch and listen, including an idiotically grinning lion. Arnie has been talking all the while and expresses his "creeping doubt" that the pair-bonding process "was ever really perfected." According to Morris, this imperfect pair-bond allowed both men and women to seek sexual comfort in the arms of multiple others.[48]

Of course, that was not to say that jealousy in the form of Ardrey's territorial (and sexual) defense would not arise. In the movie, when the hunter returns to find his partner with the interloper, he roars with rage, and a close-up of his left pupil reveals a fanged beast leaping toward the camera as he chases the other man away. A female voice-over (Principal once again) picks up the thread of the visual argument: "So there he is, our vertical, hunting, territorial, brainy naked ape. The naked ape is a new experimental departure and new models frequently have imperfections. For instance, sexually the naked ape finds himself today in a somewhat confusing situation." The film combines two images of masculinity—sexual prowess and aggression—into a single descriptor of innate human nature: the "sexual hunter." In the process of becoming fully human, the film implied, our ancestors had transformed from fruit pickers to hunters and killers.

"Part live action, part animation and all banality," one reviewer claimed. "The film has its greatest potential among the high-school set."[49] A young Victoria Principal later lamented her involvement, claiming, "That movie almost ruined my career. . . . It was worse than terrible. The popcorn was better than the picture!" She added, "Part of the deal was that I pose nude for *Playboy Magazine*. I still regret that."[50] Morris mistakenly recalled, "The film was so awful that it never received a general release."[51] After Hefner's financial loss with Roman Polanski's *Macbeth*, *The Naked Ape* drained his coffers even more.[52] Morris stopped receiving his bunny-imprinted gifts.

PART III

Unmaking Man

What happened to females when we get to humanity?

—MARGARET MEAD TO KONRAD LORENZ

IN 1972 DOUGLASS COLLEGE at Rutgers University hosted the Berkshire Conference of Women Historians at their Continuing Education Center—the first of many such conferences, although no one knew that at the time. Douglass College had been a local New Brunswick school for women that had been incorporated into Rutgers, New Jersey's state university, only two years earlier. The organizers hoped to engage an opening keynote speaker who would spark a spirited conversation about the barriers faced by women in higher education. They invited fellow Rutgers faculty member Lionel Tiger, a social anthropologist and a controversial choice because of his conviction that women seeking professional careers were fighting against their intrinsic, biological natures.[1]

Together with Robin Fox, his friend and colleague, Tiger sought to import an evolutionary perspective into the social sciences, especially anthropology.[2] Following his publication of *Men in Groups* in 1969, Tiger had found particular professional notoriety and, among other readers, devoted admirers.[3] He argued that women who sought equal opportunity in their professions bumped up against evolutionarily ingrained behaviors. Just like our hunter-gatherer ancestors, his logic implied, males in modern society bonded more strongly with each other than they did with women, in order to secure the success of the hunt. Women who threatened to challenge these age-old male bonds had failed to acknowledge their power. In *Men in Groups*, Tiger posited that even if the proportion of women in male-dominated fields increased, men would continue to see younger women as sexual objects, older women as competitors, and neither as colleagues, instinctively continuing to seek alliances with other men.

FIGURE 15. Lionel Tiger attracted particular animosity for *Men in Groups*, published in 1969. The dust jacket of the book asked, "Why do men court men? Why no women are allowed? Why wars are for men only?" A review in the *Chicago Tribune* noted that despite the rhetoric of the book, Tiger's wife was a "professional woman." The author quoted Virginia Tiger explaining her own biological irrelevance: "Studies show that more highly trained and professionally successful women have fewer children, and they have them later in life. . . . So you see, this sort of women breeds herself out of existence." Illustrator Chuck Slack's cartoon accompanied the review and echoed its evocation of a group of men shamelessly observing a solitary woman with no group bond of her own. Illustration by Chuck Slack, for Clarence Petersen, "Men Are the Spine of Society!" *Chicago Tribune*, 31 July 1969, 1. Reprinted with kind permission of Chuck Slack.

Tiger's claims infuriated women with a wide variety of political commitments, prompting him to write a short defensive essay for *Maclean's* readers: "But I Never Said Women Are Inferior" (Figure 15).[4] Putting his best foot forward, Tiger asserted that although he had claimed men and women were different, he had not intended this to be read as a statement of hierarchical value. "Difference" need not imply one sex was superior. To believe that men and women were biologically identical, he insisted, was equally disrespectful. "If we are to believe those who claim that the position of women in society results from the brainwashing they undergo, we must also accept that if the whole female sex tolerates this, then perhaps it *is* less ambitious and less discerning. This is both nonsense and unflattering to women." How much more liberating, he argued, to believe that women possessed personal warmth and a maternal skill set unmatched by men. Unless modern society recognized and supported women, he feared we would end up in a society unnaturally conditioned by test-tube babies, the socially engineered equivalent of mass produced chicken that tasted of "grubby *papier-mâché*."[5]

To the disappointment of the organizers, rather than discussing these opinions directly, Tiger devoted his address at the Berkshire conference to myths of cultural determinism, using as his key examples the Soviet agronomist Lysenko and the American behaviorist B. F. Skinner. According to one member of his audience, Tiger began by remarking, "It is idle to assume I'm happy to be here and idle to assume there is unanimity in delight that I'm here."[6] He decried the continued influence of B. F. Skinner's behaviorism on American thought. Just as the environmentalist thinking of Trofim Lysenko delayed the acceptance of genetics in the Soviet Union, his talk intimated, the continued popularity of Skinner's environmentalist thinking about behavior kept Americans from seeing the truth about the evolutionary origins of the universal human nature that constrained us all.[7]

The historian Ann J. Lane (who had taught at Douglass College until the previous year, when she left for John Jay College in New York City) provided a formal comment and took Tiger to task for his theories of male bonding. She accused him of "indulging 'in wild extrapolations from some animal behavior (for instance baboons, not gorillas) to all he-man behavior.'" Equally scathing, she suggested that at least in history "there is some reliance on evidence, some fidelity to sources." The following morning, several Berks attendees privately reprimanded Lane for the forcefulness of her attack, even as they sympathized with her intellectually.[8] After the conference, the organizers regretted inviting Tiger due to "the flap" induced by his talk. They concluded, "It was a bad idea, although we think not because Tiger has ideas repellant to most of us but because his talk was so disappointing. We expected to have someone to 'take on'—Ann Lane deserved this—and we got a mouse."[9] In general, however, they deemed the meeting a resounding success and were quite pleased with the turnout, the quality of the presentations, and especially the spirit of the discussion.

This story, in scope if not in detail, rings with familiarity: a male scientist arguing for the biological basis of differences between the sexes, a female humanist insisting his gendered logic was groundless. The familiarity reflects the frequency of tropes historians tend to marshal when telling stories about women fighting against patriarchal social structures, celebrating their critical research into the classed, gendered, and racialized basis of authority in the United States and abroad. In the early years of second-wave feminism, these commitments seemed to align rhetorically: with women, the humanities, and nurture on one side of the equation, and men on the other with technology and science.[10] These politicized conversations created professional friction for the many women who sought to join the ranks of the "establishment" themselves—especially scientists.[11]

Looking back on the legacy of the Sixties, the linguist Noam Chomsky recalled the "change in the air, young people organizing and in some cases dying in challenges to laws and institutions. And then of course there were to be the drafting of men for Vietnam and the steady rise of the women's movement. And change after change in the parts of the world addressed by anthropologists."[12] In 1972 the US Congress passed the Equal Employment Opportunities Act and the Education Amendments Act, known colloquially as Title IX. President Nixon signed both into law, mandating equitable opportunities for study and employment. This new legislation reflected a demographic shift already underway in higher education. In the United States, the number of women earning doctorates in science between 1965 and 1974 increased almost 250 percent, a dramatic change in raw numbers but also a proportionate increase from 7 to 14 percent, reversing declining trends from earlier decades.[13] As unemployment and inflation rose between 1973 and 1977, the presence of women in science faculties across the country increased almost three times faster than the total rate of growth of science faculties as a whole.[14] Work places and organizations with all-male networks of power were called into question, not only by left-leaning feminists but also occasionally by right-leaning conservatives concerned with the influence of "homosexuals" in the public sphere.[15] The visible presence of women and racial minorities at conferences and in home departments, as both graduate students and as professors in their own right, became hard to ignore.[16]

As women entered anthropology, animal behavior, and archeology in greater numbers, they asked both what our female ancestors had contributed to the origins of humanity and how their own research could make a difference in contemporary society. Such questions were especially timely as anthropologists shied away from concentrating their attention on "exotic" tribes located in remote locales (the more remote, the better) to "studying up," as Laura Nader put it. "What if," she asked, "anthropologists were to study the colonizers rather than the colonized, the culture of power rather than the culture of the powerless, the culture of affluence rather than the culture of poverty?"[17]

Male scientists, too, noticed the changing landscape of academia and strug-gled to make sense of their own experiences. Fox and Tiger wrote about the inherent difficulties women would face in a world where everyone pretended sex did not matter. They defended the normality of adult male homosocial associations against two groups they identified as combatants—the feminists and Freudians—basing their research on the ostensibly universal behaviors shared by all human cultures, as shaped by our evolutionary past. In seeking to safeguard a masculine preserve for the right kind of men, their theories also appeared to discourage women from joining the highest ranks of academia.[18] Yet these efforts to describe a past and invoke a future with men as the core instigators of intellectual progress served instead to galvanize an interdisci-plinary community of feminist anthropologists.

Such debates rendered treacherous the porous boundary between profes-sional and colloquial discourse, especially for women. As the turmoil of the revolutionary Sixties rolled into the 1970s, authors and readers often clashed in their interpretation of texts and tone. Were man-the-hunter theories mi-sogynist? Did feminists have a sense of humor?[19] The same scientific idea or passage in an article or book was read in radically divergent ways depending on who was reading.[20] If race had been the question dominating discussions of violence in the late 1960s, in the early 1970s feminists added a new facet, challenging the notion that men and women possessed different natures. In the words of Kate Millet, society had "arbitrarily assigned traits into two cat-egories; thus aggression is masculine, passivity feminine, violence masculine, tenderness feminine, intelligence masculine, and emotion feminine, etc. etc."[21] She called for nothing less than "a change of consciousness of which a new relationship between the sexes and a new definition of humanity and human personality are an integral part"—a revolution that would reshape American society and culture.[22]

For female scientists who entered academic conversations from the periph-ery rather than the center, establishing and maintaining a professional identity required more careful negotiation than demanded of their male peers. Ardrey had published The Territorial Imperative in 1966, rendering into exaggerated prose the association of aggression with cognitive prowess through the legacy of the all-male hunting group.[23] When the Animal Behavior Society invited the playwright and Hollywood screenwriter to deliver the luncheon address at their 1968 meeting, Ardrey was flattered by the invitation and warmed by his congenial reception.[24] Fox and Tiger both celebrated Ardrey openly as an honorary member of their intellectual "tribe." Elaine Morgan also ventured into colloquial nonfiction after writing radio dramas for the BBC (credentials not unlike those of Ardrey), publishing The Descent of Woman, in which she skewered man-the-hunter narratives as "Tarzanist" fantasies glorifying men and erasing women's contributions to the evolution of humanity.[25] Ardrey and Tiger merited special attention in her analysis, as did Antony Jay for *Corporation Man*, in which he adapted their arguments in a gendered reading of businessmen (and their secretaries).[26] Yet Morgan remained on the fringe of scientific respectability her entire career because she had also suggested that in our long, evolutionary past, humans had passed through a crucial aquatic phase before adapting to life on the open savannah. While Ardrey was fêted as one of the boys, feminist anthropologists could never afford to publicly align themselves with Morgan, even if they embraced at least part of her message—their own credibility would have been damaged if they became associated with Morgan's speculative reconstructions of hominid evolution.

In the new intellectual landscapes created as a result of the changing demographics of higher education, drawing lines of *hard* demarcation (between science and not-science or even pseudoscience) and *soft* demarcation (between research central to the field and tangential distractions from the real work of science) became one way of establishing professional authority. So-called pop-anthropology thus contributed fundamentally to professional controversies over reconstructions of human origins in this decade—not only as widely read accounts of evolutionary theory but also as foils against which (other) professionals could establish their own intellectual standing. Most tellingly, professional anthropologists never reached consensus on which questions, methods, or answers should form the core of their field.[27] Instead, multiple contested intellectual communities came to coexist in a single perilous academic jungle.

Thus, both masculinist and feminist theories of human evolution arose out of this same cultural moment. Just as Fox and Tiger sought to make their discipline more biological, cultural anthropologists were embracing radical critiques from the left.[28] Social changes at home and abroad highlighted the rampant inequality of economic opportunity for all people. Women in anthropology and primatology pushed against the stereotypes embedded in man-the-hunter theories, even if their reconceptualization of women's place in nature—and academia—did not come easily. Fox and Tiger, among others, stressed that feminism itself was biologically ill-conceived. In 1974 Konrad Lorenz similarly remarked, "I'm sorry to have to disappoint the ladies, but in most cases there is clear male dominance."[29] Did this "clear dominance" emerge from humanity's evolutionary legacy or from the strictures of culture? All answers to such an apparently intellectual question were deeply personal.[30]

Numerous overlapping issues worked to create professional divisions within anthropology from the 1960s and '70s, with women contributing on all sides of these discussions. In 1975 the sociologists Harriet Zuckerman and

Jonathan Cole singled out three particular barriers to women's full participation in American science: science was "culturally defined" as a male activity; if women nevertheless chose to pursue a career in science, other scientists often perceived them as less competent than their male colleagues; and some women faced "actual discrimination" from their peers.[31] Discrimination came in many forms, not all of it direct. Contemporary analyses of scientific credit indicated that awards, promotions, and prestige tended to accrue to those (predominantly male) scientists already recognized as experts in their field.[32]

Given the politics of the day, Ann Lane's response to Tiger's address at Rutgers was far from the most dramatic. As an undergraduate at McGill University in Montreal, Tiger had served as managing editor of the university's student newspaper and knew all the men who worked there, including one who eventually became the editor of *Maclean's* magazine.[33] When *Maclean's* ran its cover story about *Men in Groups*, it precipitated a three-day picket line around the editorial offices—the proximate cause of Tiger's ill-received response, "But I Never Said Women Were Inferior." Around the same time, he eagerly accepted an invitation to return to his alma mater for an invited lecture. His excitement over the opportunity quickly turned to disappointment. Writing to Ardrey just days later, he complained that "some twenty aging nymphets" had followed him onto the stage and sat at his feet throughout the entire lecture, "oohing and ahhing and in general performing the rites appropriate to an unskilled harem."[34] He discovered the reason the following day—the school newspaper had printed a full-page advertisement saying, "Women are inferior—Lionel Tiger says so, Moyse Hall, one o'clock."[35] Tiger had been looking forward to the talk and was anguished by his ridicule at the hands of undergraduate women. This anguish, in turn, was tinged by anger. He played it safe when talking to a room full of committed feminist scholars at the first Berks conference a few years later.

7

Woman the Gatherer

ONE MIGHT EXPECT that studies of primate behavior in the wild would occur naturally in zoology departments. The ethologist Robert Hinde at Cambridge University, for example, established a colony of rhesus macaques and worked with Jane Goodall and Dian Fossey, as well as with several students who also worked with Goodall at Gombe Stream Wildlife Preserve in Tanzania. In the United States, however, field primatologists often earned degrees in anthropology, perhaps working with Sherwood Washburn at Berkeley, who was fascinated by the behavioral insights that could be gleaned from careful research on wild populations. This not only gave a different flavor to their research, it meant that women entering graduate school in the United States to study paleoanthropology, cultural anthropology, or primatology could form friendships and find like-minded colleagues in each of these areas. Many anthropologists who fought to refute the picture of universal male authority implied by common narratives of human evolution were indeed women, often at the very beginning of what turned out to be long, notable careers.[1] Their research gave fuller form to a rhetorically powerful alternative to Man the Hunter in reconstructions of human origins—Woman the Gatherer.[2] Like her partner, Woman the Gatherer found intellectual support in research on long-extinct human ancestors, studies of human cultures today, and animal behavior, with a new emphasis on field research among primates.

As Mary Douglas elegantly posited, even a mistaken symbol has power, as long as the symbol is widely recognized.[3] Writing in 1957, she noted (following her reading of Konrad Lorenz's *King Solomon's Ring*) that although doves can be "relentlessly savage" and pelicans do not offer their young pieces of their own breast as food, doves serve quite well in Western cultures as a symbol of peace and pelicans of maternal devotion.[4] So feminist anthropologists mobilized not only to demonstrate that existing data could be reinterpreted but also to marshal new data against standard interpretations. Most of these now classic essays appeared in edited collections, like Rayna Reiter's *Toward an*

Anthropology of Women or Michelle Rosaldo and Louise Lamphere's *Women, Culture, Society.*[5] Such volumes illustrated the diversity of reactions among female scientists and activists and provided a common meeting ground for interdisciplinary work in anthropology that would have been difficult to publish in mainstream anthropological journals of the time.[6] The claims in these collections, and especially their argumentative frames, now read like common sense after the decades of careful inquiry they inspired into the gendering of science as both a process and a body of knowledge. There soon followed monograph after monograph by young female anthropologists, and like all first books, they necessarily took as their primary task establishing the expertise of their authors within a particular area of anthropological research.

In her now classic essay—"Woman the Gatherer: Male Bias in Anthropology"—the then graduate student Sally Slocum criticized Washburn for arguing that human sociality and civilization were made possible by hunting.[7] Channeling the wisdom of the day, Washburn had written, for example, that "the biology, psychology, and customs that separate us from the apes—all these we owe to hunters of time past."[8] Slocum objected to both their implication that men constituted the exclusive core of the evolutionary process in humans and the association of masculinity with violence. The newly minted PhD Rayna Reiter lumped professionals like Washburn together with "pop-ethology types," suggesting that their visions of human nature largely coincided when it came to their portrayal of women.[9] Feminist anthropologists, like Slocum and Reiter, worked to generate a robust set of data demonstrating the variety of gendered roles in primate and human communities around the world, thus fundamentally undermining claims for "universal" cultural behaviors shared by all men or all women, much less by all humans and, say, baboons.[10] New studies quantifying women's steady contributions to the caloric intake of hunter-gatherer societies further demonstrated that the communities' survival depended on these foodstuffs far more than on the unpredictable results of men's hunting.[11] From the perspective of many feminists inside the anthropological profession and out, these empirical findings exposed claims of universal male dominance as emerging from long-standing cultural prejudices against women rather than from our innate, biological past (Figure 16).[12]

Cultural anthropologists emphasized the diversity of social organizations and behavioral norms in human societies, especially among the hunter-gatherers they invoked when theorizing about early human culture. In Slocum's "Woman the Gatherer," she suggested that food sharing likely formed the basis of accentuated community living in humans, emerging as a function of lengthening infant dependency, as mothers came to provision infants for longer periods of time and eventually non-kin as well. According to her logic, the mother-infant bond, not hunting practices, formed the basis of family and

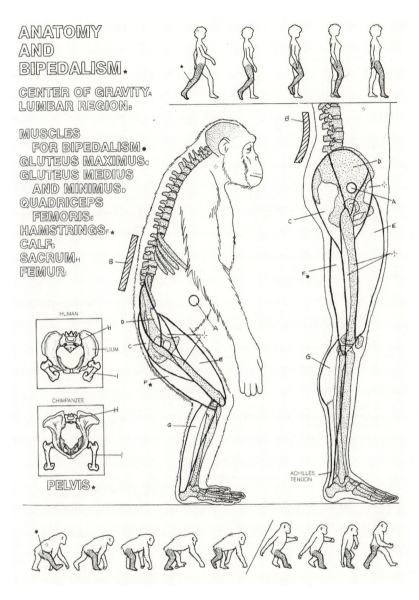

ANATOMY AND BIPEDALISM.

CENTER OF GRAVITY_A
LUMBAR REGION_B

MUSCLES
FOR BIPEDALISM.
GLUTEUS MAXIMUS_C
GLUTEUS MEDIUS
AND MINIMUS_D
QUADRICEPS
FEMORIS_E
HAMSTRINGS_F ★
CALF_G
SACRUM_H
FEMUR_I

HUMAN

ILIUM

CHIMPANZEE

PELVIS ★

ACHILLES
TENDON

FIGURE 16. Sometimes resistance to the male as universal could be countered simply by including an image of a female as representative of the species. In Adrienne Zihlman's *The Human Evolution Coloring Book, Second Edition,* she alternated between images of males and females when illustrating fundamental concepts of anthropology and biology. Here she depicts the muscles and bones that contribute to the differing gaits of humans and chimpanzees. "Anatomy and Bipedalism," image 55. Illustration by Julio C. Fernandez. Copyright © 1982, 2000 by Coloring Concepts, Inc. Reprinted by permission of HarperCollins Publishers.

community structure.[13] She further noted that among modern hunter-gatherer societies, even in difficult environmental circumstances women were able to gather enough calories to sustain their communities—so why, she asked, would we assume that human ancestors living in the Pliocene or Pleistocene would have been any different. Similarly, she argued that early tools could have been used first as "aids in gathering" rather than hunting.[14] New articles amplified this idea in the following years.

The first issue of the new journal *Signs: Journal of Women in Culture and Society*, created in 1975, contained a series of review articles, each considering the place of women in several social science disciplines, including one titled "Anthropology." The journal promised "new scholarship about women," an interdisciplinary perspective, and new truths (recognizing that "truth is never monolithic").[15] The eight authors of "Anthropology" drew on recent scholarship that reconsidered the role of women in human societies from the perspective of linguistics, archeology, biological, and cultural anthropology—all four anthropological fields. Carol B. Stack, Mina Davis Caulfield, Valerie Estes, Susan Landes, Karen Larson, Pamela Johnson, Juliet Rake, and Judith Shirek each spent time at Berkeley studying anthropology, and several were still in graduate school at the time of publication.[16] In the section "Males and Females in Nonhuman Primate Groups," they suggested that females had been observed exhibiting behavior previously ascribed only to males, and that males had exhibited so-called female behaviors. The authors also drew their readers' attentions to recent archeological and ethnographic data suggesting that women played a far more critical role in hunter-gatherer economies than had been broadly acknowledged by anthropologists. The centrality of cultural relativism within anthropology (and the social sciences generally) made it increasingly untenable to use "tribal groups" as stand-ins for early humans, according to the anthropologist Susan Sperling, writing about this moment in her anthropological dissertation on the history of human attitudes toward animals. As a result, she continued, "monkeys and apes have provided the new 'early ancestral group' from which human institutions are supposed to derive."[17] To draw humans and primates more closely together, however, required at least one of two rhetorical moves: an anthropomorphic interpretation of animal behavior or a zoocentric vision of humans. Just as Ardrey, Lorenz, and Morris invoked this logic, so too did feminist primatologists.

Like cultural anthropology and archeology, primatology provided fertile ground for exploring the vast array of models available for theorizing the human past. For example, in her dissertation research on langur monkeys in central India, Phyllis Jay (later Dolhinow) depicted a species that seldom quarreled, never defended territories, and generally led a tranquil existence.[18] Based on her research on lemurs, Alison Jolly argued the spectacular leap in intelligence that characterizes human paleoanthropological history was likely to have evolved as a result of the complexities of living in well-established social groups rather than as a function of tool use.[19] Thelma Rowell, for her part, sought to break down the picture of baboon social structure as emerging from aggressive-male dominance hierarchies by arguing that the dominance hierarchies themselves did not exist—they were merely an effect of the (male) observers' standpoint, created by the structure of the gathered data.[20] In her textbook *The Social Behaviour of Monkeys*, Rowell reviewed a great deal of the literature at the time, deliberately choosing not to place it in a singular evolutionary framework, instead emphasizing the incredible diversity of social behaviors, mating practices, trajectories of infant development, etc. When the equally young Robert Trivers reviewed the volume, he lamented this choice, writing, "The book's single, pervading weakness is that it presents no coherent view of what monkeys are up to"—which had been exactly Rowell's point.[21] Of course, the well-publicized field work of Jane Goodall, Dian Fossey, and Biruté Galdikas on "wild" chimpanzees, gorillas, and orangutans also provided evidence that the moral lessons to be learned from a primate model for human behavior depended very much on which primate model one chose.[22]

All of this new data, however, presented a problem. Why had male anthropologists not seen what these young women were seeing? Slocum, among others, attributed the phenomenon to "male bias."[23] This struck other anthropologists as problematic. When the cultural anthropologist Kay Milton analyzed the feminist scholarship of her peers, she argued they assumed that "the cultural devaluation of women exists in all societies."[24] This seemed unlikely to her. Instead, Milton suggested, male dominance as a concept within anthropology structurally echoed a cultural myth. Male bias took on symbolic importance for feminists with no grounding in reality and as a result could play no useful role in the future development of anthropology as a science. Critical of Milton's essay, Beth Elverdam wrote a letter to the editor stating that she found male bias to be a useful analytical concept and took Milton to task for what she saw as a misrepresentation of the literature in question. Milton responded by contending that "the aims of the feminist movement are in no way scientific" and therefore made "bad anthropology."[25] Whether or not feminist anthropologists saw themselves as enacting a political agenda through their research, critics ascribed such intentions to them, often expressing their criticisms in terms of preexisting fault lines between cultural and physical anthropologists, as well as between empirical and theoretical approaches within each of these fields.

Despite earlier constructive synergies between cultural and biological approaches to anthropology, conceptual space widened between them after the

Second World War as scientists in the United States and globally sought to counteract eugenicists' thrall with racialized, biological thinking.[26] By the 1970s, these tensions were solidifying into divides. In 1975 Ernestine Friedl, for example, described anthropology as consisting of two antagonistic schools— one that explained sex differences in biological terms and the other advancing social and economic causes.[27] Evelyn Kessler, writing a year later, similarly claimed that "the mass of evidence assembled tends to show that the 'environ- mentalists' have the better of the argument," again contrasting this position with biological reasoning.[28] Yet for Kessler, the environmental conditioning of women put them in a better place to act as "consiliators," enabling women to compromise and thus more easily reach agreement with others. Perhaps, she posited, women would be *better* at international relations and leadership than men.[29] Other anthropologists, like Sherry Ortner and Carol MacCor- mack, became deeply engaged in analyzing the metaphorical relationships between men and women as expressed in Western cultures, and the function of these metaphors in defining the process by which scientists framed human- ity's biological origins.[30] They transformed the science of human origins into mythological narratives reflecting the social, economic, and cultural pressures of the age in which they were created.[31]

In short, multiple centers of scholarly gravity formed within cultural an- thropology, each containing scholars who attended to the importance of in- cluding women in their analyses of cultures, bones, and primates. While some women found this work fascinating, others considered research on women's questions irrelevant to ostensibly more fundamental questions. More recently, the anthropologist Naomi Quinn has blamed the emergence of post-structuralism for marginalizing the success of many female anthro- pologists by ghettoizing their arguments as simply one of many possible readings of anthropological data. Echoing the feminist anthropologist Lila Abu-Lughod's consternation, she quotes, "Why is it . . . that just when sub- ject or marginalized peoples like blacks, the colonized and women have begun to have and demand a voice, they are told by the white boys that there can be no authoritative speaker or subject?"[32] (By the 1990s, Robin Fox also lamented that anthropology was in such "turmoil" that it was "no longer cer- tain that it can, as it has in the past, straddle both the sciences and the hu- manities."[33]) Quinn further insisted that these adversarial forces, especially as exemplified in debates over the place of women in science (as both sub- jects and theorizers), served to irreparably fracture cultural anthropology as a cohesive intellectual approach. Quinn thus laid the ultimate responsibility for the marginalization of female anthropologists within their own discipline at the feet of scholars who rearticulated theories of "ethnography-as-text" in the name of feminism.[34]

Fewer women entered biological anthropology or paleoarcheology, and those who did hewed closer to the traditional intellectual core of these fields. In 1997 the anthropologist Lori Hager looked back on the previous decades and speculated, "Fame has . . . come to three females in paleoanthropology."[35] She then named Mary Leakey (who had unearthed, among other famous finds, the Laetoli footprints of a male and female *Australopithecus* walking through sodden ash), the fossil known as Lucy (discovered by Donald Johanson and Tim White), and Mitochondrial Eve (the mother from whom the mitochondria of all humans are descended)—all associated with the late 1970s and 1980s, bridging gender as a characteristic of both scientists and their subjects.[36] When Hager wrote, she and others in her edited collection were still fighting many of the same gendered stereotypes embedded in human evolutionary narratives that had characterized anthropology twenty years earlier. Hager continued, "When a link is made from our ancestors to us, behaviors such as sex and gender roles become credible and justifiable since they are seen as 'inherent' or 'natural' to the species. These features become identifiable as part of 'human nature,' and because of this it is inferred that they cannot be easily modified or altered."[37] She worried that such evolutionary reasoning reified damaging stereotypes about men and women in the guise of science. For this reason, Hager bemoaned the continued underrepresentation of women in scientific discussions about the evolution of humanity.

Following this logic, it is perhaps unsurprising that women succeeded most easily in the disciplines and departments where they built dependable networks of support across all fields of anthropology. The Committee on the Status of Women in Anthropology (COSWA) was founded in 1969 to investigate the effects of sex discrimination in the profession, and advocate for equality in the hiring and promotion of women in academic institutions.[38]

One of the most successful nexuses of female graduate students in anthropology turned out to be the University of California, Berkeley, where a substantial number of women earned their PhDs under the guidance of Washburn and Dolhinow.[39] Mary Haas taught linguistics with the same empirical bent that Washburn applied to physical anthropology.[40] Laura Nader advocated turning the gaze of cultural anthropologists on cultures of power.[41] Gerald Berreman and Elizabeth Colson both called for greater attention to social responsibility among anthropologists.[42] By the mid-1960s, Berkeley had developed its reputation as a hotbed of radicalism. Student protests throughout the decade reinforced this impression.[43] The research agendas of many members of the faculty there began to reflect the climate in which they lived and worked.

When asked to reflect on her experience as a graduate student at Berkeley, the physical anthropologist Adrienne Zihlman responded, "Washburn was incredible." Yet she also remembers being mocked by her male peers. "Washburn was made fun of for having a harem." Male graduate students and even faculty members would intimate that Washburn liked women and that was the only reason he agreed to work with so many female students. It made her furious. She noted that when a friend got married and had a child, Washburn had encouraged her to take a little more time but to stick it out and finish her dissertation—which she did. Zihlman added that although she had not been aware of it as a student, Washburn had also been incredibly supportive of his colleague Nader. It was only at Washburn's memorial service that his students and peers began to share these stories with each other, and only then that Zihlman truly appreciated the extent of his behind-the-scenes efforts.[44]

Once she left Berkeley to join the faculty at the University of California, Santa Cruz, Zihlman found a like-minded colleague in the cultural anthropologist Nancy Tanner, who had also finished her PhD at Berkeley a few years earlier.[45] Zihlman and Tanner started teaching a senior seminar, The Biological and Cultural Basis of Sex Roles, and planned to write a book synthesizing the class. She recalls distributing mimeographed copies of Sally Slocum's essay before it was published and the students' great interest in civil rights, the women's movement, and activism on campus.[46] Zihlman also co-organized with Alex Alland a symposium at the American Association for the Advancement of Science (AAAS) annual meeting in 1971. (The following year Alland published *The Human Imperative*, in which he tried to "reinstate the dignity of man after his sufferings at the hands of latter day evolutionists and ethologists."[47]) Other speakers included the primatologist Suzanne Chevalier-Skolnikoff (whom Zihlman had met at Berkeley), the cultural anthropologist Janet Siskind (who had earned her PhD at Columbia with Alland but by then worked at Rutgers University), Lionel Tiger, and Margaret Mead.[48] In other words, quite an interdisciplinary and young group of scholars—with the exception of Mead, who had earned her doctorate more than thirty years before anyone else in the session. As a sign of the times, the AAAS meeting in Philadelphia that year also featured the symposium "The Role of Aggression in Human Adaptation," featuring a talk by Lionel Trilling on aggression in the arts and another on "Women in Academia" more generally.[49]

Tanner and Zihlman's areas of expertise complemented each other well. In Zihlman's doctoral research, she explored the question of whether *Australopithecus* could have preferentially walked on two legs (rather than four), by analyzing available examples of the pelvis, leg, and foot bones.[50] She traveled to London, Nairobi, and Witwatersrand, examining fragments of *Australopithecus* specimens recently unearthed in Kenya and South Africa and comparing them to the skeletons of gibbons, chimpanzees, and humans. Whereas humans balance by rotating their pelvis when walking (which allows the trunk of our body to move in a relatively straight line forward), Zihlman concluded that Australopithecines walked with their knees together but their feet angled outward. Given the position of their pelvis and hip joints, they would not have been able to internally rotate their body efficiently. Instead, they likely balanced while walking with a great deal more arm motion and side-to-side swaying of the body. Australopithecines, in short, had been effective but not efficient walkers. Zihlman further speculated (following Washburn) that the evolution of a bipedal way of life may have been one of the most important factors in the evolution of humanity, because it allowed our ancestors to keep their hands free for carrying and using tools. As a cultural anthropologist, Tanner's doctoral research had taken her to western Sumatra, where she lived with the Minangkabau for several years and documented their overlapping religious, matrilineal, and national legal codes.[51] Although nonevolutionary in focus, Tanner wrote about legal systems as a form of conflict management. She questioned the assumption that a society necessarily shares a single culture and therefore functions according to a single legal tradition. Anthropologists of legal systems, she wrote, typically sought to piece together the remnants of "native" or "customary" law onto which colonial systems had been imposed. Rather than seeing the resulting legal and linguistic amalgamation as artificial, however, Tanner built on recent scholarship to suggest that human cultures are more like a patchwork quilt. Overlapping legal traditions thus allowed diverse societies to negotiate a fragile but hopefully persistent peace. Despite dramatically different research interests, Tanner and Zihlman found common ground in their desire to rethink theories of human origins across the four fields of anthropology.

Although Tanner and Zihlman's collaboration dissolved in the process of writing their planned book, they did coauthor a paper in the new journal *Signs* entitled "Women in Evolution. Part I: Innovation and Selection in Human Origins."[52] *Signs* planned four issues a year, the first number would explore the social sciences, the second would concentrate on the humanities, the third on the natural sciences, and the fourth would concern questions of work and the professions. Tanner and Zihlman's paper appeared in the third issue. They submitted it to *Signs* because they did not think they could get it published in an anthropology journal.[53] Yet publishing it in a feminist journal also distanced their research from the anthropological mainstream. The primatologist Sarah Blaffer Hrdy, whom we will meet again, lamented publishing one of her articles in an edited feminist collection because she thought it never received attention from biologists until it was reprinted in a more conventional volume years later.[54] On the one hand, by publishing in edited collections or journals devoted to the topic of women, female anthropologists strengthened their social and intellectual ties to each other and found ways of putting their research ideas into print, particularly when other options were unavailable to them. On the other hand, this protective self-segregation also separated feminist perspectives from the central intellectual issues of the field.

In their paper, Tanner and Zihlman argued that females played a crucial role in the transition from ape to human in evolutionary history. Out on the dry savannah, where food was scarce, natural selection favored gathering as a dietary specialization, along with tool use and bipedal behavior.[55] Females, they suggested, had been crucial to prehuman adaptation to the savannah. Although Zihlman had started her training in physical anthropology, she was becoming interested in cultural anthropology, too—Tanner followed the opposite path. So they combined their interests in locomotion, chimpanzee behavior, sexual dimorphism, fossils, and human cultures. But what started out as fun ended up as trouble. Within a couple years of writing that first paper, they had a "big falling out."[56] Zihlman authored "Women in Evolution. Part II" alone.[57] In it, she posited that just as females had played an important role in the evolution of humanity, the social organization of women in early human groups made possible the rapid transformations of human culture that have characterized our history. Tanner, for her part, published *On Becoming Human*, again combining evidence from primatology, physical, and cultural anthropology to argue for a more gender-balanced reconstruction of human evolution.[58]

If advancing feminist arguments and publishing in nonscientific journals carried a degree of professional risk, then both Zihlman and Tanner chose to mitigate that risk by building from empirics, sticking to their training, and deriving new theories according to the same logic they had learned in graduate school.[59] (Doing this would have driven them even farther from the postmodern critiques of anthropological methods that so occupied the 1980s.) Like Zihlman and Tanner, other women may have steered clear of trying to write for an elusive "cross-over" market between academic and trade audiences, at least until after tenure. For the most part, science journalists, too, remained largely uninterested in feminist anthropologists' critiques of man-the-hunter theories of human origins, perhaps viewing work in feminist anthropology as a niche academic market.

If, instead of reveling in the historical complexity of this moment, we in the twenty-first century were to quickly glance back, we would catch sight of a mass of female anthropologists joining forces to pit themselves against select high-profile men who advanced what the former deemed overly simplistic theories of innate, universal differences between the sexes. Who better, then, to spark conversation as a keynote speaker at the Berks than Lionel Tiger?

8

The Academic Jungle

FOX AND TIGER met in London in 1965. According to Fox they started talking after one of Desmond Morris's seminars on animal behavior at the London Zoo—in front of the gibbon cages they discussed human and animal instincts and the importance of male bonding and competition in controlling human social relations.[1] (Tiger insists there were no gibbons.) Fox "was taken immediately by this funny, smart, talkative, small but confident son of the Montreal ghetto" and they bonded "fiercely."[2] Both eventually moved to Rutgers, and there they made quite a pair. Fox was lanky, charming, and British and chose to settle in the farmlands of rural New Jersey in a house outside of Princeton. Tiger was shorter and steelier, preferring the urban sophistication of New York City. They commuted to Rutgers from different directions, literally and figuratively. Fox and Tiger, both at the time and retrospectively, described their collaboration and the dynamics of the academic culture in which they worked as a function of theories about men in groups.

Tiger found his way to animal behavior as a result of his long-standing fascination with the logic of leadership. In his sociological dissertation on Kwame Nkrumah, the leader of Ghana who oversaw its transition from a colonial to an independent state in the late 1950s, Tiger picked up Max Weber's writings on charisma. As he read, he wondered about the source of Nkrumah's personal power. In a recent interview, he recalled one phrase that really struck him at the time: "Charisma is only understandable with an imperceptible transition to the biological."[3] He added a section to his thesis in which he wrestled with the overlapping jurisdiction of sociological and biological explanations of human behavior. When he submitted his thesis, someone (possibly his supervisor, Tom Bottomore) objected to the "offensive" pages and forced him to delete them.[4] Thereafter Tiger paid attention to research on social hierarchies in primates when he encountered them in the writings of John Crook, Desmond Morris, and many others on hominid evolution. He came to believe that standard sociological factors (inequality of access or prejudice against women, for example) were entirely insufficient to explain the dominance of men in politics.[5] Zoology provided a crucial new resource.

Fox came to animal behavior through a different route entirely, sparked by meeting the "Darwinian psychoanalyst" John Bowlby.[6] Before Bowlby, Fox had considered himself an anthropologist—a social anthropologist, really—and remembers Bowlby was the first person to tell him that "all social structures are supported by emotional systems. People have got to be devoted to their social institutions. . . . They can collapse overnight if people lose interest or lose faith in them, and so on. Emotions are a part of it."[7] The clincher, from Fox's perspective, was Bowlby's ethological insight that the origins of emotions lay in our evolutionary past. Bowlby gave him an article by Michael Chance on the sociality of monkeys to read on his trip home. Around the same time, Fox also heard Noam Chomsky lecturing at MIT on the deep structures within human language and started rereading Sigmund Freud's *Totem and Taboo*.[8] With Bowlby, Chance, and Chomsky rolling around in his head, Fox felt as though he appreciated Freud's theory for the first time—perhaps bands of males *would* murder their fathers, if not for the posited reasons. Surely, he thought, Freud's primal horde and Chance's hordes of macaques connected somewhere in the middle.

M.R.A. Chance, as his favored byline read, started his professional career in what was once called psychobiology, or the psychological study of animal behavior (predominantly in laboratory settings). He quickly formed an interest in understanding the "natural" behavior of animals, partly as tool for diagnosing humane laboratory conditions and techniques.[9] The more natural the animals' behavior, the better the conditions of the lab. By the mid-1950s, Chance had become entranced by the social behavior of monkeys and argued that aggression or threat, rather than male-female relations, was the primary factor structuring social relations in animals and people.[10] Chance also hypothesized that "attention" could be used as a proxy for understanding the cohesiveness and rank of the members of a small community. For example, if one male monopolized 80 percent of the combined attention of the females, this would mean he was of much higher rank than a more subordinate male who could command only 10 percent and devoted over half of his own attention to the dominant male.[11] This could be more than an *effect* of the group's internal hierarchy, Chance postulated, since differential attention might help to *cause* social hierarchies. This insight formed a key component of Tiger and Fox's argument in their later work that males born to high-ranking lineages were, as adults, able to command more females than those from lower-ranking lineages.[12]

FIGURE 17. Konrad Lorenz leading Lionel Tiger, Robin Fox, Robert Ardrey, and Desmond Morris down the garden path of human ethology. The original caption read, "The man/animal bandwagon, undoubtedly on its way to the bank." From Judith Shapiro, "I Went to the Animal Fair . . . the Tiger and Fox Were There," *Natural History*, October 1971, 90–98.

Tiger and Fox published the first in a series of collaborative efforts in 1966—just after they met, and several years before they both landed at Rutgers: "The Zoological Perspective in Social Science."[13] Tiger later referred to it as their "brief and impudent article on the deadness of most social science and the vitality of contemporary biology."[14] Social scientists ought to pay more attention to recent advances in the biological sciences, they argued, especially those emerging from the study of animal behavior and genetics. By understanding humans as cultural animals "with an as yet insufficiently explored repertoire of genetically programmed behavioural predispositions," anthropologists and sociologists could better understand how biology constrained the variability of human social action that typically occupied their research.[15] Tiger and Fox hoped that the object of study in the social sciences would remain the same but could become more nuanced as a result of insights from ethology (Figure 17).

Of particular concern to Tiger was the question of homosocial association—how and why groups of men function the way they do.[16] Fox, on the other hand, had been rethinking notions of kinship in human societies in a new lecture course. Although he did not apply the lessons of an evolutionary perspective to his work on kinship directly, he emphasized the mother-child bond as the primary basis for understanding kinship patterns.[17] If one looks at their work side by side, a sexual division of labor characteristic of anthropological theories of human nature in the 1960s comes into stark relief—men hunted and women reproduced.

When importing an evolutionary perspective into the social sciences, Tiger and Fox noted two subsidiary scholarly arguments worth particular attention. First, both made academic names for themselves by suggesting that human societies should be studied as a function of social relationships—the friendships uniting men and the parental love of women toward their children. These sympathetic bonds seemed to require a sexual division of labor that worked against the claims of contemporary feminists.[18] (They rarely named *which* feminists in their texts, but the names Simone de Beauvoir, Betty Friedan, and Kate Millett recur in both their footnotes and reminiscences.[19]) In biological terms, male bonding had been crucial in driving the intellectual evolution of humanity, while female-child bonding had perpetuated the species. Following Chance, Fox added that as a result, female-male bonding was irrelevant to overall social stability of the population.[20] Both located the origins of the imbalance between male and female contributions to human evolution in the development of early man's hunting practices.[21] When hunting in groups (the origins of social cooperation), men exhibited higher levels of testosterone and competitive behavior and later built these qualities into the foundations of life in the corporate (and academic) jungle.[22] Tiger argued that feminists, by ignoring these scientific facts, made achieving their goal of social equality more difficult.

Second, Tiger and Fox worked self-consciously to replace a Freudian psychological conception of homosociality—as either a juvenile phase or associated with homosexual tendencies—with a biological understanding of homosocial association as fully adult and normal for all men.[23] Whereas Family of Man theories of human evolution that characterized the 1950s had similarly assumed complementary roles for men and women within a conjugal pair, Tiger and Fox often portrayed these male-female partnerships as temporary, potentially agonistic alliances rather than as essentially cooperative. Ardrey (who had similarly attacked Freudian psychology as so much nonsense in *The Territorial Imperative*) enthusiastically agreed. In his review of *Men in Groups*, he wrote, "Men want, need and must have the opportunity of exclusive association. And the all-male group, bonded by long familiarity, furnishes society with its spine."[24]

The sympathy and mutual regard worked both ways. Tiger attributes part of his initial interest in biology and paleoanthropology to picking up Ardrey's

African Genesis. "Sometimes a book makes the difference," he said. "I thought to myself, this, this is something."[25] They, too, struck up a correspondence, finally meeting in London when Ardrey invited him to dinner at the Savoy. Conversation naturally turned to animal behavior. Fox first encountered Ardrey through a chance recommendation from his cousin, who had called *African Genesis* "quite dramatic" and "all about" Fox's "latest enthusiasms for Zoology and early man."[26] When he got around to reading the "remarkable" book, he was impressed by its basic argument, that "society was older than man; we did not invent it, we inherited it. This animal heritage could only be understood by putting together the knowledge of animal society (territory, mating, dominance etc) with the knowledge of primate and human evolution."[27] Fox finally met Ardrey while in Bristol for a televised joint interview and found him "instantly likeable in his no-nonsense, tell-it-like-it-is fashion . . . He had an easy manner and a witty delivery, with a wicked line in sarcasm."[28] Fox continued, "Bob (he was immediately Bob)" and he became "firm friends forever."[29]

Fox wrote his first book, *Kinship and Marriage* (published in 1967), for use in university courses and it received no attention in the mass media.[30] His next book, however, was an introduction to anthropology designed to appeal to the elusive "general public" and consisted in large part of republished essays adapted, translated when necessary, and compiled for the book, written in a "personal, autobiographical" tone.[31] In the last of these, "The Cultural Animal," Fox juxtaposed his position that human cultural patterns were limited by biology with the opinions of earlier cultural anthropologists who argued that all human behavior and actions were determined by nurture, not nature.[32] By attributing these claims to an older generation rather than to his peers who held the same beliefs, Fox avoided attacking his colleagues directly but painted their perspective as hopelessly outdated. He suggested that the "'no links between biology and culture' argument" was understandable given the social concerns of the 1950s, when anthropologists had needed to attack "racists who wanted to explain seeming inequalities between cultures as a result of biological differences." In keeping "the world safe for humanity," he posited, they had also "kept the gap between us and the brutes nicely wide."[33] The time had come for a new perspective, a perspective Fox was happy to provide.

Fox argued that culture and nature were inextricably intertwined—just as early humans produced culture, it in turn produced us—much as Chance had posited with reference to his macaques.[34] But culture, in Fox's vision, could never be expanded to the infinite possibilities of habits and traditions that we might intellectually conceive. All humans possessed "the *capacity* for culture," he continued, but were simultaneously bound by "the *forms* of culture, the universal grammar of language and behavior."[35] In his argument, Fox invoked Chomsky's differentiation between highly variable "surface grammar" and the "deep structures" underlying all human languages as analogous to the relationship between biologically irrelevant cultural variation and true behavioral universals. Even if the laws, rules, and customs differed, every culture possessed "laws about property, rules about incest and marriage, customs of taboo and avoidance, methods of settling disputes with a minimum of bloodshed," etc.[36] Fox proclaimed that all human social systems performed two basic functions—they defined kinship (Fox termed this "descent") and defined who counted as an eligible mate ("alliance"). In so doing so, social systems established traditions ensuring that members of the same kinship group could not mate, thereby preventing inbreeding without the need for a psychologically imposed "incest taboo."[37]

Fox's larger point was that evolution acted to modify human behavior, just as it had altered our anatomy.[38] A couple of years earlier, Fox wrote an article for the *New York Times* that began with the same sentiment but quickly expanded to talking about the evolution of differences in male and female sexual behavior.[39] Some of his lessons echoed those of Ardrey: males competed with each other for access to reproductively available females, and females fought for status within the social hierarchy to ensure the survival and health of their offspring. To succeed evolutionarily, he posited, a male had to be smart, able to defer gratification (sexual or otherwise), be socially graceful and cooperative (with larger, more important males), and acceptable to females. Most important, "he must also be tough and aggressive in order to assert his rights" within the hierarchy. Control over such emotions turned into the capacity to use tools, wield weapons, and ultimately shape his environment. Of course, none of this would have been possible without the capacity to communicate. So what made a man? Control. Control over oneself and control over others. Females, on the other hand, lived within the same social hierarchy, and their rank seemed to affect the status of their sons, Fox suggested. There might be other qualities, too, that contributed to their high status, some of which they shared with dominant males; yet unable to escape another quip, Fox added, "with the exception, perhaps, of bitchiness and bossiness." Given these disparate goals (and traits), Fox described marriage in human society as a contractual affair. "Love and marriage may go together like a horse and carriage," he wrote, "but let us not forget that the horse has to be broken and harnessed." In Western cultures, harems were out of vogue, making it difficult to gauge male status according to the number of women they "controlled," but Fox observed that a "big man" in the office may have a number of women at his beck and call: "one wife, two full-time secretaries, 20 typists, and a girl who comes in to do his manicuring." Here was evidence of females as status symbols. Here too was evidence that "monogamous nuclear families" were merely cultural fictions, perpetuated by (among others) the *Saturday Evening Post*.[40]

Yet Fox attributed the most carefully elaborated example of a deep cultural structure to Tiger: "Whatever the overt cultural differences in male-group behavior ... in society after society one thing stands out: men form themselves into associations from which they exclude women."[41] Tiger intended his *Men in Groups*, published in 1969, to reach a wide audience.[42] He succeeded, receiving two reviews and two interviews in the *New York Times* alone. The interviews described him as a "puckish, 33-year-old Canadian" and as "tailored in London, shod in France and automotively equipped by Alfa-Romeo."[43] He appealed as a new, hip man of science.

As a representative of this new breed of scientific men, Tiger sought to make two points. The first, like Fox, was to explore the possibility of uniting biological and sociological analyses of human behavior. Second, in his case study of such an approach in action, he wondered, "Why do human males form all-male groups? What do they do in their groups? And, what are the groups for?"[44] He proposed, in answer, that all-male groups are ubiquitous in human societies and require some kind of initiation ceremony to demarcate members from nonmembers. These ceremonies, in turn, reflected a form of "unisexual" selection "for work, defence, and hunting purposes" that paralleled sexual selection for reproductive purposes. In unisexual selection, male bonding and aggression were intimately linked. "Aggression," Tiger wrote, "is both the product and cause of strong affective ties between men."[45] He further posited that unisexual organizations, like all-male prep schools and universities, promoted conceptions of masculinity that arose out of these institutions. The net effect kept males from high-ranking families dominant and subordinated males from families without access to such resources; it also automatically excluded females from the upper echelons of political and social power.[46] If he were right, Tiger reasoned, then "modifying the dynamics and repercussions of the male bond may be a crucial feature of altered attitudes to power, to the value of destroying other communities' people and property, and to the concept that manliness is strength rather than flexibility and authority rather than attentiveness to others."[47] Coming from a working-class background, Tiger was acutely aware of how class was perpetuated through codes of masculinity defined by youthful experience with exclusive all-male institutions.[48] Tiger combined sociological and evolutionary arguments in fascinating ways but despaired of changing the resulting social dynamics that characterized modern life. His reluctance to admit that the system could be changed left him open to charges of biological determinism.

Reactions to *Men in Groups* were mixed. One reviewer argued that Tiger's theory of male bonding dominated discussions of human social interactions with almost no attention paid to female-female relations.[49] Another called Tiger's theory "necessarily tentative and tenuous, based as it is on analogy, speculation and a staggering variety of scholarly sources," and quoted "one of his more infelicitous phrases, 'Now let me softly blow my sheepish horn.'"[50] In *Sexual Politics*, Kate Millett called *Men in Groups* "a genetic justification of the patriarchy" and claimed he had misrepresented Konrad Lorenz's work.[51] Tiger fared slightly better at the hands of the physical anthropologist Sherwood Washburn, who nevertheless emphasized the book's "deep problem . . . the nature of the evidence; how can such an interesting point of view become more than a very tentative hypothesis?"[52] Ardrey, unsurprisingly, penned the most positive review and emphasized the book's scholarly nature. In his usual style, he wrote, "Footnotes buzz by like June bugs in a summer cottage. But let the reader be tolerant . . . predators are waiting. And in the academic jungle a footnote can be a man's best friend. Let the reader likewise endure a 'paradigm' or two as scholarly decor. He may read with assurance that no cloistered jargon will muffle the explosion let loose in these pages."[53] In Ardrey's reading of *Men in Groups*, two factions would find their cultural assumptions smashed by his findings: the Freudians, who posited that male friendship resulted from latent homosexual attraction, and the feminists who sought to challenge the "age-old male bond."[54] All-male groups were not the result of psychological deviance or cultural prejudice, Ardrey gleefully concluded, but of our human biological heritage. Tiger's book sat uncomfortably at the intersection of academia and the mass media, providing different messages to both sets of readers.

Given Tiger and Fox's earlier publications, *The Imperial Animal* (1971) begins much as you would expect. They drew evidence from biology, history, and genetics in order to describe the evolution of human behavior.[55] The opening chapter outlines their approach in terms of "biogrammar," again echoing Chomsky. Their attention then turned to the political nature of human social interactions, highlighting Tiger's work on competition and dominance hierarchies. "Sex and politics," they asserted, "are two sides of the same evolutionary coin"—reproduction was fundamentally a political process, and (reflexively) politics were driven by the "ancient impulses of sexual competition."[56] Sections on the importance of the mother-child bond and of male-male competition in forging social order followed next. Tiger and Fox then devoted the second half of *The Imperial Animal* to more traditional social scientific topics, illustrating how evolution had probably influenced the origins of economic systems, educational practices, and efforts to maintain the health of the social body. In writing the book, Tiger and Fox fulfilled what they saw as their scholarly duty of diagnosis and analysis—it was now up to their colleagues and politicians to embrace and learn from their biological past.[57]

Fox later recalled that *The Imperial Animal* attracted reviews by "hostile feminists, turf-defending social scientists, snotty humanists and friendly laymen—and women for that matter."[58] Elizabeth Fisher questioned the "clear case" for dominance hierarchies in chimpanzees but quipped, "That they exist among academics I have no doubt—one feels that no woman can advance in those circles without making certain ritual submission gestures."[59] Zihlman's review was anything but submissive. She deemed the *Imperial Animal* confused and marred by "glaring shortcomings."[60] Tiger and Fox, in her estimation, conflated predation and aggression, overemphasized male contributions to humanity's past "almost to the exclusion of the female role," and wrote in a tone too tedious for lay readers and too glib for professionals. Fox chose to see negative reviews as the result of infighting among social anthropologists, describing their attacks as territorial defense. Nowhere was this allusion clearer than in a conversation where Fox remembers Ardrey gleefully claiming that his antagonists were "being consistent with their own Darwinism: they struggle for existence, for the reproductive success of their own ideas."[61]

Both Fox and Tiger attributed the "hostile" reaction their work received to the clouding effects of feminism and a knee-jerk "Ardrey reflex." In an autobiographical essay written in 1973, Fox suggested that critics equated their work "lock-stock-and-barrel" with the writings of Ardrey, Lorenz, and Morris. Feminist scholars were "applying the same blanket criticisms—including the usual string of insults: racism, fascism, sexism, reductionism or whatever other-ism happens to be on the unwanted list. Along with this would go the imputed motives: male chauvinism, reactionary politics, cupidity and general meretriciousness."[62] Tiger was especially disappointed by feminists' lack of apparent interest in *Women in the Kibbutz*, which he coauthored with his former student Joseph Shepher in 1975 and in which they described the persistence of sexual division of labor even in self-consciously egalitarian kibbutzim.[63]

As they built their professional identities, Fox and Tiger worked to include the playwright Robert Ardrey in their professional networks. At one conference (possibly "Man and Beast," held in Washington, DC in 1969), Fox defended Ardrey against attacks by scientists who lambasted his (and Konrad Lorenz's) characterization of aggression as "instinct."[64] According to Tiger, someone made a nasty comment about Ardrey, "holding him responsible for the Vietnam War and for everything." Tiger still believes that many of Ardrey's professional difficulties stemmed from the lack of academic accreditation that marked him as an outsider. In Tiger's retelling, Fox first came up with the idea of granting Ardrey an honorary degree, thus making him "a member of the group"—perhaps then people would have left him alone.[65] Yet, at least according to Fox, Ardrey was happy with life outside the academy. After discussing the faults of work by a fellow Darwinian, Fox remembers Ardrey questioning whether the point was worth a big fight. "Remember . . . you are on the same side in the great struggle: the struggle to get the evolved, biological aspects of behavior recognized and incorporated into a scientific view of human behavior." Try at all times, he encouraged Fox, "to boost and encourage and promote" people on your side.[66]

Rather than creating a new field of study, however, Fox and Tiger envisioned returning anthropology to what they identified as its true mandate. Once anthropology reclaimed its evolutionary roots, they firmly believed, other behavioral sciences would naturally follow suit. Darwin, too, had been criticized and in the end emerged victorious. "That was the plan."[67]

9

The Edge of Respectability

IN CHAPTER 4 of *The Descent of Woman*, published in 1972, Elaine Morgan asked her readers to take science into their own hands. "Try a bit of fieldwork," she suggested. "Go out of your front door and try to spot some live specimens of *Homo sapiens* in his natural habitat. It shouldn't be difficult because the species is protected by law and in no immediate danger of extinction." After completing observations of twenty random people, substitute them when you are reading statements about universal human nature. The result?

> 'That window cleaner is one of the most sophisticated predators the world has ever seen.'
> 'The weapon is my grocer's principal means of expression, and his only means of resolving differences.'
> 'The postman's aggressive drive has acquired a paranoid potential because his young remain dependent for a prolonged period.'

Morgan then added that you might imagine you were observing the wrong species. "But," she continued, "if you're going to be any good as an ethologist, you must learn to trust the evidence of your own senses above that of the printed word and the television image. Remember, you have been living among thousands of these large carnivores all your life, on more intimate terms than those on which Jane Goodall lived among the chimpanzees."[1] In positing such a scenario, Morgan was engaged in serious work. All theories of human evolution to date, she insisted, were based on a male-centered notion of human evolution. Where were the evolutionary scenarios that began, "When the first ancestor of the human race descended from the trees, she had not yet developed the mighty brain that was to distinguish her so sharply from all the other species"?[2]

Morgan's *The Descent of Woman* was one of the earliest public rejoinders to masculinist narratives of human ancestry. An Oxford-educated Welsh writer for BBC Radio, Morgan chose as her target several of the most prominent books on human evolution circulating through the UK. Unsurprisingly, Ardrey's *The Territorial Imperative*, Lorenz's *On Aggression*, and Morris's *The Naked Ape* topped her list. So, too, did Tiger's *Men in Groups* and Antony Jay's witty *Corporation Man*. By the early 1970s, in other words, evolutionary approaches to human nature were sufficiently current for other writers (and publishers) to deem them worthy of pillory. Yet humor was a double-edged sword. Sarcasm might garner readers but for authors with no formal training in anthropology or zoology, like Morgan, maintaining a respectable public persona as a popularizer of science proved difficult.

The situation was slightly easier for male humorists. Published in 1971, Jay's *Corporation Man* valorized the work of office men as heroic hunters. A British author and broadcaster—he worked for the BBC until 1964, when he resigned to go freelance—Jay appropriated a Man the Hunter perspective for a new look at corporate culture.[3] The cover copy on the paperback announced that it "probes beneath the 'civilized' trappings to reveal the primitive instincts that really run the business world. Status display, aggression, appeasement, comradeship, tribal gatherings . . . these are the powerful emotional reflexes and rituals that govern the modern nine-to-five day as profoundly as they did the dark, terror-filled nights of the Stone Age" (Figure 18). Jay's previous book, *Management and Machiavelli*, had argued that the rules for efficiently running modern corporations could be found in Machiavelli's writings on political power.[4] Together these books give voice to Jay's studied frustration with institutional life and reveal the Paleolithic corporation man as his analytical flavor of the day.[5] He counted on readers' familiarity with evolutionary tropes and deployed them vividly to recast the world of sales as a cut-throat business in which only the strong could survive.

The gendered stereotypes embodied in masculine theories of human nature were part of a larger reimagining of the roles men played in American social life. David Riesman's keen sociological analysis in *The Lonely Crowd* and the immense popularity of Sloan Wilson's *Man in the Gray Flannel Suit*, for example, spoke to widespread anxieties about the soullessness and behavioral conformity of middle-class suburban life.[6] By the close of the 1960s, both corporate culture and masculine ideals were rapidly changing. *Playboy*, in promoting itself as a magazine for men who preferred to stay single, depicted secure family life as a trap to be avoided for as long as possible and included articles on the "sexy hunter" as an embodiment of masculine prowess.[7] Greater numbers of women entered the work force as "career girls," creating office spaces potentially fraught with sexual tension and fluctuating power dynamics.[8]

Like Morgan, Jay emphasized that interpreting human behavior was something his readers already did on a regular basis. Although he admitted that his

FIGURE 18 (*above and following pages*). Beneath the surface of every businessman
lurked a cave man, or so Antony Jay suggested in his semi-satirical look at company
life in the early 1970s. Photographs by Steve Horn and Norman Griner, for Antony
Jay, "Corporation Man," *New York Magazine*, 20 September 1971, 31.
Courtesy of Steve Horn/Norman Griner.

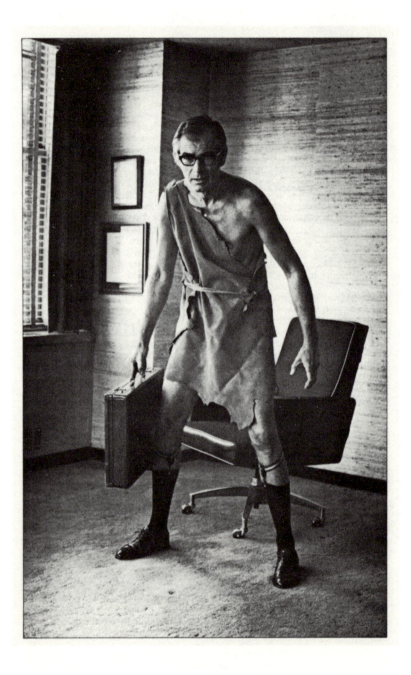

own fieldwork was surely less systematic than that of professional scientists, he added it was also "a good deal more comfortable, since the species I have been studying—*Homo sapiens corporalis*, Corporation Man—flourishes in a much more accessible and agreeable habitat." He lauded the well-kept records of the species (stretching over thousands of years) that made his job easier. Yet, the most convenient aspect of devoting himself to the study of human corporate culture he described as his ability to blend. Jay wrote that he was "accepted so completely by the objects of my study that my presence in no way inhibited or modified their behaviour. Indeed, there were times when I could honestly say that I felt I was one of them myself."[9]

In his analysis of modern corporatism, Jay divided companies into two halves: the hunting bands and the camp. Men in hunting bands—those who made or sold products for the company—were, for Jay, the "heart" and the "ruling circle" of the corporation, "broadly charged with ensuring the firm's long term survival." All other employees formed the camp: members of finance departments, planning, personnel, maintenance, and public relations. Nothing, he believed, irritated hunters more than members of camp pretending they carried just as much weight or were as important to the company's survival as the hunters themselves. Once he understood this delicately balanced system, Jay noted, he was able to derive a number of important rules for maintaining the vibrancy of a corporate environment. All companies, he suggested, needed to employ women. Women were less aggressive than men. Left to their own devices, all-male groups might go on strike, leaving the company in tatters. Just like on the savannah, he continued, corporate hunting groups were territorial and competitive. They should succeed (or fail) as a unit, with reward or punishment meted out equally among the members, in order to enforce a strict cooperative spirit within the group and fierce competition between groups. Jay noted, too, that all company employees interacted within a hierarchical system of status, demarcated for all to see by the size of offices, the quantity of rubber plants, the number and location of secretaries, and the cost of the cars they drove. Companies were cultures, too, each characterized by folklore that functioned to unite disparate members into a single cohesive community with shared norms of conduct. Although members of one company's hunting band routinely went to war with men from other companies, Jay contended that their aggressive interactions took the "ritualized" form of changing a product's advertising, packaging, and accompanying frills. No one stooped to price-cutting, as "a fight to the death would upset the delicate ecology of the industry."[10] How very Lorenzian!

Both Jay and Morgan built their narratives in reaction to the same set of colloquial science books. According to Jay, he obtained his insights about corporate culture by following the principles of a "revolution" he called "the New Biology," a phrase he borrowed from Ardrey.[11] Jay suggested that well-known authors like Ardrey, Lorenz, Morris, and Tiger had "begun to arrange all the separate mosaic pieces into a new picture of the origins and nature of man" (as a nonscientist himself, he especially appreciated Ardrey's captivating presentation of evolutionary ideas). The fundamental element of this New Biology, he suggested, was its appreciation for the evolution of social behavior. To Jay, these men had dramatically extended Darwin's original principles (which he inaccurately described as a theory of evolution and natural selection accounting for merely *physical* characteristics). In the New Biology, he wrote, "status-seeking emerges not as an unworthy failing of jealous executives, but as an immutable ingredient in man's make-up—an ingredient he shares with many other species including the jackdaw, the baboon, and the domestic hen."[12] Humans exhibited aggressive, exploratory, and defensive behaviors, Jay believed, because early humans who lacked these qualities had died—our ancestors came from heartier stock and survived. Morgan's frustration with the "Tarzanists," as she dubbed this same growing circle of male writers, is equally palpable when reading *The Descent of Woman*. She began by noting, "The legend of the jungle heritage and the evolution of man as a hunting carnivore has taken root in man's mind. . . . He may even genuinely believe that equal pay will do something terrible to his gonads." The scientific facts used to buttress such arguments were unassailable, she insisted, but not the interpretations of those facts.

Morgan's *The Descent of Woman* and Jay's *Corporation Man* fit well into the contemporary genre of colloquial science that aimed to present scientific arguments in ways accessible to people with common sense (but little scientific training)—much like the Tarzanist books Morgan critiqued and the New Biology Jay lauded. Primarily Morgan sought to call attention to the sexism inherent to contemporary savannah-based theories of human evolution. She noted that the term "man" was ambiguous, denoting both the entire species and also the males of the species. The trick was not to confuse the two. Adding that although this observation might sound like "a piece of feminist petulance" to some, Morgan wished to convince her readers that her semantic point vitiated much of the "speculation" concerning the evolutionary origins of humanity.[13] An evolutionary theory that purported to explain all human behavior needed to account for the roles of both men *and* women in human history. She also explored what she considered an equally plausible hypothesis: that humans had undergone an aquatic phase in their evolutionary past. When Morgan turned her attention from fiction to anthropology, it was largely out of irritation with books that emphasized the evolution of man.[14]

The aggressive masculinity they advanced left little room for women, she argued, and in stark opposition to these volumes, Morgan was read as a radical feminist with a sense of humor.

Morgan contended that, based on the available evidence, it was impossible to distinguish weapons from tools and impossible to know who invented them, man or woman. "A knife is a weapon or a tool according to whether you use it for disemboweling your enemy or for chopping parsley." In fact, she reasoned, it was likely that early humans used tools for both purposes—the strict dichotomy between herbivores and carnivores in our past was never clear-cut.[15] Morgan also noted that although male hunters were widely acknowledged as being the inventors of tools and pottery, there was no solid evidence supporting this assumption. She characterized the dominant theory of pottery invention as follows: "One day he noticed with a secret chuckle that the little woman was wearing herself out trotting to and fro carrying seed home by the handful. He quietly laid aside his beautiful symmetrical weapons and forsook his male-bonded companions for a few weeks while he devoted himself to the problem, and finally invented the pot. He gave her a few prototypes and a crash course of instruction, patted her on the head, and sped away across the savannah to rejoin the hunting party." That was equally likely, she contended, as assuming that a woman invented her partner's weapons, saying, "Play quietly among yourselves today, children: I'm busy inventing the bow and arrow for your father." Both tales were equally plausible and equally unsubstantiated.[16]

Why, then, were such assumptions about the importance of men to the evolutionary progress of humanity so widespread? Morgan chalked this up to male pride. She quoted Ardrey's *The Territorial Imperative* extensively in order to rebut his characterization of men as baboons. A male baboon, he had written, "is a born bully, a born criminal, a born candidate for the hangman's noose. He is as submissive as a truck, as inoffensive as a bulldozer, as gentle as a power-driven lawnmower. He has predatory inclinations and enjoys nothing better than killing and devouring the newborn fawns of the delicate gazelle." Morgan imagined that a male reader would naturally find this characterization appealing. She pictured him polishing his glasses, thinking: "Yeah, that's me all right. Tell me more about the bulldozer and how I ravaged that delicate gazelle." The reason such theories continued to be popular was because Ardrey's typical reader, and the author, got "no end of a kick out of thinking that all that power and passion and brutal virility is seething within him, just below the skin, only barely held in leash by the conscious control of his intellect."[17]

It was high time, she thought, to expose these arguments for what they really were—myths, yes, but politically useful myths. Morgan argued that man-the-hunter theories were used to "bolster up with pseudo-history and pseudo-anthropology the belief that it is 'against nature' for women to play a part in economic life; that 'from time immemorial' men have said 'she shall have no other food and that will make her my slave'; and that we are descended from females whose sole function was to placate the hunters and keep them happy and mind the babies."[18] In particular, Morgan cited the wide acclaim and recognition that Tiger's *Men in Groups* had received. His idea of "male-bonding" dominated discussions of human social interactions with almost no attention paid to female-female relations.[19] Based on recent evidence from modern hunter-gatherer societies, she also noted that the bulk of the total diet of early humans was probably the result of gathering vegetable matter, not hunting meat.[20] Women had been important to the evolutionary history of humanity but were ignored by the "blood-and-thunder boys."[21]

Morgan then conjectured that perhaps a period of aquatic adaptation had preceded life on the savannah. In our semi-adaptation to a watery world, she suggested, we would have lost our body hair, gained a layer of subcutaneous fat to keep us warm, learned to walk upright (keeping our head above water while foraging for tasty snacks in the shallows), learned to use stones and manufacture tools for breaking open shells, and developed the ability to control our breathing when diving beneath the surface—a precondition for symbolic language and an obvious boon to any individual trying to communicate with most of her body submerged. These activities, Morgan noted, were associated with gathering, not hunting, an activity in which women's contributions were already acknowledged. (Margaret Mead called Morgan's assertion that women had been excluded from the anthropological profession "pure rubbish."[22]) In short, she believed we acquired precisely those traits that distinguish us from the rest of the animal world while living around water, not in the arid grasslands.

Morgan first came across the idea while reading a scant two pages that Morris devoted to a possible aquatic phase in human history in *The Naked Ape*.[23] She contacted the theory's originator—the professor of zoology Alister Hardy at Oxford University—and asked whether he would mind if she worked on a book responding to the contemporary theories of human evolution and developing his idea. Like Ardrey, she may have believed that artists and writers were well-developed students of human nature, despite their lack of anthropological training.[24] Hardy had begun considering an aquatic theory of human evolution more than thirty years earlier but initially refused to publish anything on it lest he ruin his nascent academic career (at the time, he had plenty of experience at sea and in marine biology but lacked a permanent post).[25] In 1960, as a recently knighted and well-established professor of zoology at Oxford, he agreed to write a brief article in the *New Scientist* proposing an aquatic past for humanity.[26] It was to be one of Hardy's only publications on the topic, and he was more than happy to let Morgan run with his idea.[27] At the time, he had planned to publish his own book and, after consulting with his editor, believed that a more popular account of the aquatic ape could only help his later sales. (His book never appeared, however, a fact Morris attributed to a lack of material evidence.[28])

Originally titled *The Evolution of Eve*, Morgan renamed her book just before publication. Foremost among her concerns, the argument had nothing to do with Eve, and she worried that as a symbol of womanhood, "Eve" had become "old-fashioned, especially to the younger generation." The "Women's Lib" movement on the other hand, had made "'woman' . . . a highly topical and commercial word." She hoped her book would appeal to these readers, who, she reasoned, would be especially put off by biblical references. The argument that cinched the title for the publisher, however, was its self-conscious echo of Charles Darwin's *The Descent of Man* (which enjoyed a minor centennial moment of its own in 1971).[29] Also, Morgan insisted.[30]

Everyone expected *The Descent of Woman* to sell. The British publisher with whom Morgan worked—Souvenir Press—sold the rights of publication to Stein and Day in the United States and to publishers in nine other countries. *McCall's* magazine paid $10,000 for a two-part serial excerpt, and Bantam books $100,000 for the paperback rights. When news broke that the journalist Clifford Irving had faked all of his interviews with the millionaire recluse for his highly anticipated *Autobiography of Howard Hughes*, the Book of the Month Club upgraded *The Descent of Woman* from an "alternate" to a "main" selection. Even Ardrey, who likely read this excerpt sent to him by a colleague in Hollywood, wrote back that, "the anthropology may be appalling but it sounds like fun. Bet it sells like mad."[31] Yet in the United States, *Descent of Woman* made the *New York Times* best-seller list only once (the week of June 25, 1972). According to her American publisher, the book sold "steadily, but not spectacularly."[32]

Reviews of the book quickly picked up on her revisionist agenda. *Playboy*, for example, hailed it as a "stunning tour de force." The review hypothesized that "even the most militant male chauvinist will find it difficult to cling to all his prior convictions in the face of the evidence marshaled here."[33] Similarly, *Life* magazine described *The Descent of Woman* as a lively "women's-lib prehistory."[34] Several reviewers missed that Morgan did not have any anthropological training, describing her as "a female anthropologist" and "a scholarly woman, educated at Oxford in the fields of paleontology, ethology, and anthropology."[35] Yet even when people recognized her outsider status, they still acknowledged the force of her arguments against the standard evolutionary history of humanity. Her reviewer in the *New York Times*, for example, characterized *The Descent of Woman* as "a potent commentary on the state of the social sciences in general, and anthropology in particular."[36]

By the late 1970s, feminist anthropologists were already citing Morgan's *The Descent of Woman* as an early critique of Man the Hunter theories of human evolution.[37] Yet although Morgan's book was useful, it also contained her dalliance with Hardy's aquatic ape. As a young professor, Adrienne Zihlman argued that Morgan's sensible and "substantial critique of existing evolutionary dogma did not get the attention and credibility it deserved." It had been "contaminated by Morgan's own elaboration and support of a very dubious theory of human origins, the 'Aquatic Ape' hypothesis."[38] Zihlman regretted Morgan's advocacy. First, the myths against which Morgan had been fighting were still present more than a decade later.[39] Second, readers might get the mistaken impression that Morgan's book was the best feminist anthropology could offer. "A feminist revision of human evolution," Zihlman insisted, "does not require life in the water."[40]

Morgan invested in the idea that humans had an aquatic past because she needed a viable theory to fill the hole left by her insistence that Man the Hunter could never be the full story. She insisted that in concentrating solely on the males of the species, evolutionary narratives like those advanced by Ardrey, Jay, and Tiger overstated the capacity of females (and therefore the species as a whole) to survive. From a female perspective, without protective weapons or easy hiding places and weighted down with a nursing infant, "the only thing she had going for her was the fact that she was one of a community, so that if they all ran away together a predator would be satisfied with catching the slowest and the rest would survive a little longer." In other words, when this sweet "generalized vegetarian prehominid hairy ape" experienced the "first torrid heat waves of the Pleistocene," she would have been unable to avoid being eaten by predators.[41] Both our putative ancestor and Morgan needed an escape route, and that is what the water provided.

A watery environment, Morgan intimated, could explain a wide variety of our uniquely human characteristics. As our putative ancestor ran into the water up to her waist or even her neck to escape predators, she was forced into upright, bipedal walking. Lakeshores and seashores provided much easily accessible food, but hard shells needed to be broken open to access the tasty tidbits they contained—the use and eventual manufacture of tools thus started with gathering food and were only later adapted as weapons for hunting. Living in caves along the seashore would have provided shelter and also an explanation for the origins of family structure. The loss of our body hair would then be the result of the first stages of adaptation to an aquatic environment.[42] Even our capacity for speech could have emerged as a result of learning breath control for swimming and diving. Communicating while wading can be difficult because our limbs would be covered with water (preventing active gesturing), and the water would additionally mask the chemical particles that form our olfactory communication.[43] After this aquatic phase of our prehistory, she posited, the rains once again returned to Africa, the Pliocene merged into the Pleistocene, and Man the Hunter learned to roam the savannahs in search of prey.[44]

Morgan, as had Jay, presented evidence for her claims that nonscientifically trained readers could easily understand.[45] She suggested that the incredible sensitivity of our fingertips resulted from the need to grope for food under the water. She noted that our heartbeats slow down when we dive to great depths and further pointed out that newborn babies can float. She also explained long hair on the heads of women and their "pendulous, dollopy breasts" as adaptations allowing babies to firmly grasp their mothers in the water and providing easy handholds for breastfeeding rather than traits controlled by the sexual preferences of men.[46] In a broad survey of the animal kingdom, she argued, the only animals that exhibited traits like these were aquatic mammals.

Morgan remained aware that her interest in the aquatic ape might be perceived as outside the normal bounds of science. In an interview, she remarked that perhaps it was easier for her than for an established scientist—one can only assume she meant Hardy—because she had "nothing to lose, no high academic position to think of." She added, "If you talk about flying saucers you're branded a kook. I don't believe in flying saucers but I suppose this kind of thing looks flying-saucerish to the Establishment."[47] In fact, that was exactly what "Establishment" science thought. Her reviewer in the *New York Times*, for example (although he had lauded her critique of contemporary anthropology), suggested that her characterization of human ancestors as including "a breed of sea beasts" should be consumed along "with a grain or two of salt."[48] Although Ardrey thought her book likely to sell, he also remarked that "the more the [Germaine] Greers and the [Kate] Millets and the Morgans present us with the intellectual rationale for Women's Lib, the more I realize why the female is the second sex, and is going to stay that way."[49] This was the kind of reaction professionals like Zihlman sought to guard against.

Several years later, together with her husband, the physician Jerold Lowenstein, Zihlman cowrote a brief article for *Oceans* magazine comparing the Aquatic Ape Theory to "the existence of bigfoot" and "visitors from outer space."[50] After analyzing key aspects of the evidence, they concluded, "the Aquatic Ape Theory does not hold water, anatomically, biochemically, behaviorally or archaeologically. With a similar combination of imagination, a grab bag of unrelated 'facts' and a popular literary style, one could make an equally convincing case that our ancestors evolved in the air—as von Däniken has

FIGURE 19. In their critique of Elaine Morgan's *Descent of Woman*, Jerold Lowenstein and Adrienne Zihlman proposed wrapping together several of the fringe countercultural stories concerning the origin of humanity: "Intelligent apes from outer space land in the middle of the ocean, evolve into amphibian creatures with huge footlike flippers and . . . Finish it yourself," they added. "It is sure to sell." Note, too, the eye chart in the illustration that accompanied their article, which reads: "Piltdown Man is an ethnic slur." Illustration by Bill Prochnow, for Jerold Lowenstein and Adrienne Zihlman, "A Watered Down Version of Human Evolution," *Oceans* 13 (May–June 1980): 3–6.

more or less done in his cult book *Chariots of the Gods?*" (Figure 19).[51] Despite this dismissal, and with some urging from an American fan, Morgan wrote a second book, *The Aquatic Ape*, in 1982, in which she re-presented her evidence in more scientific terms.[52] This time, she included illustrations visualizing her points, and notably, her prose no longer contained the biting feminist critique that had been so well received in her first book. Scientists remained skeptical. One reviewer noted, "Until some hard evidence is found though, I fear we are left with several equally convincing theories floating in a sea of speculation."[53] Lowenstein, too, remained unconvinced, remarking, "It is fun to make up evolutionary fables."[54] Unfazed by such criticisms, Morgan published *The Scars of Evolution* in 1990 and *The Descent of the Child* five years later, each time updating her evidence to reflect more recent findings.[55]

Jay, on the other hand, never published another book on evolution after *Corporation Man*. Even if it reads now as a dry send-up of evolutionary reasoning, at the time his American readers either did not find it funny or felt they could not afford to take the chance that it might be considered serious scholarship. There were any number of issues embedded in his light-hearted methodological account that would have alarmed professional scientists if they had taken him seriously, and I have no evidence that they did.[56] Ardrey, of course, favorably reviewed the book in *Life* magazine, suggesting it was "a revolutionary work . . . as entertaining as it is profound."[57] Yet Jay's approach differed in degree and sophistication from the arguments advanced by professional scientists, even in their most popular writings, and for women invested in professional careers his assertions seemed downright hostile. When *New York Magazine* published an excerpt of *Corporation Man* in one of their issues, several women wrote letters to the editor protesting Jay's analysis. Judith Shapiro, then an assistant professor of anthropology at the University of Chicago called it "fashionable," "half-baked ethologizing." She posited that behavioral biologists could offer an illuminating perspective on human nature, but she also urged caution. "First of all," she declared, "the fact that man is part of the animal kingdom was not discovered by Robert Ardrey or Desmond Morris any more than sex was invented by Norman Mailer." Additionally, she insisted that neither "genetic determinism" nor "self-conscious rationality" could solely account for the "special and unique attributes" that make man human.[58] Susan Kreisler, identified only as residing in Manhattan, took issue instead with the basic facts of corporate life. She pointed out that many of the people working in "camp" (secretaries, personnel assistants, cleaners, and coffee ladies) were actually men and recent social changes had lead women to "hunt" as well.[59] Jay wrote to Ardrey as reviews started to appear, wryly noting, "I perceive much more clearly the degree of resistance you encounter when you talk in evolutionary terms about human behaviour."[60]

Historians of science have spilled much ink on the demarcation question, asking how scientists and philosophers have tried to cleanly differentiate legitimate scientific inquiry from nonscience, pseudoscience, pathological science, or simply bad science.[61] In contemplating Morgan's aquatic ape theory, we can see that in addition to these hard demarcations, scientists in the 1970s also used the label "feminist science" as a tool of soft demarcation. By categorizing *The Descent of Woman* as feminist science, they sidestepped Morgan's critique of "Tarzanist" anthropological narratives as male biased. Emulating the efforts of many fringe scientists, Morgan responded by rearticulating her argument and incorporating new evidence in an attempt to present her claims in a style and language scientists would take seriously.[62] Notably, as the first step in this process, she removed her humorous critique of other theories of human evolution.

Female scientists reacted strongly to Morgan's *The Descent of Woman* because the implications of the book's arguments being taken seriously mattered to them. Zihlman had worried that male readers who picked it up might think, "Is that the best you can do about the role of women?" When later recalling her frustration, Zihlman (ever the scientist) focused on Morgan's lack of evidence for her theory. Morgan relied on common sense and experiences her readers might have shared, which made her feminist arguments vulnerable but did not dissuade those devoted few who found themselves transfixed by the aquatic ape. Zihlman remembers her first encounter with one of Morgan's American disciples who contacted her after her review in *Oceans* magazine. Zihlman took the time to explain why fossils are generally found near water and why that did not constitute evidence for the aquatic ape theory—in short, bones from individuals that die near streams are preserved in alluvial deposits at higher frequencies than bones from individuals who die in a field or forest. She also added that if he were to take into account the variety of ethnic groups in the world, he would quickly realize that not all women have long flowing hair to which babies could cling while their mothers waded into the shallows, as Morgan had suggested. He replied stridently and without responding to her arguments. In the end, Zihlman gave up. What was the point of arguing with someone who *believed*, she told me. "It's like a religion, so it was ridiculous to even try to be logical. It was a waste of time."[63] Although scientists easily swept Elaine Morgan and her supporters off the edge of respectability, the fate of academic scientists who actively courted colloquial readership was not yet clear.

PART IV

Political Animals

Modern man has brought this whole world to an awe-inspiring threshold of the future. He has reached new and astonishing peaks of scientific success.... We have learned to fly the air like birds and swim the sea like fish, but we have not yet learned the simple art of living together as brothers.

—MARTIN LUTHER KING JR.

JOHN B. CONLAN had his eye on a long career in Washington. He styled himself a sharp and polished politician attuned to the concerns of conservative Christians. He kept his dark hair perfectly trim and wore (in the summer months at least) white patent-leather shoes and white socks.[1] On April 9, 1975, he took the floor of the US House of Representatives and railed against the educational philosophy and damaging content of the Man: A Course of Study (MACOS) program. For progressive evolutionists, including MACOS designers, the cultural relativism of the program and the ascent of humanity from our animal heritage had embodied their hopes for an egalitarian future. Conlan's objections to the program came from directions they had never anticipated while building the curriculum: the moral depravity of the Netsilik, the violence of their lifestyle, and the secularized vision of humanity as united by a common, amoral nature. Concerns over violence in contemporary society provided a seed, crystallizing Americans' fears across the political spectrum, and biological accounts of human nature were caught in the lattice.

Conlan spoke with the clipped cadence of an experienced congressman, his rapid patter sprinkled with ineradicable relics of his Illinois roots like "fellas" and "Warshington." MACOS, he asserted, called for a "radical break from traditional loyalties." No wonder, he continued, "parents to 10-year-olds in scores of American communities find MACOS and its Federal backing repugnant and threatening to their way of life." Conlan called particular attention to the theme of aggression in MACOS materials and the way the program contrasted peaceful societies lacking an "ideal of brave, aggressive masculinity" with the warlike tendencies of technologically sophisticated, power-seeking nations. He insinuated that the Netsilik people used as the primary ethnographic case study in the course exhibited totally immoral behavior: wife-swapping, cannibalism, divorce and trial marriage, female infanticide, murder, and senilicide. He suggested that children in the MACOS program were called upon to engage in mutual surveillance of their peers. Worse still from Conlan's perspective, MACOS films contained "many lurid and gory scenes of Eskimos killing and butchering animals they hunt. Children are shown scenes of Eskimos eating the eyeballs and other organs of slaughtered animals, and drinking warm blood." He called the films "repugnant and vulgar" and insisted that grade-school children were "impressionable and sensitive" and thus "should be shown the beauty and wonders of life, not just its seamiest and most uncivilized aspects."[2] Conlan feared the captivating medium of film would normalize these (un-American) behaviors for the students who watched them.

Worries over the influence of violence in film had led to the establishment of the Hays Code in the 1930s. For decades, film studios had believed that any movie produced without the stamp of approval from the Hays Office (technically, the Production Code Administration within the larger organization that comprised the Motion Picture Producers and Distributors of America) would result in financial failure. As a result, studios would typically submit movie scripts before filming even began, carefully removing profanity, violent acts without moral retribution, sexual embraces deemed too risqué for innocent viewers, any mention of abortion, drugs, or miscegenation, and even explanations of evolution as a demonstrable theory.[3] In monster movies of the 1950s, the Hays Office allowed depictions of evolutionary reversion, like *I Was a Teenage Werewolf* (1957), or spontaneous mutations due to atomic explosions, including everything in the *Godzilla* franchise, but only if the films did not include any description that would make the transformations seem scientifically plausible.[4] For years, that responsibility had fallen to Joseph Breen, a tough-minded Catholic who stringently enforced Hollywood's morality. Breen once declared to a new colleague, "I am the Code!" and was known around Hollywood (despite his responsibilities) as profane, belligerent, and anti-Semitic.[5] When Breen retired in 1954, his more laidback second-in-command, Geoffrey Shurlock, took Breen's place as the administrator of the Hays Code. Under Shurlock, studios and independent producers fought for greater space to raise previously banned topics in their movies, effectively undermining the power of the office to strictly regulate the content of Hollywood's films.[6] The voluntary self-censorship of Hollywood thus faltered in the

1960s as film studios began to take risks releasing films without the Hays Office imprimatur.

After the final demise of the Hays Code in 1968, Jack Vizzard (who had worked with Breen and Shurlock for decades) published a memoir about his life as a Hollywood censor. Reviews of *See No Evil* emphasized Vizzard's own fall from grace—as a young boy he planned to become a priest but at the age of twenty-nine left the Jesuit seminary where he had been studying, got married, and found a job in the censor's office, then he "liberate[d] himself" from that, too.[7] The simultaneous dismantling of the Hollywood studio system made moving pictures into a veritable cornucopia of consumable violence, profanity, miscegenation, and other previously banned content. Gone were the "ten-second kiss rule" and "the double-bed decree," replaced by the unruly sexuality of *Lolita* (1962), *Blow-Up* (1966), and *Midnight Cowboy* (1969). The demise of the Hays Code exemplified how much American society had changed since the 1930s. Although liberals celebrated an end "to the abiding hypocrisy" of the old system, conservatives continued to worry about the ability of film to normalize immoral behavior and the liberal bias of the industry.[8] Emphasizing the importance of freedom of expression, the Motion Picture Association of America designed a new ratings system that they hoped would protect children from exposure to the more bestial aspects of human nature by allowing their parents and the movie theaters to regulate their visual consumption of social ills.[9]

Violence was certainly not new to a country that had mobilized against Japan and Germany in the Second World War or in Korea during in the 1950s. Ever larger numbers of young men ventured to Vietnam, and images of their lives and deaths were projected back into homes across the nation.[10] During the long, hot summer of 1967, a series of urban riots in Cleveland, Newark, Detroit, and surrounding areas gripped communities already reeling from news of violence in Los Angeles, Chicago, and other cities in previous years. These conflicts led to further loss of jobs, white flight to the suburbs, depopulation of many inner cities, plummeting property values, and ultimately, commercial decline.[11] Concerns about the spread of violence in American society were further heightened by a series of high-profile assassinations. At 12:15 a.m., on June 5, 1968, Senator Robert Kennedy was shot in the kitchen of the Ambassador Hotel in Los Angeles as he sought to exit a party celebrating his victory in California's Democratic primary that day. Only two months earlier, on April 4, Dr. Martin Luther King Jr. had been killed while catching a breath of fresh air on the balcony of his second-floor room at the Lorraine Motel in Memphis. When he learned of King's assassination, Robert Kennedy had spoken briefly but passionately, urging his listeners to "make an effort, as Martin

Luther King did, to understand, and to comprehend, and replace that violence, that stain of bloodshed that has spread across our land, with an effort to understand with compassion, and love." He empathized with listeners who were angered by the news, mentioning for the first time in public the assassination of his brother, President John F. Kennedy, five years earlier.[12] As news of King's assassination spread, another wave of violence broke out in many American cities, most devastatingly in Washington, DC, Baltimore, and Chicago, which intensified politicians' anxieties about urban violence even more.[13] In these contexts, bright lines between heroes and criminals, between just and wrongful violence, became hopelessly muddied.

Less than twenty-four hours after Robert Kennedy was shot, President Lyndon B. Johnson announced the establishment of a National Commission on the Causes and Prevention of Violence. Johnson charged the commission with investigating "the causes, the occurrence, and the control of physical violence across this nation, from assassination that is motivated by prejudice and by ideology, and by politics and by insanity; to violence in our city streets and even our homes." He continued, "What in the nature of our people and the environment of our society makes possible such murder and such violence?" This broad mandate was constrained by preexisting commissions and reports, such as the Kerner Commission, which had explored civil disorder and urban violence (and was published in February 1968, selling more than two million copies) and the Crime Commission (released a year earlier), which investigated the criminal justice system.[14] The struggle for civil rights looked dramatically different than it had a few years earlier. Increasing urban violence had led President Johnson to charge the National Advisory Commission on Civil Disorders to answer the questions, "What happened? Why did it happen? What can be done to prevent it from happening again?" The Kerner Commission, under the chairmanship of Otto Kerner, then governor of Illinois, summarized their findings with a soon to be oft-quoted line: "Our nation is moving toward two societies, one black, one white—separate and unequal." What had started as a conversation about violence ended as a study on race relations.[15] The solutions they offered, including an end to segregated housing, were politically dead on arrival. President Johnson had blocked several social scientists from participating due to their left-wing political views, and others refused to join because they objected to the Vietnam War. The public controversy over the Kerner Commission's conclusion—that white racism was the root cause of black violence—steeled Johnson's refusal to move forward with any of their recommendations.[16] The National Commission on the Causes and Prevention of Violence established seven task forces that would not, at least in theory, duplicate these previous efforts: historical and comparative perspectives; group violence; individual acts of violence; as-sassination; firearms; law enforcement; and media.[17]

By the early 1970s, the lion's share of public outrage was directed at media that seemed to promote violence as a necessary feature of human existence—in both the fictions of Hollywood and the reality of the nightly news.[18] Several contemporary movie critics associated this rise in violence on the screen with changing scientific conceptions of humanity and the violent masculinity em-bodied in killer ape theories of human nature. Professional scientists who disliked this caricature of depraved humanity in turn worried that Americans encountering such ideas solely through mass media, like Conlan, would be-lieve that most anthropologists and zoologists agreed that humans, at their core, were little more than aggressive, competitive animals.[19] At stake, too, was the continued concern that children learned about sex and violence from the media they watched—lessons their parents had been striving to keep from them. Back in the 1930s, the Payne Fund had financed thirteen studies under the title "Motion Pictures and Youth," arguing that children learned a great deal about sex and violence from the movies and from radio.[20] These studies had been crucial to establishing the Hays Code and forty years later remained the most significant body of research on the topic—even if professional social scientists by then deemed the studies' conclusions suspect on methodological grounds.[21] As the code crumbled, the disturbing possibility remained that children would alter their behavior in the face of their enthusiastic consump-tion of movies and television programs.

The son of well-known baseball umpire "Jocko" Conlan, John B. Conlan had spent his undergraduate years at Northwestern, earned his JD at Harvard, and then studied international law at the University of Cologne for year on a Fulbright scholarship. He joined the US Army, serving as a captain from 1956 to 1961, after which he moved to Arizona. Within a year he was attending Re-publican conventions and was elected to the Arizona Senate in 1965, around the same time his father retired to the area. Eight years later he left for Wash-ington to serve in the US House of Representatives. In 1975 he had just begun his second term in the House. It was a well-manicured career, and he made waves as a politician who brought Christian values back to national attention. Conservatives like Conlan considered both the evolutionary logic of biology and anthropological cultural relativism to be symptoms of the increasing secularization of American public culture. MACOS, to Conlan, not only threatened the morality of otherwise virtuous children, but it did so with funds provided by the federal government—he would not stand for either.

Conlan's attack on the anthropological curriculum struck a variety of chords, from his lamenting the social changes currently underway in the country to laying the blame for this moral decline squarely at the feet of left-leaning academic elites out of touch with regular families. He also attacked MACOS from the perspective of fiscal conservatives who, after years of rising inflation and the more recent energy crisis in the United States, sought to re-duce federal spending and prevent the country from slipping into an even deeper economic recession.[22] As a result of the Arab-Israeli War of 1973, oil prices had ratcheted quickly upward, and Americans formed long lines at gas pumps waiting to buy as much as they could afford. By 1975 the economy was recovering, but calls for financial responsibility still resonated strongly with politicians who sought to curb what they saw as unnecessary excesses in gov-ernment spending. In this climate, the funding for the social sciences fared especially poorly.[23] Social scientific research had been used as evidence in recent legal decisions of the Supreme Court under the leadership of Chief Justice Earl Warren, regarding desegregation, the rights of criminals, school prayer, and welfare—decisions considered inappropriate "legal activism" by many conservatives.[24] When Conlan did not deem social science research dangerous, he condemned it as a "squanderous waste."[25]

As New Lefts and New Rights in American politics were forged in the same fire, evolutionary theorists who posited a universal human nature were caught betwixt them.[26] Leftist activists worried that a biological lens could be used to dismiss the politics of protests as due to unavoidable circumstances. They saw in evolutionary theories of human behavior a denial of self-determination and political power to members of minority communities. When conservative Christians encountered evolution, they were reminded that national morals appeared to be spinning out of control. Just as Jerome Bruner, Peter Dow, and the other designers of MACOS had sought to transform the United States by providing children with a set of intellectual tools that would help them to be better citizens in the future, so did Conlan and others seek to remake the country according to their own aspirational values.

Popular depictions of humanity's innately violent nature infused these dis-cussions with urgency. Although MACOS designers had anticipated difficul-ties with students (and teachers) reluctant to recognize the equality of all cultures, they were shocked by criticisms that the program was amoral or pro-moted violence in society. The power of the progressive, evolutionary vision behind the program's creation had begun to break down with resistance from both ends of the political spectrum. On the Left, reduction of human experi-ence to any biological explanation (whether environmental, genetic, or evo-lutionary) appeared to deny personal agency. This proved extremely problem-atic for activists seeking to redress gendered and racial discrimination in the coming decade. On the Right, conservative Christians identified evolutionary theories of humanity as a fundamental component of secular humanism, which was characterized as one of the most terrifying threats to moral order of the twentieth century. Popular films of the era further amplified concerns about the social implications of evolutionary accounts of humanity's past for making sense of the present. Scientists feared that growing environmental destruction or overpopulation could bring an end to human life on Earth as they knew it, and greater numbers of Christians planned for the coming Armageddon.[27] These secular and religious specters of apocalypse no longer resonated with the illimitable future of humanity posited by progressive evolutionists in earlier decades. The Family of Man, a resonant vision only a handful of years earlier, now looked like a utopian dream divorced from the political reality of the present. MACOS became a moral lesson in and of itself—another cultural controversy to be added to the ranks of the Vietnam War, urban violence, and the sexual revolution.

10

The White Problem in America

THE ANTHROPOLOGIST Colin M. Turnbull and Joseph A. Towles penned an intervention into the politics of racialized urban violence that appeared in the pages of 1968's June–July issue of *Natural History* magazine.[1] Turnbull was white, Towles was black, and they met at a gay bar on the Upper West Side of New York City in the fall of 1959, less than two months after Turnbull started working as a curator at the American Museum of Natural History.[2] As an interracial couple immersed in the world of anthropology, they were perhaps acutely aware that the question of violence in American cities was impossible to disentangle from preconceptions about race. At issue were the root causes of urban unrest, why it had started, and ultimately who should be held responsible. Analogical reasoning from anthropological studies of other cultures and zoological explorations of violence in animals played an important role in these debates as this research proffered a variety of models with which to make sense of the social chaos of the time.[3] In their article, Towles and Turnbull echoed the conclusions of the recent Kerner Commission (released just months earlier), arguing that present social policies were dividing the country into ever-more separate societies. They agreed as well with the commission's conclusion that white society itself was "deeply implicated in the ghetto. White institutions created it, white institutions maintain it, and white society condones it."[4] In short, Towles and Turnbull endeavored to convince the relatively affluent, predominantly white readers of the American Museum of Natural History's magazine that urban violence stemmed ultimately from the white problem in America.

African American intellectuals had been wrestling with these issues for a long time. In "Letter from a Region in My Mind," James Baldwin's essay published in the *New Yorker* in 1962, he too had located the fundamental difficulty with race relations in the hearts of white Americans. "White people in this country," he suggested, "will have quite enough to do in learning how to accept and love themselves and each other, and when they have achieved this—

which will not be tomorrow and may very well be never—the Negro problem will no longer exist, for it will no longer be needed."[5] In captivating prose, Baldwin explored the tensions between his own conviction regarding the promise of integration in the United States and the beliefs of black nationalists that true understanding between blacks and whites in America was all but impossible. Baldwin recognized the power of the religious leader Elijah Muhammad's arguments and witnessed firsthand their effect on his followers in the black separatist Nation of Islam. In the coming years, many black nationalists identified with the colonized peoples of foreign empires, from the Algerians to the Vietnamese, characterizing both as forms of unjust imperialism and hailing the rights of independence and self-governance.[6] Baldwin, however, continued to believe that the only hope for the future of the country lay in blacks and whites working together. He called for all his countrymen not to falter in their duty to "end the racial nightmare, and achieve our country, and change the history of the world."[7]

Self-described "good liberal" Norman Podhoretz responded controversially in the pages of the Jewish magazine *Commentary*: "We have it on the authority of James Baldwin that all Negroes hate whites. I am trying to suggest that on their side all whites—all American whites, that is—are sick in their feelings about Negroes."[8] Podhoretz doubted that integration could succeed in the face of such personal animosity, even as activists struggled, no matter what the color of their skin, to break down the institutionalized structures arrayed against African Americans economically and socially. Perhaps the only real solution, he proposed, was total miscegenation. His essay, not only for these claims, inspired over three hundred people to write to the magazine—a few lauding his bravery, but most not.[9]

Two years later Lerone Bennett Jr. opened his introduction to *Ebony* magazine's first special issue by claiming "there is no Negro problem in America."[10] Bennett continued, "The problem of race in America, insofar as that problem is related to packets of melanin in men's skins, is a white problem." In closing he suggested, "by asserting that the Negro problem is predominantly a white problem, this issue summons us to a new beginning and suggests that anything that hides the white American from a confrontation with himself and with the fact that he must change before the Negro can change is a major part of the problem."[11] The special issue generated such excitement among the magazine's readers that *Ebony*'s parent publishing company reprinted the collected essays that composed "The White Problem in America" as a book. For many of the authors, "race" was an idea that physically and culturally demarcated some groups of Americans from others and, it seemed, an idea that could twist itself into misunderstanding. Turnbull and Towles's essay in *Natu-*

ral History drew on this rich conversation for their argumentative thrust, as well as for their title.

Given the weight of these issues, tracing personal views about race among historical figures, especially among authors who wanted to maintain the appearance of scientific objectivity, can prove quite difficult. Not everyone was as willing as Baldwin or Podhoretz to put their private thoughts into print. The well-known advocate of the killer ape theory Robert Ardrey, for example, confessed to W. H. "Ping" Ferry in a letter that he was extremely sensitive about what he said or recorded publicly on the subject of race.[12] Too many left-leaning anthropologists were waiting in the wings, Ardrey thought, eager to label him a racist, just as they had Carleton Coon for his *Origin of Races*— Kenneth Boulding, Sol Tax, Ashley Montagu, Geoffrey Gorer, and others of their ilk.[13] Ardrey did not know Ferry, but based on a recent article he had read, Ardrey deemed Ferry likely to be sympathetic to his own perspective. In typical fashion, he dashed off a long letter of praise.[14] In Ardrey's reading, Ferry had asserted that "Negro and white are sub-species of Homo sapiens" and that "there exists almost no natural precedent for the integration of sub-species." With the institutionalization of slavery in the United States, Ferry suggested and Ardrey agreed, the country had initiated an exceptional situation of two races living in intimate proximity. To correct this situation, Ferry had contended that the United States ought to establish a "system of black colonies" to be managed by a Department of Colonial Affairs based in Washington, DC.[15] He imagined this would provide a legal opportunity for self-determination among blacks without interference from the opinions of whites who did not (he worried, never would) comprehend the difficulties confronting African Americans in the United States. Ferry's proposal drew fierce criticism and struck those who learned of it as an unthinkable parallel to apartheid in South Africa.

In the early spring of 1968, Ferry's suggestion resonated with Ardrey, who agreed that there was a "high likelihood that integration is impossible and unwanted by either people." Although on the surface Ardrey's contention may appear to resonate with the claims of black nationalists, his underlying logic differed significantly.[16] Recall, for example, Ardrey's friendship with Raymond Dart and general sympathy for apartheid laws in South Africa. Ardrey's analysis was antithetical to the racialist politics of black nationalists, which became clear as his letter continued. Ardrey may have been joking when he wrote that "one might regret that more white men through history didn't rape more Negro girls, by this time accomplishing total hybridization and absorption of the Negro into the white population genetically."[17] He nevertheless drew a sober conclusion from the example: "On the whole I don't think much of hybridization as a solution."[18] He complained that a kind of "albo-centrism" gripped his colleagues and prevented them from realizing the hatred that hybrid peoples in other parts of the world had to endure. In closing Ardrey noted, "You have said it: the black will shortly hold our cities, and our years of grace are few."[19] Ferry asked whether Ardrey would mind him quoting an excerpt of his letter in the *Center Magazine*.[20] Despite Ardrey's stated hesitancy over discussing race in print and the content of the letter itself, he raised no objections. He even added that "one of the things that my African experience has taught me is never to generalize about Africans. . . . It's quite probable that a portion of the American problem AND DON'T QUOTE ME is that the slaves were universally selected from docile tribes by the fact of their capture, and by the fact that dealers wouldn't accept potential headaches. Natural selection took place, of a backwards order."[21] When Ferry responded to Ardrey's letter, it was on April 5. He typed hastily, with wide margins and hand-penciled corrections. "M. L. King was murdered last night & I don't have much enthusiasm for writing." He speculated that King would become "an authentic martyr" even among "the nationalists" and noted "there is a rumor that Stokeley [sic] C.[Carmichael] cried when he heard the news."[22]

Across the political spectrum, then, Americans believed the country's "racial problem" had started in the deep entanglement of the nation's history with slavery—even if they disagreed about whether those effects still continued in the present and what, if anything, still needed to be done.[23] Towles and Turnbull reinforced the Kerner Commission's account of the causes of recent urban violence by asking, "What is the point, or purpose, of the violence? And above all, the ultimate anthropological question, what is the function of this violence?"[24] Their answer emphasized violence's symbolic power. "The Negro youth who runs from a burning store holding a child's doll is clutching far more than fifteen dollars' worth of merchandise: he is clutching a symbol of all that he is deprived of in this anything but equal world."[25] Violence was a symbol of defiance. These were not "riots," they constituted a form of collective action.[26] Even the magazine's layout was arrayed against Towles and Turnbull's argument. Inserted into the article were two ads for safaris. The first, to Uganda, Tanzania, and Kenya, promised "21 beastly days in Africa." "We're going on safari," the ad continued, "Not with guns or other lethal weapons. But if you're deadly with a camera, bring it along. There'll be lots to shoot." The second proffered a "civilized safari" in the Amazon.[27] Towles and Turnbull rolled on without regard to the silent interruptions they could not have anticipated.

The outbreaks of violence in cities the previous summer appeared to Towles and Turnbull to be a reaction to a broader crisis that had befallen American society, a response to the country's "total failure . . . to behave in what is considered a correct manner." Because everyone was guilty, the violence was indiscriminate. "Houses are torn down, property destroyed, people beaten, animals slaughtered. This is not primitive, barbaric savagery; it is a very sensible and necessary catharsis. It purges society, which then returns to the lawful way—at least for a time."[28] This almost ritualistic violence, they suggested, had the potential to become much worse. In fact, they wondered why there had been so little. If the government reacted by suppressing the violence, by treating the outbreaks as merely lawless acts, they believed the country risked blocking a "vital safety valve" that guarded against "a much greater disaster."[29] By this Towles and Turnbull meant nothing less than a complete loss of human kindness. They had seen it happen, they thought, in the mountains of northern Uganda, among the Ik.[30]

In his recent book *The Mountain People*, Turnbull wrote a first-person account of the massive cultural changes the Ik had experienced in the years before his arrival. Formerly a hunter-gatherer society, according to Turnbull's reconstruction of their history, the Ik had been excluded from their major hunting grounds on grassland savannahs by the creation of Kidepo National Park and forcibly relocated to the precipitous and rocky terrain of nearby foothills, where they were supposed to become farmers. Then the rains failed and famine set in. Written with engaging intimacy, the book documents Turnbull's journey and stay with the Ik.[31] After his earlier research on the Mbuti, with whom he had found a safe haven, he had been excited to embark on a new adventure. These expectations quickly turned to horror when confronted with the starvation, illness, and death of many members of the community. In the face of their powerful hunger, Turnbull surmised, Ik society was transformed from cooperative and friendly to highly individualistic and devoid of mutual compassion—each man, woman, and child had to fend for him or herself. "There is simply no community of interest, familial or economic, social or spiritual," he despondently related.[32] Even violence no longer existed, except among children as a form of play.[33] Struggling against impossible odds, the Ik survived, but at the cost of their very humanity, Turnbull suggested. "There is no goodness left for the Ik, only a full stomach, and that only for those whose stomachs are already full. But if there is no goodness, stop to think, there is no badness, and if there is no love, neither is there any hate. Perhaps that, after all, is progress; but it is also emptiness."[34] Turnbull feared the social disorder would spill into his own camp and so erected three gates between himself and the Ik.[35]

In their desolation, the Ik provided a moral lesson for Turnbull about the impending dangers of nuclear holocaust or an almost universal famine as a result of world population growth—now was the time to change, before it was too late for humanity.[36] The United States was not Uganda, not yet, Towles and Turnbull seemed to warn in "The White Problem in America." American blacks had not been deprived of so much that cultural decay had set in, not yet. Urban violence was therefore a hopeful warning, a psychological safety valve allowing denizens of inner cities to retain their dignity. The violence indicated that residents of Atlanta, Newark, Detroit, and Baltimore still cared enough to protest, to push back. The Ik exemplified how much worse things could become if the social causes that led to the violence remained unaddressed. Several years later, Turnbull perhaps came to see an optimistic message in Ik society as well.[37] Writing in 1973 he suggested, "Our basic nature, whether we be hunters or computer technicians, insofar as it is inherent in our humanness is probably not so different from any animal nature." But, he continued, "what is special about humanity is not what man *is*, but what he *can* be. . . . Our nature is determined by what we do with that potential, not the other way around."[38]

No readers missed that Turnbull intended *The Mountain People* to be a cautionary tale about the possible future of any human society, but they took away radically divergent morals. The *Book World* supplement of the *Washington Post* hailed it as "a brilliant, terrifying book."[39] The *Chicago Tribune* warned, "Don't read this horror story unless all the lights in your soul are on."[40] Peter Brook, who had directed the film version of William Golding's *Lord of the Flies*, turned the book into a play emphasizing the fragility of the human social contract and the glimmer of humor and lasting social bonds that prevailed despite the famine of the Ik.[41] Clyde Kuhn, in the pages of the *Insurgent Sociologist*, condemned it as "a poor ethnologic study" that said less about the Ik than how Turnbull had become "impotent with frustration."[42] Kuhn insisted that the Ik would have fared far better if the Ugandan government (with Turnbull's assistance) had helped them socially transform by introducing modern agricultural technologies into their lives. The Norwegian anthropologist Fredrik Barth felt compelled to pen a vicious attack in *Current Anthropology* because, he wrote, "it exhibits a number of anthropological difficulties and failings in such a crass form that it deserves both to be sanctioned and to be held up as a warning to us all."[43] Not only did Turnbull reveal various illegal activities (like cattle stealing), he also identified the perpetrators by name in his book. Equally bad, Turnbull had lost all perspective, letting his emotional reaction to the Ik "distort his judgment, erode his integrity" until he developed a "paranoid hate towards the people he lived among so that all genuine anthropological ballast is lost."[44] And yet, Barth fumed, the blurbs on the cover attested to the book's congenial reception by many people he considered colleagues: Desmond Morris called it "beautifully observed and beautifully written," Ashley Montagu urged "we would do well to read it," while Carleton Coon hailed it as a "masterpiece . . . a magnificent if ghastly tale." The controversy over the book has never disappeared, reemerging in anthropology classrooms each fall.[45]

At the time, Ardrey, like Morris and Coon, was delighted with the new perspective on violence provided by Turnbull's research. Asked to blurb *The Mountain People*, Ardrey passed along his congratulations through Turnbull's editor, Michael Korda, at Simon and Schuster. Ardrey said he had read the book in two sittings—"I cannot conceive of the concerned, informed citizen who can read THE MOUNTAIN PEOPLE without having his wits scared out." He agreed with the moral of Turnbull's book because it caused him for the first time to face the real world consequences of his own philosophy: "that humanity can vanish like a candleflame, meaning in less than a generation." He added, "How can one read this book and not think of all one's sons? (Please don't quote that: I have enough problems.)"[46] Turnbull wrote to Korda about the prepublication buzz, excitedly noting the approval of animal behavior experts. The editor at *Smithsonian* magazine had "tried to get the rat man (the name has completely gone from my mind Custer, Cuddles, Calgon . . .) to review the book." Although "C . . ." said no, Turnbull's letter related, the book had inspired him to write a short article about its implications. Turnbull facetiously worried that if the "animal behaviorists and psychos" were to take it up as enthusiastically as it seemed they might, he would be "out of work as an anthropologist for keeps!"[47]

The C . . . in question—the experimental psychologist John B. Calhoun, renowned for his work on the social behavior of rodents—published an article in the very next issue of the *Smithsonian* and titled it "Plight of the Ik and Kaiadilt Is Seen as a Chilling Possible End for Man."[48] He gushed with enthusiasm for Turnbull's book. Here, at long last, was evidence that his own theories of behavioral breakdown in rodents were applicable to humans. Just as mice in his starving rodent cities lost their "mousity," the Ik lost their humanity and turned into a "village of mutual hatred." What gripped Calhoun most in Turnbull's account of his eighteen-month stay with the Ik was the cruel laughter that characterized their social interactions with one another. Rather than reacting to personal calamity with empathy or at least mild concern, the Ik had laughed. They laughed at an old man when he cried out while being pelted with rocks; they laughed at a baby who burned himself on a hot coal near the fire; they laughed when stealing food from family members too sick to resist. By the end of the book, Turnbull laughed with them. Calhoun grimly recounted the difficulty Turnbull had extricating himself mentally from the Ik's worldview even after he physically left their dry mountain village. The dissolution of family life and total valuelessness that beset their culture might not be reversible, Calhoun feared.[49] When Turnbull revisited the Ik a few months later, the rains had returned but not yet human caring. Calhoun suspected the same was true of his experimental mice—if deprived of love and affection during a critical phase in their social development, the harm could never be undone. He emphasized the lessons for American social scientists: "Taking a child from a slum environment and giving him some sort of ideal life is never going to repair the damage done in his earliest years."[50] If urban ghettos were a microcosm of overcrowding, Calhoun speculated, world population increases would soon push the rest of Earth's cultures to the brink of disaster, too.

Before the burst of interest in man as a killer ape, overcrowding had formed the most common naturalistic nonmedical explanation for violent behavior.[51] In the mid-1940s, Calhoun had built what would become the first of a series of experimental rodent cities in which he provided plentiful food, protection from predators, and much-reduced exposure to disease—a seeming utopia. That experiment ended when Calhoun accepted a position at the National Institute of Mental Health (NIMH had been funding his research for years) and moved to the suburbs of Washington, DC. There he constructed new rodent enclosures in, initially, a barn and then in more formal laboratory space (Figure 20). In these man-made structures, the rodent populations skyrocketed, leading to a host of behavioral problems: roving males with no territory of their own lashed out at unprotected females and young; the mating behaviors of some males turned hypersexual (they would mount both males and females), while others became exclusively homosexual; pregnant females built shoddy nests, then abandoned or even attacked their offspring; most of the dead were cannibalized by their cell mates; many rats were psychologically traumatized, unable to defend themselves or engage in normal behavior.[52] For his admirers, the implications of Calhoun's artificial rodent colonies for the overcrowded cities of human construction could not have been clearer.[53] Naturally peaceful species could become violent when forced to live at higher-than-normal concentrations of individuals.[54] These two tendencies—to see violence as caused by social circumstance and as a biological characteristic of humanity—coexisted in public discourse throughout the 1960s and 1970s.

In his preoccupation with overpopulation, Calhoun's voice merged with other doomsayers who feared the growing population bomb, including the ecologists Paul Ehrlich (Stanford University) and Garrett Hardin (University of California, Santa Barbara).[55] In the 1960s Ehrlich made his reputation as an outspoken expert on the consequences of unmitigated population growth, including war, pestilence, famine, and new social problems. He worried that "crowded cities seem inevitably to increase aggressiveness, which manifests itself even now in general disorder and steadily soaring crime rates."[56] Hardin similarly stressed the unforeseen consequences of population growth, but from a different angle. Americans, he submitted, had been conditioned to

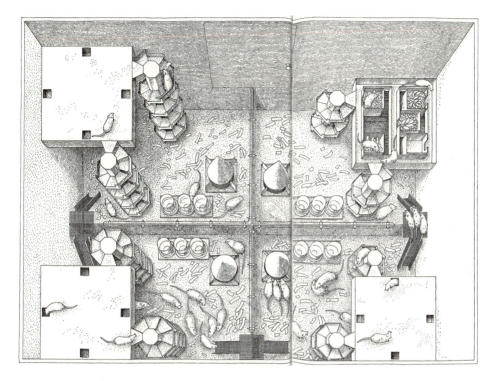

FIGURE 20. John Calhoun's experimental laboratory setup to demonstrate the effect of population density on rat behavior. Calhoun reasoned that because the lower two pens were accessible from both sides, they were harder to defend, which explained why the rat populations in those areas tended to swell while the densities remained lower in the more defensible territories. Illustration by Bunji Tagawa, for John B. Calhoun, "Population Density and Social Pathology," *Scientific American*, February 1962, 140–141. Courtesy of Don Garber/Bunji Tagawa estate.

think according to economies of scale—as demand for a product increased, it became easier to produce and cheaper to produce. "But," Hardin argued, "this economy does not hold when it comes to relational matters. The size of your phone bill goes up with the size of your city. You have to pay for the privilege of talking with more people."[57] As the population of cities in the United States and abroad continued to grow, he posited, so too would the systems on which they depended become more complex, more vulnerable, and more costly. Both Hardin and the Ehrlichs (Paul and Anne; married, longtime intellectual collaborators, although only his name appeared on the cover of *The Population Bomb*) believed humanity's ruthless exploitation of Earth's resources threatened the very life support system on which humans depended.[58]

Barry Commoner, a plant physiologist at Washington University in St. Louis, similarly fretted about the ecological future of the world but sincerely doubted that population growth could be identified as the primary (far less the sole) cause of the environmental crisis.[59] For Commoner, the origins of the environmental crisis lay instead in the free-market system. In *The Closing Circle*, published in 1971, he outlined four general ecological laws that governed the biotic and abiotic interactions on Earth (including humans): "Everything is connected to everything else.... Everything must go somewhere.... Nature knows best.... There is no such thing as a free lunch." These laws integrated the ecological networks of Earth into a single, delicate whole. Commoner insisted that the consumption patterns of industrial society, not the birth rate, lay behind the ecological crisis of the planet. Capitalist greed and growth, he opined, needed to be replaced by a more "socialist relationship with technology" (a stance largely ignored by even those policy makers who took his ecology seriously).[60]

Ehrlich and Commoner—dubbed dual "Cassandras of ecodoom"—came to loggerheads over how to best prevent the imminent demise of man's environment.[61] Ehrlich's preferred remedy was to slow population growth (in concert with reducing pollution and consumption), while Commoner instead favored dramatic changes in US economic policy, insisting that any effort to lower the birth rate was a dangerous canard distracting lawmakers from the real causes of the crisis. The battle became personal as well as public, spreading over the pages of the *Bulletin of the Atomic Scientists* in 1972.[62] An editorial in *Science* suggested that perhaps a fundamental difference in argumentative strategy might explain their continued dissent; whereas Ehrlich had attempted to argue on purely scientific grounds, Commoner considered politics part of the equation, too.[63] Hardin approached their debate from a different angle, giving a brief nod to the weight of Ehrlich's scientific acumen, calling Commoner a "great popularizer of science," and maintaining that both were necessary contributions to intellectual discussions about the ecological health of the planet.[64] To observers less personally involved, however, all three men appeared to be environmental "doomwatchers" who shared a set of beliefs about the future of humanity. Most important, they agreed that the fate of humanity was closely tied to Earth's ecosystem. Because of this tight connection, humans were systematically wrecking the environment on which their survival depended. Unless policy makers mounted a "wholistic, urgent, and radical attack" on the status quo, they all feared the same outcome—destruction of Earth's fragile ecosystem and its capacity to sustain human life.[65] Ehrlich, Commoner, and Hardin just could not agree on what that action should be.

Few scientists read such environmental explanations of violent behavior as antithetical to innate behavioral tendencies in either animals or people, as even normally peaceful animals could be made to kill each other under extreme circumstances, like those created by Calhoun in his rodent hell.[66] But much of the reasoning from overcrowding experiments in animals to lessons for humanity struck scientists as tenuous. So the Ehrlichs collaborated with their Stanford colleague the psychologist Jonathan Freedman to test directly whether population density on a microscale could affect humans' tendency to cooperate.[67] In one experiment, they assembled men and women in groups of four or eight to play a modified prisoner's dilemma game.[68] These groups were placed either in small rooms or large ones to see if crowding altered players' level of cooperative or competitive behavior. The results were ambiguous— although male-only groups did become slightly more competitive in the smaller rooms, female-only groups became more cooperative. The idea of sex differences in violent tendencies would continue to crop up in public and academic conversations, but because both of these effects disappeared in mixed-sex groups, the experimenters concluded there was no consistent connection between crowding and aggression for either sex.

More metaphorically, in *The Human Zoo* Desmond Morris compared human cities to animal cages. (It was published two years after *The Naked Ape*, in the middle of his time in Malta.) He contended that in overpopulated urban centers normal dominance relations governing human social interactions broke down.[69] Morris also suggested that as a result of living in such cages, people naturally developed neuroses and stereotyped behavior patterns, like caged panthers daily pacing the same worn path. The human equivalent of Calhoun's rodents eating their abandoned young was, for Morris, the "battered baby" syndrome of modern cities.[70] Child abuse had leapt to national attention in both the US and the UK during the previous decade, and Morris capitalized on the immediacy of the problem.[71] So, too, the statesman of science Julian Huxley described population as "pressing on man's psychological stability and satisfactions." The stress that afflicted animals living in overly dense environments could produce deranged behaviors, suppressed immune systems, and increased delinquency.[72] "It all boils down to one insistent question," he wrote. (Actually, it was three.) "Do you, we, mankind, want more people at a lower level of existence and a higher risk of disaster, or fewer people at a higher level, with more opportunities for fulfillment? Do we want man's possibilities smothered or cherished? Do we want mere quantity of human units—or better quality of life?"[73] Konrad Lorenz also joined the fray. In an interview in *Psychology Today*, he worried, "Man is a very conditionable animal, and we don't know just how far you can train man to fit another environment. But I strongly doubt whether you can condition man so that he does not become nervous and neurotic when he is crowded."[74] In a separate article, Lorenz more vividly announced that humanity was "in danger of being killed by its own excretion. . . . Humanity will be forced to invent some sort of planetary kidney—or it will die from its own waste products."[75] *Playboy*, too, chimed in, wielding the dangers of overpopulation as a justification for the widespread use of birth control—a strategy with fringe benefits.[76]

Nor did such wanton ecological destruction strike scientists as entirely new. Humans, it seemed, had been wreaking havoc on the environment for a long time.[77] "Preindustrial man" came under suspicion for his destruction of the ecological balance that purportedly existed before his arrival on previously uninhabited lands, like Australia, or South America, where human migration coincided with the retreat of glaciers, resulting in the mass extinction of many species of large herbivores on the continent.[78] Contemporary anthropological analyses of human history suggested that indigenous Australian cultures, for example, played a role in dramatically altering the flora and fauna of the continent. According to a 1966 article in *Scientific American*, "Man and the dingo together represented a scourge to the prehistoric fauna; the two were virtually the sole predatory carnivores on the continent."[79] Ehrlich used similar reasoning to dismiss Commoner's overwhelming concern with modern technology.[80] "The earth," Ehrlich reasoned, "has been badly scarred by the results of ecocatastrophes which predated by centuries the faulty technologies that have attracted Commoner's attention."[81]

The respected English zoologist Vero Copner Wynne-Edwards dated the origins of humanity's problematic relationship with nature a little later.[82] To his way of thinking, only with the agricultural revolution had humans begun to reproduce out of control, destroying the ecosystem in their wake.[83] Traditional hunting and gathering lifestyles, he believed, involved cultural taboos and customs that "kept the population density nicely balanced." Then, overgrazing of domesticated animals turned "rich pastures into deserts," while overhunting led to the extinction or near extinction of countless animal species for food, clothing, and even aphrodisiacs. In fact, Wynne-Edwards insisted, the inability of modern humanity to regulate our population growth rates placed us "conspicuously out of line with the rest of the animal kingdom. Man is almost alone in showing a long-term upward trend in numbers; most other animals maintain their population size at a fairly constant level."[84] Only humans treated their natural resources with such imprudent disregard.[85]

For Towles and Turnbull, however, writing in "The White Problem in America," each explanation of urban violence that relied on theories of aggression, overcrowding, or even rising unemployment effectively amounted to the same thing: comfortable answers that fundamentally missed the cultural function of the violent outbursts witnessed across the United States in recent years. "Comfortable, above all," they reasoned, "because they avoid the nagging possibility that the violence might stem from something a great deal more serious and basic—the white man himself and his myopia."[86] To them, any answer rooted in natural causes, whether environmental or evolutionary, foregrounded the explanatory power of nonpolitical factors and therefore undermined the potential for meaningful social action.

For social conservatives, on the other hand, an even darker lesson emerged from these discussions of violence at the home front—the possibility that anthropologists considered all cultures "equal" whether accounting for differences between whites and blacks or between Americans as a whole and the Ik, Netsilik, or the Soviets. Such cultural relativism implied a dangerous moral equivalency. From both perspectives, and those in between, the urban unrest of the late 1960s and early 1970s required explanation and provided evidence of a moral peril facing the country. Where had their fellow citizens learned to commit violence on such a massive scale?

11

A Dangerous Medium

THE HOLLYWOOD DIRECTOR Sam Peckinpah dismissed concerns that the violence in his films could be causing aggressive outbursts among his audiences as a reaction confined to overly sensitive feminists not in tune with the brutal reality of modern male life. In a profile in *Playboy* magazine, his interviewer asked him about the well-known film critic Pauline Kael's review of *Straw Dogs* (1971)—she had claimed Peckinpah "discovered the territorial imperative and wants to spread the Neanderthal word."[1] Not swayed in the least, Peckinpah responded, "More, more, I love it!"[2] He explained that watching a violent film served as a physical release. It allowed men to experience violence safely within the movie theater, much like sports fans.[3] "Do you think people watch the Super Bowl because they think football is a beautiful sport? Bullshit! They're committing violence vicariously." When pushed even harder about the role of violence in films and its relation to human nature, Peckinpah explicitly invoked arguments circulating in colloquial scientific publications at the time: "I think it's wrong—and dangerous—to refuse to acknowledge the animal nature of man. That's what Robert Ardrey is talking about in those three great books of his, *African Genesis*, *The Territorial Imperative* and *The Social Contract*. Ardrey's the only prophet alive today." Peckinpah had first come into contact with *African Genesis* while working on *The Wild Bunch*, which was released to critical acclaim in 1969. A friend had passed it along, suggesting that Peckinpah and Ardrey were "both on the same track." After finishing the movie, Peckinpah finally had time to read it. In recalling the experience, he described Ardrey as a kindred spirit. "I thought, wow, here's somebody who knows a couple of nasty secrets about us."[4] Ardrey, for his part, was so taken with at least this portion of Peckinpah's interview that he wrote to "Dear Playboy," saying, "Like him, I believe that until we have the courage to grasp the whole of human reality—namely, our propensity for violence— we possess small hope for improvement of our lot."[5]

When *Playboy* interviewed Clint Eastwood two years later, the question of violence in Hollywood again arose. Eastwood's reaction was similar to Peckinpah's. He, too, argued that fictional on-screen violence would reduce aggression in viewers. Eastwood admitted that he knew he produced "tough films" but reckoned the violence was sufficiently satirical that the films could still serve as a catharsis for viewers. He continued, "I'm not a person who advocates violence in real life, and if I thought I'd made a film in which the violence inspired people to go out and commit more violence, I wouldn't make those films." When asked to elaborate, Eastwood added that the movies were a form of escapism, and he related a story he remembered reading in the *Los Angeles Times* in which a journalist reported that inmates at San Quentin said they loved Clint Eastwood westerns. According to Eastwood, "Any pent-up emotions they had were released when they saw those films. After they'd see one, everything would be very calm in the prison for the next few weeks."[6] (In the original article, the journalist's inmate informant added, "One year they showed seven comedies and eight guys got stabbed."[7]) Eastwood, however, dismissed Peckinpah's *The Wild Bunch* as part of a recent snowballing of excessive violence in films. Peckinpah, he suggested, simply "wanted to make a superviolent flick" to one-up other directors, to show "how *beautiful* it [violence] is, with slow-motion cameras and everything."[8] Eastwood hoped that members of the audience would not be permanently brutalized by the violence of such recent films through becoming inured to it.

Hollywood depictions of violence on screen, almost always performed by men, perpetuated popular conceptions of human nature as inherently brutal. They also served as a focal point for the concerns of parents, psychologists, and educators about the potential effects of violence in the media on sensitive viewers like children or adults with emotional disturbances.[9] Renewed debates over what constituted "appropriate" content on television and in the theaters quickly followed.[10] Coupled with a commitment to freedom of expression, activists ensured that when the Hays production code lost its power, an MPAA rating system took its place in November of 1968. Studios submitted a finished film to a seven-person rating board that evaluated it according to four criteria: theme, language, nudity and sex, and violence. The board then classified the movie as G (General Audiences, "without consideration of age"), M (Mature Audiences, "parents should exercise their discretion"), R (Restricted, viewers under the age of sixteen admitted only if accompanied by an adult), and X (no one under sixteen admitted, due to the film's "treatment of sex, violence, crime or profanity").[11] All ads, trailers, and radio and television copy would exhibit the rating. This allowed viewers (or their families) to decide in advance what level and kind of violence or sexual activity they were willing to consume. Some parents, following the logic of Peckinpah and East-wood, argued that violence was an inevitable aspect of modern life and there-fore children should be taught to understand and deal with it. Film provided one means of doing so. A great many more (or at least a more vocal subset) insisted that through a steady consumption of shoot-'em-up westerns and other tales of revenge and mayhem, innocent children could turn into trou-bled youths. Yanks could beat up Nazis, and Batman could deliver solid right hooks to criminals, as long as the violence carried symbolic meaning.[12] Of all the concerns parents and teachers raised about the subtle lessons children would learn through violence in film and on television, two repeatedly recur: that the line between criminality and justice was nebulous, and that all men had the potential for violent action when subjected to great pressure.

In 1968 the Task Force on Mass Media and Violence of the National Com-mission on the Causes and Prevention of Violence chose to focus primarily on the depictions of violent behavior in television programs, including both journalistic coverage and entertainment. Much of the concern with journalis-tic depictions of violence concerned the explicit and shocking footage broad-cast on evening news programs covering the war in Vietnam and civil rights protests at home. The commission worried, for example, that because both news and entertainment were delivered to American homes through the same medium of television, the effect might be to blur the boundary between truth and imagination. They further feared that families watching the news might view the coverage as they would a grisly "spectator sport."[13] Following the assassinations of King and Kennedy, networks promised to reduce the quan-tity of violence they aired—in fact, they had made (and neglected) much the same promise after President John F. Kennedy's death five years earlier. If viewers could switch the channel and see "citizens and the police locked in a bloody real life no-holds-barred conflict" on the news, the networks reasoned, there was little point in reducing the violence of cartoons or the frequency of shootings in Westerns or pistol-whippings in crime dramas. "No network," the commission reported, "would have dared stage in make-believe anything as violent as the battle in Chicago" at the Democratic National Convention in 1968, which was marred by unruly demonstrations and police brutality.[14]

The commission argued that the underlying problem of violence in visual media was that "television does teach." Children, they suggested, "mimic the aggressive behavior of adults, whether they observe this behavior in the flesh or on film, but . . . this imitation was drawn equally from realistic and cartoon-like films."[15] To demonstrate the magnitude of the problem, the commission amassed statistics on the frequency of violence on the three major television networks (ABC, CBS, and NBC) for the 1967–68 season. The shows that year included, for example, on ABC, *The Guns of Will Sonnett, Custer, Batman,* and

The F.B.I. (which often ended with a "most wanted" segment); on CBS, *Lost in Space, Mission Impossible,* and *Hogan's Heroes;* on NBC, *Bonanza, Star Trek, Daniel Boone, Dragnet,* and *Get Smart.* According to their analysis, 58 percent of leading men in television dramas committed some violence (in comparison, only a third of leading female characters did so). Men were also more likely to kill or be killed than were female characters.[16] Reading between the not-so-subtle lines, the researchers held westerns and science fiction especially responsible for the degradation of America's youth. Their report cited both of these genres as depicting male-dominated societies in which the rule of law was ineffectual (where it existed at all), leaving questions of justice, often retribution, in the hands of the protagonists themselves.[17] Television also brought images of unattainable affluence into the slums and ghettos of American inner cities.[18]

Defenders of violence on television and in the movies usually invoked at least one of two strategies. They often suggested that contemporary children were already living in a society infused with violence. Even if adults and children delighted when watching violence on TV or in the movies, the psychologist Bruno Bettelheim argued that people still feared real-life violence in big-city schools, when they walked home at night, and sometimes even in the bright light of day.[19] Television, he proposed, provided an opportunity for parents to discuss violence with their children and to help them develop strategies for coping with it. Without discussing violence at home or in school, Bettelheim worried, children would learn about it the hard way. Additionally, as with Peckinpah and Eastwood, defenders invoked the possibility of a "catharsis effect," in which audiences were cleansed of their aggressive urges and became less likely to commit real violence. The authors of the *Mass Media and Violence* report of the National Commission on the Causes and Prevention of Violence argued there was no evidence supporting the existence of a catharsis effect and that vicarious violence probably served instead as a triggering mechanism.[20] Researchers had not yet proved this relation, the report admitted, but if they were right, the burden of risk (and presumably blame) lay in the daily airing of violence-inflected news and entertainment shows on television.[21]

Predictably, the report and its conclusions failed to meet with universal approval. Michael Couzens of the influential Brookings Institution objected to the basic framework of the reports. According to Couzens, the commission began by recognizing that "some violence is necessary and inevitable." The directors of the program, he insisted, believed that "violence inheres in human nature, and it is requisite to the maintenance of a community in which large numbers of people live together."[22] This meant that the commission had been placed in the awkward position of demarcating legitimate from illegitimate violence, an impossibly tricky task. Additionally, he complained that the report's social scientific leanings led to determinist thinking about violence (either because the commission sought to classify a unitary vision of a personality type shared by all presidential assassins, or because they believed society inexorably drove communities of citizens to act out) rather than investigating the role of personal choice.[23] Violence as a tool for gaining an audience *worked* according to the rules of contemporary mass media, and Couzens worried that this would serve to radicalize political groups, forcing them to ever greater acts of horror in order to gain the attenuated notice of journalists and, through them, a hearing with millions of Americans.[24]

Perhaps ironically, two films whose subjects were most directly associated with scientific depictions of humans as primates were less violent than contemporary television shows: *Planet of the Apes* and *2001: A Space Odyssey*, both released in 1968. At least one reviewer, Judith Shatnoff of *Film Quarterly*, lauded these films for once again asking the big questions, "Who is man? From whence cometh and to what destiny goeth?" albeit using Darwin, rather than religion, as "a launching pad to the millennium."[25] Starring the "perfect, lean-hipped, powerful body" of Charlton Heston, *Planet of the Apes* turned the animal-human relationship on its head.[26] The movie was based on Pierre Boulle's 1963 novel, *La Planète des Singes*, originally translated into English as the far less mellifluous *Monkey Planet* the following year.[27]

The vision of humanity in *Planet of the Apes* was equal parts bestial origins and American spirit. When three US astronauts emerge from deep hibernation to discover they have crashed on an alien planet, they encounter an ape civilization in which humans lack the capacity to speak and are considered mere animals.[28] The script makes a point of noting that of all primate species, only humans had killed for sport, lust, and greed, and of all the animals, only humans murdered their own.[29] These sentiments echo the conception of man as a killing ape popularized by Ardrey, Lorenz, and Morris between the time Boulle penned his novel for French readers and Franklin J. Schaffner adapted the story for an American audience.[30] In her review, Pauline Kael noted that Heston embodied the physical "archetype of what makes Americans win"— the "beauty of strength" and a "moral revulsion one feels toward the ugliness of violence." She added, almost as an afterthought, that he had "the profile of an eagle."[31]

Planet of the Apes asked viewers to directly engage with questions about the future of humanity and our role in the nuclear and ecological destruction of the planet. Consider, for example, the simian commandment distilling their fear of humans, "Let him not breed in great numbers, for he will make a desert of his home and yours. Shun him! For he is the harbinger of death." Remember, too, the shocking climax of the movie, when Heston, riding bareback on the beach, discovered the Statue of Liberty embedded in the sand and realized that the alien world on which he thought he was trapped was the future Earth itself, where human social regression had been caused by our violent battles. Remarked Judith Shatnoff, "What is said is very moral, full of concern about the evil nature and ways of man once thought due to original sin, but now attributed to man's evolutionary warp."[32] Her favorite scene of the movie came rather early—the hunting of savage humans by civilized gorillas on horseback. The gorillas gathered specimens for scientific research, caged them for display at zoos, and posed, as have humans past, with their trophies. The luckless astronauts, now subject to their own treatment of other species, appreciated its cruelty for the first time. The last of these, hunting for sport, was a particularly contentious issue in the 1960s. Although many Americans (if they were not hunters themselves) thought of hunting for food as part of a valorous national heritage, the concept of killing for sport evoked images of privilege, class, and conquest at odds with ideals of political and racial equality.[33] The continued success of the franchise, in four more films and two television series released over the next seven years, effectively popularized the idea of humanity's evolutionary origins in an animalistic past, and caused great consternation among some conservative Christians (Figure 21).

In the first scenes of *2001: A Space Odyssey*, on the other hand, audience members relived "The Dawn of Man" as a group of *Australopithecus africanus* encountered a glossy black alien obelisk. Collaboratively produced with Arthur C. Clarke, Stanley Kubrick's *2001* mixed theology and science.[34] The first hominid brave enough to touch the obelisk received, as if by divine inspiration, the knowledge that bones could be used to kill prey and each other. As the strains of Richard Strauss's *Also sprach Zarathustra* swelled through theatres, a lone figure picked up a bone, repeatedly smashing it into the bleached skeleton of a tapir, seeing in his mind's eye his prey falling. Later, after his tribe fought another tribe to claim rights to a favored watering hole, the leader in exultation threw his bone-club into the air, where a split screen transformed it into a spaceship, transporting audiences from their furry past to a bright technological future. Inspired by contemporary popularizations of humans as born in a moment of violence, Kubrick worked with paleoanthropologists who instructed the actors in the film how to move and act like Australopithecines.[35] Reviewers at the time noted the film's fidelity to current science. Shatnoff, for example, remarked that audiences were "watching a recreation of the newest candidate for 'missing link'—the small, meat-eating warrior ape whose remains anthropologist [Louis S. B.] Leakey discovered in Tanzania, thereby suggesting man is 4,000,000 years old."[36] Another reviewer described *2001* as a quintessentially American nightmare in comparison to Kubrick's next film, an equally violent, dystopian future envisioned by Anthony Burgess

FIGURE 21. *Primal Man?* formed the sixth of eleven full-color comics published by Chick
Publications in the Crusader comics series (1974–1985). In the comic, Tommy, the star
of a film about humanity's gruesome evolutionary past, visits with an anthropologist who
calls evolution "one of the cruelest hoaxes ever invented." With a little divine help, the
anthropologist manages to convince both Tommy and the movie's director, Dr. Finlayson
from the "International Geographic Commission," that evolution "is a lot of *bunk*." Finlayson
even admits that his films were "brainwashing and damaging" children in audiences and that
"many will lose their souls because of these films." Despite these revelations, on the last page,
Finlayson decides to continue making his films. Image: Jack T. Chick, *Primal Man?*,
The Crusaders, vol. 6 (1976), 32. Copyright 2007 by Jack T. Chick LLC.

in *A Clockwork Orange*, which reflected common British preoccupations with
class, socialism, and manners.[37]

In the coming years, critical attention to violence in popular media turned
to the far more graphic New Hollywood films of urban jungles and rustic wil-
dernesses in which heroic men found the very essence of their humanity challenged. Independent movies made significant money at box offices around the

country and often included far more violence and brutality than their studio competitors.[38] Stripped of the civilization that usually protected them from the harsh reality of bare existence, the heroes of these "ultraviolent" films, as they came to be known, used every means possible to survive. In 1971 Clint Eastwood starred in *Dirty Harry*, Sam Peckinpah directed *Straw Dogs*, Roman Polanksi released *Macbeth*, and Gordon Parks directed *Shaft*. Kubrick's *A Clockwork Orange* appeared in theaters the following year, along with John Boorman's *Deliverance*. Under the MPAA rating system, Kubrick, Peckinpah, and Polanski all received an X rating the first time the board viewed their films, but Peckinpah and Polanski reedited *Straw Dogs* and *Macbeth* to achieve an R for theatrical release; they protested the changes but feared the X rating would scare away viewers who assumed any film meriting an X must be pornographic.[39] Months after the theatrical release of *A Clockwork Orange*, Kubrick also produced an alternate R-rated version by replacing thirty seconds of footage with a less explicit outtake of the same scene—a decision that satisfied no one. Kubrick's fans thought he had caved to commercial interests and critics accused the MPAA of softening its principles.[40] As in Calhoun's rat cities or Turnbull's depictions of the Ik, the true horror of these movies was not that the protagonists were extraordinary characters, but rather that they were normal people yanked from their average lives and placed in bad situations.[41] Heroes shot first. Villains were sympathetic. Clear demarcations between legitimate and illegitimate violence quickly vanished.

Contemporary film scholars argued that these movies reflected the existing state of society and, more dangerously, contained the seeds of social change in their power to mold (or warp) the minds of adolescents. It therefore seemed inauspicious that violence on television and in films appeared to be on the rise in popularity.[42] Given the commercialism of the film industry, critics and scholars read movies as channeling the social concerns of the communities that produced and enjoyed them.[43] When contemplating the financial successes of ultraviolent films in the very late 1960s and early 1970s, one reviewer despaired that the major issues of the day must therefore include themes like corrupt power, racism, drugs, and brutality.[44]

Pauline Kael had initially railed against the idea that movies could harm society. When she began publishing reviews in the *New Yorker*, she suggested that the real horror lay in the select audiences who loved and reacted to the violence in movies. "It's hysterical to blame the violence in the world on American movies," she argued.[45] Because Kael found the increasingly brutal scenes of recent movies revolting, she believed the "head-movie audiences" who enjoyed such moments had dissociated pain from violence so that graphic killings and gratuitous acts of cruelty became the "richest ingredient of fantasy."[46] In early 1971 she wrote, "It is not uncommon now for fights and semi-psychotic episodes to take place in the theatres, especially when the movies being played are shockers," and with movie houses in then seedy locales such as Times Square and Greenwich Village. "Whether the movies bring it out in the audience or whether the particular audiences that are attracted bring it to the theatre," Kael wrote, "it's *there* in the theatre, particularly at the late shows, and you feel that the violence on the screen may at any moment touch off violence in the theatre. The audience is explosively *live*. It's like being at a prizefight."[47] New releases, like *The French Connection*, seemed to go out of their way to give the audience "jolts" she began to worry, and when watching them in the theater she felt a "raw, primitive response."[48]

Kael had initially been sympathetic, too, to the idea that filmmakers could use graphic violence to make a political point. In the wake of Peckinpah's *Wild Bunch* (1969), she had believed the director when he said "that by making violence realistically bloody and gruesome he would deglamorize warfare and enable the audience to see how horrible it is."[49] But audiences had misunderstood this point, she reckoned, and instead of taking his film as "an attack on violence," "simply enjoyed it as a violent Western."[50] As moviegoers continued to flood box offices to see New Hollywood films, Kael feared that even normally passive audiences were "gradually being conditioned to accept violence as a sensual pleasure." These movies had no intention of condemning the actions of the characters, and Kael became angry. "You don't have to be very keen to see that they are now in fact desensitizing us. They are saying that everyone is brutal, and the heroes must be as brutal as the villains or they turn into fools."[51] Kael no longer blamed audiences for loving violence in films; she instead held directors responsible for inspiring a newfound appreciation for blood and gore.

Widespread audience enthusiasm for violent movies clashed with the skeptical reception of liberal professional critics, who like Kael began to associate filmic violence with reactionary politics. Science fiction appeared an especially "pessimistic" genre, with its stalwart reliance on technology and science in a godless future devoid of Judeo-Christian traditions and hope—"The overwhelming tone is despair; the overwhelming emotion is fear," remarked one observer.[52] Another compared Kubrick's *A Clockwork Orange* to the voyeurism of pornography and claimed it heralded a "new fascist mood in American society."[53] Given the sexual nature of much of the violence perpetrated by Alex DeLarge and his band of droogs, critics reported that female viewers were especially shaken by the film.[54] Kael had read and enjoyed Anthony Burgess's book that Kubrick adapted for his production, describing it as a "parable about the dangers of soullessness and the horrors of force, whether employed by individuals against each other or by society in 'conditioning' "—a message to which she was entirely sympathetic. Kubrick's film, on the other hand, she decried as a vindication of Alex's actions. Kubrick had made the victims un-sympathetic and less human than their attackers, and therefore Kael feared viewers would empathize with the villains. She accused Kubrick of promulgat-ing the "deformed, self-righteous perspective of a vicious young punk who says, 'Everything's rotten. Why shouldn't I do what I want? They're worse than I am.'"[55] In defending his artistic vision, Kubrick (like Peckinpah) invoked Ardrey's books on human nature, arguing that Rousseau's conception of soci-ety corrupting man, rather than the other way around, was nothing but a "ro-mantic fallacy."[56]

Ardrey's status as a Hollywood insider possibly leant his books visibility among directors and actors. These connections did not help, however, with a favorite pet project—a feature-length film based on his first three anthropol-ogy books.[57] Specifics about "The Ardrey Papers" are difficult to reconstruct. The character actor Anthony Zerbe signed on to play a young Ardrey.[58] Other actors dressed up as prehuman man-apes. Mel Stuart and Walon Green di-rected.[59] The producers planned to film on location in East Africa, Germany, and Japan.[60] Filming started on June 9, 1972, in Kenya.[61] The film cost around $800,000 to make and lasted for 84 minutes.[62] In early 1974, Wolper changed the name to "Up from the Apes."[63] In an interview with *Variety* magazine the following year, Ardrey reported he was "physically and emotionally ex-hausted" from working on the film (once again renamed) "Animal Within" and his publication of a fourth anthropology book with a planned forty-eight full-color stills from the film.[64] Conceptualizing the same material as both a screenplay and a book had proved "particularly tiring," he claimed.[65] "I don't intend to do any writing for a while." As late as December of 1974, Ardrey still hoped for a Chicago release, timed for his return to his alma mater.[66] Instead, the film officially premiered as a four-wall release in Winnsboro, a small town in northeast Texas, in January of 1975.[67] It was also screened at the Dallas Film Festival a couple of months later, which Ardrey attended.[68] Then the film sank into oblivion.[69]

Meanwhile, *Dirty Harry* opened in New York City the same weekend as *A Clockwork Orange* and was pegged by Kael as having similar "fascist poten-tial."[70] Kael argued that societies cause crime. She refused to believe that vio-lence resulted from the innate evil of individuals who therefore had to be stopped by any means necessary. Movies like *Dirty Harry* not only violated the facts of reality as she saw them, they also reflected an ideological position that the ends always justified the means. Kael described the director Don Sie-gel as a former liberal who had "put his skills to work in a remarkably single-minded attack on liberal values, with each prejudicial detail in place. *Dirty Harry* is a kind of hardhat *The Fountainhead*."[71] She was only slightly gentler when discussing his lead character, Harry Callahan, played by Eastwood.

Callahan, to Kael, was "a free individual, afraid of no one and bowing to no man." He could have been the quintessential law enforcement officer, a "Camelot cop, courageous and incorruptible, and the protector of women and children," except for the corruption of the system.[72] In the movie, Callahan set out to stop a serial rapist and murderer. Dubbed the Scorpio Killer by the press (and described by Kael as a "hippie maniac"), Charles Davis knew how to work the system, and although Callahan arrested him halfway through the movie, he was released because Callahan had illegally broken into his home and impounded his rifle. The Scorpio Killer also managed to remove Callahan from the case by paying someone to beat him up and then claiming the injuries were the result of police harassment. This series of events led Callahan's commanding officer to yell at him: "Where the hell does it say that you've got a right to kick down doors, torture suspects, deny medical attention and legal counsel? Where have you been? Does Escobedo ring a bell? Miranda? I mean, you must have heard of the Fourth Amendment. What I'm saying is that man had rights." (These legal reforms had just been secured by recent Supreme Court decisions—more evidence, for critics of the Warren Court, of its misguided efforts in "legal activism."[73]) In the film's climax, Callahan nevertheless tracks down the Scorpio Killer and shoots him. Tossing his badge into the water, Callahan renounces the rule of law in the service of true justice. Audience members loved it. Just before the release of *Dirty Harry*, Judy Fayard wrote in *Life* magazine that "the character Eastwood plays is inevitably a man in total control, able to handle anything. He is his own law, and his own morality—independent, unfettered, invulnerable, unfathomable and unbelievable."[74] For Fayard, this made him a "superstud," not a crypto-fascist. Fayard also noted her approval at the pattern established several years earlier in *Fistful of Dollars*—Eastwood shot first. Kael predictably disagreed. She argued that *Dirty Harry*'s simplistic juxtaposition of good versus evil could only be plausible when surrounded by the natural beauty of San Francisco, where "crime can be treated as a defiler from outside the society." The film could never have been set in New York, she posited, where "crime is so obviously a social outgrowth" that the narrative itself would ring false.[75] More than that, she derided *Dirty Harry* as "deeply immoral" for insinuating that the recent Supreme Court decisions upholding the rights of suspects charged with crimes were unnecessary because crime was caused by "super-evil dragons" rather than "deprivation, misery, psychopathology, and social injustice."[76] A reviewer in the conservative pages of *Human Events* agreed that *Dirty Harry* might indeed be "the first right-wing melodrama," yet he cheered Siegel for embracing the political views of Americans "long forgotten" by liberal Hollywood. For a change of pace, he wrote, "It is the *liberal* who is the villain," who "wears a peace emblem on his belt-buckle."[77]

In films about urban, black communities, the newly murky line between law enforcers and lawbreakers became positively nebulous. Taking part of their inspiration from Melvin Van Peebles's *Sweet Sweetback's Baadasssss Song* (1971), "blaxploitation" films, including *Shaft* (1971), *Super Fly* (1972), *Blacula* (1972), and *Coffy* (1973), justified the violence inherent to the characters' lives as an appropriation of white violence against blacks, turning it against the original aggressors.[78] Black characters now fit into new roles as heroes, including private detectives, cocaine dealers, pimps, and sex workers, all fighting to improve their daily lives by outsmarting a system stacked against them.[79] Film historians have argued that despite their embrace of potentially damaging stereotypes, blaxploitation films problematized questions of police brutality and racism, often more so than self-styled "serious" African American movies of the same era.[80] Of his hero, Van Peebles remarked, "He's 'a brown Clint Eastwood.'"[81] Taking taciturn to a new level, Sweetback uttered only forty words in the entire film. Long atmospheric scenes—of Sweetback running, of an urban ghetto church meeting, of a sex show, of a confrontation with a biker gang—occupied much of the film and rendered it far more psychedelic than the action-oriented films it would inspire. *Sweetback* was exceptional in another way, too. As the director, writer, star, and producer, Van Peebles created a film packaged, financed, and distributed solely by African Americans. The involvement of whites in the production and marketing of other black action films, argued Kael, undermined any "therapeutic" function these later films might have had through the creation of black heroes.[82] She suggested this was exactly why "so few whites go to black-*macho* movies: they know the show isn't meant for them, and it's very uncomfortable to be there."[83]

African American communities reacted to black action films with deeply divided sentiments. The cofounder of the Black Panthers Huey Newton crafted a review of *Sweetback* for the Panthers' newspaper in which he lauded the movie for its revolutionary message and its hero's resistance to the violence inherent to white hegemony in the United States.[84] Only a few months later, *Ebony*'s Lerone Bennett published a scathing critique of both the movie and Newton's analysis.[85] The publicity surrounding ongoing trials of Black Panther members and the Attica uprising, in which inmates gained control of the New York prison and were violently quelled, further complicated their reception.[86] Looking back on the early success of these films from 1975, the film scholar Brandon Wander wrote that on the one hand the artists who made blaxploitation films chronicled their personal experiences, including the Watts rebellions of August 1965 and the emergence of the Black Panthers in Oakland, and in so doing, "distilled their own self-made stereotypes, most notably the All-Powerful Black Stud." On the other, he worried that depictions of black culture and masculinity came "full circle to approximate initial white stereotypes." (The ouroboros of cultural stereotypes continued in the James Bond franchise's appropriation of blaxploitation with Roger Moore in *Live and Let Die* [1973], when Bond pitted his wits against those of an urban drug lord known as "Mr. Big."[87]) Thus, Wander noted, "it remains to be seen whether they are used ironically or unwittingly."[88] In fact, criticisms of blaxploitation emerged quickly from the ranks of middle-class liberal leadership in the NAACP and the National Urban League, who contended that the stereotypes of the films might hamper their quest for equal civil rights.[89] In the face of these criticisms, enthusiasm for blaxploitation films proved to be short-lived, and by the mid-1970s the number of new releases had dropped considerably.

Rather than seeing aggression as an essential component of human identity or as the result of psychological disturbance, blaxploitation films and revolutionary organizations like the Black Panthers conceptualized violence as a powerful political tool whites had used to oppress blacks, and that blacks could in turn mobilize in defense of their own justice.[90] This vision of everyday crime—as a phenomenon caused by economic realities beyond the control of any one individual—sat at odds with the depiction of violent criminality in *Dirty Harry*. Even so, social scientists and the National Commission on the Causes and Prevention of Violence interpreted the revolutionary rhetoric of active resistance used by the Black Panthers and in Frantz Fanon's influential *Wretched of the Earth* as evidence of the ubiquitous presence of violence in modern society.[91]

The masculine heroes of New Hollywood or blaxploitation films not only took justice into their own hands, the internal logic of the films made them sexy for doing so. In her article on Clint Eastwood, Judy Fayard lingered on his appearance: his "egg-beaten" hair and his "eyes sparkling with exhilaration." He was tall, she continued, "with chiseled features and long, fine bones. At 43 he still looks draft-age, but his deep lines and creases have begun to strengthen his once almost pretty face."[92] *Shaft's* theme song, for which artist Isaac Hayes won the film's only Academy Award (he also won a Grammy for the soundtrack as a whole), described the hero as a "black private dick, that's a sex machine to all the chicks."[93] In Peckinpah's *Straw Dogs*, David Sumner (played by Dustin Hoffman) traveled to England with his new, beautiful British wife, Amy (played by Susan George), to escape the pressures of academic physics and write his book (Figure 22). Over the course of the film, David proved incapable of protecting his home from invasion or wife from rape until, in the final climax, he shot an intruder in self-defense. The locals who tried to murder him and steal his wife were vanquished and David triumphant. Pauline Kael was furious. She protested, "The goal of the movie is to demonstrate that David *enjoys* the killing, and achieves his manhood in that self-

FIGURE 22. *Newsweek* claimed that Sam Peckinpah's *Straw Dogs* (1971) "flawlessly expresses the belief that manhood requires rites of violence." American Broadcasting Companies, Inc., enthusiastically embraced the assessment, incorporating it into movie posters advertising the film. Author's collection. © American Broadcasting Corp.

recognition." Kael was appalled by Peckinpah's desire to make audiences "dig the sexiness of violence" and David's lack of remorse. "There is even the faint smile of satisfaction on the tarty wife's face that says she will have a new sexual respect for her husband." Kael insisted that the movie implied that "pacifism is unmanly, is pussyfooting, and is false to 'nature.'"[94] In learning to kill, Kael argued, David "lost his intellectual's separation from the beasts," a victory for Peckinpah's obsession with cowboy machismo and a loss for his viewers.[95] It was, she declared, "no more than the sort of anecdote that drunks tell in bars."[96] Albeit in less accusatory prose, other reviewers agreed, connecting Peckinpah's conception of masculinity with indecent brutality.[97] In conquering other men, in learning to fight and ultimately to kill, David Sumner, John Shaft, and Harry Callahan proved their mettle as men and became objects of women's sexual desire.

For concerned citizens from across the political spectrum, the increase in violent content in Hollywood and on television powerfully mirrored the changing times in which they lived.[98] If popular films and movies were passively educating young, eager audiences, these mixed-up amalgamations of gendered stereotypes and the normalization of violent masculinity were lessons neither liberals nor conservatives wanted amplified. When Peckinpah and Kubrick justified the violence of their films by referencing Ardrey's books, they gave weight to conservative fears that evolutionary conceptions of humanity reinforced a moral relativism that twisted notions of right and wrong. As refracted through contemporary cinema, evolutionary connotations of a violent human animal traveled far beyond the halls of academic institutions and presented a revolting vision of human nature. In the wrong hands or burdened by inappropriate content, film was a dangerous medium.

12

Moral Lessons

CONSERVATIVE CHRISTIANS numbered among those who worried about violence in the streets, the breakdown of "traditional" family values, the lawlessness of Hollywood, the horror of mainstream comic books, the licentiousness of the counterculture, government intervention in the education of their children, and more—so they sought to promote more wholesome messages within each of these genres. Throughout the 1970s, alternative Christian radio and television shows gained in popularity, as did Christian movies, sex manuals, textbooks, and universities.[1] As a young ambitious lawyer, John Conlan sought to channel this energy, transforming his constituents' collective outrage into political action. When he entered the fray against Man: A Course of Study in the early spring of 1975, trouble had already been brewing for the curriculum. The intellectual battleground chosen by critics of MACOS bore little resemblance to the concerns anticipated by the scientists and educators who had worked to bring the program to fruition. The shifting political landscape of the early 1970s caught MACOS designers by surprise. They had taken the progressive nature of humanity's deep history for granted, but this was precisely how critics of MACOS attacked the curriculum's sincere embrace of anthropological cultural relativism and its secular undertones. Even in the thick of the debates, the three major scientists involved in MACOS—Jerome Bruner, Asen Balikci, and Irven DeVore—never testified in defense of their creation, convinced that this was a fight over politics, not science, and that therefore their involvement would have little effect on the unfolding drama.

MACOS became commercially available to public schools in the fall of 1970 and represented the National Science Foundation's first venture into the social sciences. By then, almost one-quarter of American high school students were enrolled in science classes designed and made available to local schools thanks to an influx of funding to the National Science Foundation in the wake of the Soviet launch of the Sputnik satellite.[2] By the mid-1970s, NSF geared its curriculum projects to accomplishing two goals: training the next generation of scientists and increasing the scientific literacy of everyone else. A multimedia science curriculum that could hold the attention of "culturally-deprived minorities," like MACOS, looked especially attractive.[3] This came with a significant price tag: acquiring all of the films, filmstrips, booklets, games, and teacher training books totaled nearly four times the cost of a contemporary social studies course. (The comparison to contemporary natural science courses was far more favorable.) The cost, the unfamiliar format of Jerome Bruner's approach to social studies, and the inclusion of potentially controversial materials made education experts wonder how readily school boards would adopt the new program. Yet their skepticism was balanced with praise for the dramatic changes MACOS wrought in the organization of the classroom. Rather than relying on "authoritarian" teaching, MACOS's "student-centered" style "opened" classrooms and helped teachers reach difficult or withdrawn students.[4] According to all of these desiderata, MACOS gleamed with promise.

The first inkling of a problem with MACOS immediately arose in Lake City, Florida, in 1970.[5] MACOS organizers had attempted to minimize controversy by including anthropological materials based only on the light-skinned Netsilik. When criticisms of the program came, they were from unanticipated intellectual directions—that it propagated cultural and therefore moral relativism, that it embodied the humanistic religious convictions of its designers, and especially that it introduced children to violence through its sustained attention to Netsilik traditions.[6] A group of concerned citizens (none of whom had children enrolled in the program), led by the Baptist minister Don Glenn, approached the local school board and tried to ban the teaching of MACOS. Outrageous rumors circulated through the community—students were asked to get naked and touch each other, reincarnation was being taught, and the course itself reflected a "sensual philosophy."[7] Rev. Glenn argued, "The basic objection we have is that the entire course is predicated on the philosophy of humanism," which was itself based on an evolutionary conception of humanity.[8] According to its mandate, the school board assembled a panel of four teachers and four townspeople and asked them to report back in two weeks. After extensive research, the panel recommended that the course be continued but added that the following year parents should be given the option of placing their children in a separate, non-MACOS class.[9] The school board made the course optional immediately. Of the 360 children enrolled in MACOS, only 45 transferred to another course.[10] Wary of controversy, the following fall the school nevertheless dropped MACOS from their curriculum entirely.

Peter Dow, curriculum director of the Education Development Center (the educational nonprofit that had developed MACOS and several other post-Sputnik science curricula), took to heart his conversations with Glenn.[11] Times had changed since the early 1960s when a small, idealistic group of scientists and teachers began work on what would become MACOS. In the intervening decade, Dow reflected, "dramatic political events and social changes of considerable magnitude" highlighted the continued importance of education in helping students wrestle with the cultural tumult of the era.[12] Equally important, the pastors and parents who objected to MACOS had likely not read the sympathetic reflections of Loren Eiseley or Sherwood Washburn on humanity's evolutionary past. In a memo after events unfolded in Florida, Dow wrote to a colleague reflecting on what had happened. Glenn had asked him whether or not "the human hand was a product of natural selection," and Dow wondered what would have happened had he "replied more directly and honestly." He understood that Glenn was really asking, "Where do you stand, Dow? I need to know if I am going to put my child in your hands." Or more fundamentally, "How does the function of the school differ from the function of the church?" These questions belied any universal answers, Dow reasoned, and were ultimately something the local community needed to settle for itself. "We say in effect that knowledge rather than blind faith is our best hope, that mankind has a chance, and that in particular an intelligent understanding of his condition and his origins will improve his chances." Yet distinguishing between school and church was something Dow found difficult. He had taught for years at Germantown Friends and wrote, "I know how schooling can become a profoundly religious experience, far more so than that which passes for it on Sundays."[13] The difficulty proved unresolvable.

The language with which Lake City denizens expressed their anxieties with MACOS echoed the rhetoric of similar disquiet with the grade-school curricula in California. Max Rafferty titled his 1969 screed against secular humanism in education *Guidelines for Moral Instruction in California Schools*.[14] Rafferty served as the superintendent of public instruction of the State of California and became close with Ronald Reagan during the latter's first years as governor of the state.[15] Rafferty made a name for himself in the early 1960s when he blamed public grade schools for the declining moral vigor of American youth, those "youngsters growing up to become booted, side-burned, ducktailed, unwashed, leatherjacketed slobs, whose favorite sport is ravaging little girls and stomping polio victims to death."[16] Toward the end of his seventy-four-page *Guidelines*, Rafferty outlined the worst culprits advancing secular humanist thought in the country: advocates of progressive education, like John Dewey; sex education; psychological behaviorism, as advanced by B. F. Skinner; the social sciences; Marxists; and evolutionists. Teaching a naturalistic basis to the "origins of man" was tantamount to teaching atheism and, he suggested, could be challenged legally.

Rafferty's vehement opposition to the "rising Humanist movement" in the United States came from reading an odd array of sources: an article in *Time* magazine entitled "The Supreme Being—Man," a brief description in the *Brooklyn Tablet*, and most crucially he hearkened back to a 1933 treatise, the "Humanist Manifesto."[17] By excerpting portions of these documents, he painted a picture of humanism as essentially European, with strong roots in German philosophy, and preoccupied with the physical nature of Man to the exclusion of interest in his spiritual nature. Perhaps worse, he argued, humanists were committed to birth control and had created a religious doctrine that stood in opposition to traditional faith. In short, humanism constituted "activity alien to our [American] heritage" and likely "contrary to public policy."[18]

Concerns like Glenn's and Rafferty's were repeated in Phoenix the following year. Conlan may have first noticed the curriculum as a state senator in Arizona that fall, when the program caused a stir among local conservative families.[19] Course critics suggested that MACOS students were introduced to "a steady diet of blood-letting and promiscuity." As the course continued, they believed, "the children lose touch with good and reality." They suggested that in recent years, there had been too many "innovative" curriculum changes in the school, that the lesson that "man is an animal and nothing more" was tantamount to "teaching about the existence of God and religion," and that the school ought to return to "basic education."[20] Critics also raised concerns with the inclusion of "nightmarish films" portraying Netsilik culture, asking "Why is such a lawless culture portrayed as desirable to copy? For dessert the students study how to skin a daughter-in-law and masquerade in her skin."[21] Events unfolded much as they had in Lake City—concerned community members wrote to the school board, which in turn convened an investigatory panel and eventually voted to keep MACOS in the curriculum. MACOS critics tried to take their fight to the Arizona State Board of Education, but it refused to hear the case, stating that the matter was a local, not a state, concern.[22] This time, however, the MACOS team reacted quickly by sending a small crew to record the public discussions and gather materials. Transforming this footage into a film called *Innovation's Perils*, they were determined to learn from the events how best to deal with such local concerns if they came up in the future.

In preparation for his trip to Phoenix, Peter Dow talked with the guru behind MACOS's design, Jerome Bruner, to practice his talking points and get Bruner's perspective on the surprising turn of events. Bruner expressed optimism at the possibility of a real conversation. "I am very moved by the amount of discussion going on in Phoenix," he told Dow. "In one sense, it's a great thing that a community is that much concerned about the kind of course that they are putting into their schools." He wanted Dow to convey two things to the parents. First, that the course was designed to give students a way of "comparing their own plight as human beings in a culture with those in another culture sufficiently different from our own so that they could see the difference and yet sense the similarity." Bruner invoked the French proverb that the fish will be the last to discover water: to find out about ourselves, he continued, comparison with another culture was therefore necessary. Second, he wanted parents to know that although the course was geared to identifying continuities between primates and man, it also posited that there is "something distinctively human about human beings." Bruner confessed that he found himself "shocked and taken aback" by the argument that this could be seen as "somehow an un-Christian way of going about the study of man." He insisted that the films that Asen Balikci and Father Rouselière produced provided an "intensely pure, honest" and "extraordinarily compassionate" perspective on the Netsilik. More people should see them, Bruner believed. The films belonged on local television, and although he admitted it was a "terribly vain thing to say," he thought they were better than anything else people might watch.[23]

Guided by Bruner's comments, and using the material gathered in Phoenix, Dow encouraged prospective school administrators and teachers to think broadly about the kinds of questions that might be raised by MACOS in their communities and helped schools develop answers *before* trouble started rather than after.[24] A MACOS representative in Phoenix—Edward Martin—suggested the attacks could be divided into two broad arenas: values and politics. Difficulties arose, Martin reasoned, because teachers and those who criticized the program had different answers to values questions such as, "Whose values are the children exposed to in school? What values should they be exposed to? How should children be exposed to these values? Who is really responsible for teaching and maintaining values? How do children learn values? What kind of people should schools help children become?" Similar disagreements stemmed from conflicting answers to questions about the politics of education.[25] Until educators and local communities could come to some kind of rapprochement regarding these questions, Martin worried, MACOS would continue to stir up trouble. When ripples of local discontent multiplied, the MACOS team hoped to use *Innovation's Perils* to prepare administrators for what might come. The downside of this strategy, felt even by curriculum designers at the time, was that it led to the impression that MACOS was already in trouble.[26] The film gathered dust as opposition mounted.[27]

Dow slowly began to realize that these seemingly local protests were linked. Conservative education experts like Onalee McGraw, who worried about the "secular humanist" influence of MACOS and other curricula, spread the word through regional and national education organizations, hoping to stop the program's adoption in yet more cities.[28] In April of 1972 news reached Dow that the social studies coordinator in the Bellevue Public Schools, Washington State, had been receiving concerned letters from community members who claimed that MACOS was teaching local elementary school students about sex, human evolution, and other cultures. Dow immediately replied, "As you know we have encountered this sort of problem before. . . . My suspicion is that there is a small group of extremely conservative and well organized people who are behind this kind of opposition and that they are determined to prevent significant innovation from taking place in the schools." He added, however, that any of these debates, once they became public, inevitably drew into the fray parents with "honest concerns about what is being taught to their children." These parents deserved an "honest discussion" about the "real issues," no matter who had started the trouble. Concerns over sex education were easy to dismiss, Dow noted—MACOS simply did not cover the topic. When it came to human evolution, he argued that MACOS largely avoided the question by adopting a "comparative approach" to human nature rather than a developmental, historical one. (Recall that the MACOS team had considered developing a paleoanthropological component to the course but ultimately decided against it.) One booklet for students did discuss natural selection explicitly, although even that omitted the word "evolution" (Figure 23). The materials for teachers included several readings on evolutionary theory, but students would never encounter those readings directly, Dow noted. Trying once again to address the problem behind objections to "teaching another culture in depth," Dow admitted that many of the materials describing Netsilik myths were "strongly worded." Through these legends and heroic tales, students encountered the "deepest fears" of the Netsilik. "Like our Bible, these stories are not always pleasant, but they carry the fundamental ideas and beliefs which strengthen the culture's capacity to deal with its most difficult problems."[29] His diagnosis proved prescient—Christian parents were concerned with exactly this: that MACOS led students to believe that biblical stories were just another form of cultural myth-making and therefore represented a fundamental attack on religious truth more generally.

In Burlington, Vermont, the following year, an organization called Citizens for Quality Textbooks brought the Catholic conservative John Steinbacher to join the fight against MACOS.[30] Mrs. Norma Gabler traveled from Longview, Texas, to warn local audiences about the program, too.[31] Steinbacher had published two books railing against secular forces conspiring to bring about the

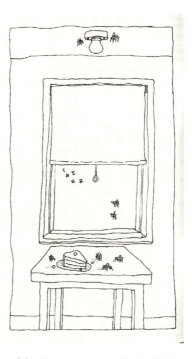

Some scientists believe that Try

All the different animals in the world today
are like they are because of natural selection.
Natural selection has brought about
the baboon and the herring gull, the house
fly and the hippopotamus. To understand
the many different animals of this world, we
must understand what natural selection is
and how it works.

2

FIGURE 23. As a result of the MACOS scandal in Congress, some schools added a
stamp to their booklet that described the process of natural selection for students:
"Some scientists believe that . . ." MACOS, *Natural Selection* (Washington, DC:
Curriculum Development Associates, 1970). This version is available online
from the (Hu)mans: A Course of Study website: http://www.macosonline.
org/course/booklets/Natural%20Selection.pdf. Developed by Education
Development Center, Inc. under grants from the National Science Foundation.

fall of freedom and democracy, one aimed specifically at failures of the public
education system.[32] Like McGraw, Steinbacher sought a national audience.
He, too, saw the controversy over MACOS as a proxy for a long battle between
the truth of Christianity and the ideology of science. On one side, he described a religious worldview in which the forces of Christ fought with the
forces of Satan, a world in which evil was real and humanity had fallen. On the
other, he saw a secular humanistic vision in which humanity was endowed
with the capacity for self-improvement and destruction, in which humans determined their own fate. Whereas a Judeo-Christian perspective provided a
concrete basis for absolute truth, Steinbacher fulminated, MACOS taught
that truth and morality were relative, contingent on culture. In the secular relativism of psychologists and anthropologists lay mortal peril for children (Figure 24). From Steinbacher's perspective, teachers should be providing students with answers, not encouraging them to ask questions of their own.

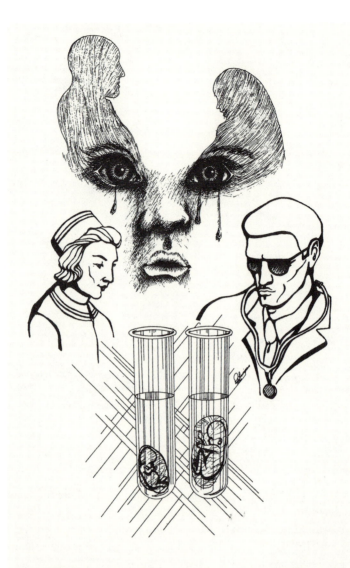

The children of America weep — and the behaviorists callously look upon America's children as experimental animals for their test tubes.

FIGURE 24. In Steinbacher's *The Conspirators*, issues of atheism, humanism, and behaviorism were all closely intertwined. He blamed the "Humanistic takeover of the school system" on "the godless disciples of Moloch, the children of John Dewey and his missionaries." The images of babies in test tubes echoed fears over artificial reproduction. Image: John Steinbacher, *The Conspirators: Men against God*, illus. Ramon Garciaparra (Whittier, CA: Orange Tree Press, 1972), 37–39, image on 39.

MACOS captured the attention of Christian textbook watchers like Stein-bacher and McGraw because they found traction in MACOS designers' educational philosophy as represented in the teacher-training materials. They reacted viscerally to the MACOS films but vocalized the lion's share of their intellectual disagreement by quoting from the teachers' handbooks in which the intent behind the design of the course was clearest. When presented with resistance from school communities, Dow appreciated the deep moral conviction underpinning their concerns and countered with his own. On a local, personal scale his strategy worked brilliantly.[33] Moments of community outrage over the new social sciences curriculum remained sparse. By 1975, seventeen hundred school districts in forty-seven states had adopted MACOS for either late-elementary or middle-school students.

National politics worked differently. In addition to resisting the adoption of "secular humanist" courses locally, conservatives attacked these new curricula in the courts. Until Chief Justice Earl Warren retired in 1969, it seemed doubtful that the United States Supreme Court would recognize their concerns. The Supreme Court had already ruled in two freedom-of-religion cases that public schools could not require students to participate in school prayer or Bible devotionals. Then in 1968, the court declared unconstitutional Arkansas's law banning the teaching of evolution in public school classrooms. *Scientific American* promptly announced "The End of the Monkey War," a quiet denouement to the long aftermath of the 1925 Scopes Trial in Tennessee.[34] Using this ruling as precedent, Mississippi repealed the last remaining anti-evolution law in the country the following year. *Scientific American's* glee was matched by conservative discontent. Taking a different tactic, in 1972 William Willoughby, an editor for the *Washington Evening Star* on religious issues, sued H. Guyford Stever, director of the National Science Foundation, on behalf of "forty million evangelistic Christians in the United States."[35] He alleged that in creating the Biological Sciences Curriculum Study (another of Zacharias' educational projects), the NSF had violated the first amendment by using taxpayer dollars to establish "secular humanism" as the "official religion of the United States," as evidenced by the inclusion of evolutionary theory as a cornerstone of the program.

Building from her early antipathy to the MACOS program, in a 1973 letter to the *Washington Post* (coincident with the program appearing in local suburban schools), McGraw outlined the logic of Willoughby's challenge.[36] Two recent Supreme Court decisions could provide the basis of a legal strategy to confront secular humanism more generally, she reasoned. In 1961 the court ruled that the state of Maryland could not prevent someone from holding the office of notary public if he or she refused to declare a belief in God. The decision contained a footnote listing a series of religious beliefs such an individual might hold that do not generally include arguments of the existence of God: "Buddhism, Taoism, Ethical Culture, Secular Humanism and others."[37] Citing this footnote, the Supreme Court ruled in 1965 that atheists could not claim conscientious objector status in order to escape the draft.[38] Taken together, McGraw believed these cases established precedent for thinking of secular humanism as a religion. And if secular humanism was indeed a set of religious beliefs recognized by the Supreme Court, then their recent decision requiring public schools to maintain complete neutrality when it came to religion implied that publicly funded schools should not be able to teach secular humanism.[39] In the foreword to the first edition of his provocative *Scientific Creationism*, Henry Morris expressed similar frustrations: "In the name of modern science and of church-state separation, the Bible and theistic religion have been effectively banned from curricula, and a nontheistic religion of secular evolutionary humanism has become, for all practical purposes, the official state religion promoted in the public schools."[40] This argument in turn provided an entrée to a second strategy, that if evolution *were* to be taught in schools, teachers should also be required to spend "equal time" on other explanations of biological diversity, including creationism.[41]

Yet after being dismissed by the US District Court in Washington, DC on the grounds that the biological curriculum, by its secular nature, could not violate the Establishment Clause (that "Congress shall make no law respecting an establishment of religion, or prohibiting the free exercise thereof . . ."), the US Supreme Court refused to hear the case. Willoughby's logic built on the conservative evangelicals' conviction that secular, evolutionary humanism constituted a religious doctrine, but they needed more direct evidence to win in court.

In response to MACOS, then, activists additionally tried to stop the production of new science curricula at their source—funding from the National Science Foundation. As a second-term member of the US House of Representatives, Conlan served on the House Subcommittee on Science and Technology that oversaw approval of the NSF's annual budget. Together with Bill Bright's Campus Crusade for Christ and Howard Kershner's Christian Freedom Foundation, Conlan helped to produce books like *In the Spirit of '76: The Citizen's Guide to Politics*, a "handbook for winning elections" designed to increase the number of conservative Christian politicians.[42] The cover featured a stained glass window from the Congressional Prayer Room at the Capitol depicting George Washington kneeling in prayer at Valley Forge and a quote from Billy Graham: "If America is to survive, we must elect more God-centered men and women to public office; individuals who will seek Divine Guidance in the affairs of state. Christians should get involved in good government—not to conform, but to transform."[43] Conlan considered the evangelist

Billy Graham a model Christian statesman and earlier in his career had urged him to run for president.[44] In the name of fiscal responsibility, Conlan began to dismiss numerous zoological and social science projects funded by the agency as "absurd expenditures," including one study on "sexual behavior of Polish frogs" alongside another on "the smell of perspiration from Australian aborigines."[45]

With these ideals at his back, in 1975 Conlan attacked MACOS with a three-pronged strategy demonstrating his conservative credentials and Christian commitments. He denounced the program as "secular humanist," a dog whistle signaling grassroots activists that he was aware of MACOS's evolutionary premises.[46] He suggested that the curriculum materials encouraged cultural relativism in the minds of its students and forced them to interpret the moral values of their parents as simply one system of belief among many. He also insisted that the purportedly scientific content of the program used the suggestive medium of film to expose previously innocent children to violence, promiscuity, and a litany of other evils. As the director of the NSF, H. Guyford Stever attempted to get ahead of Conlan's objections. In March of 1975, Stever wrote to Congressman Olin E. Teague, chair of the committee, that no more funds would be made available to MACOS and that he would additionally allocate no money to any curriculum implementation activities until he had time to conduct a thorough internal review of the Directorate for Science Education.[47] Conlan nevertheless continued his attack in April, in front of the entire House of Representatives: "Mr. Chairman, MACOS materials are full of references to adultery, cannibalism, killing female babies and old people, trial marriage and wife-swapping, violent murder, and other abhorrent behavior of the virtually extinct Netsilik Eskimo subculture the children study. . . . The course was designed by a team of experimental psychologists under Jerome S. Bruner and B. F. Skinner to mold children's social attitudes and beliefs that set them apart and alienate them from the beliefs and moral values of their parents and local communities."[48] (For the record, Skinner was never associated with the program.) The MACOS directors included a two-paragraph quotation from Skinner's *Science and Behavior* on "education as the acquisition of behavior" in the teacher-training materials.[49] They also included passages on the goals of education by Bruner, John Dewey, and Lawrence S. Kubie, which together formed the basis for a seminar discussion among teachers. By referencing Skinner's behaviorism, Conlan provided support for his argument that MACOS was a program bent on modifying the behavior of American youth and echoed earlier complaints against the program by Steinbacher and McGraw. McGraw testified before Congress, repeating her by-now well-rehearsed attack on MACOS.[50] Conlan further claimed that only a monstrous lack of oversight within the NSF could have allowed the program to be funded in the first place, reflecting a deep disconnect between government spending and public utility. (None of this excitement prevented Conlan from slipping away from Washington in May to testify at Billy Graham's crusade in Jackson, Mississippi.[51])

Conlan objected to the violence and sex he saw as inherent to the Netsilik way of life, and a key part of his argument hinged on his perception of their hunting practices as unnecessarily cruel. Conlan and his press secretary repeatedly cited filmed sequences of seal and caribou hunting as a key cause of student distress.[52] Such traumatic scenes of violence, Conlan claimed, left at least one boy unable to sleep for days.[53] He rejected Bruner's notion that MACOS provided a set of tools that students could use to negotiate their lives, seeing the program instead as delivering a world of violence to the doorsteps of good families who had so far been successful in protecting their children from the horrors of the changing world in which they lived.[54] Unlike Hollywood films of the era, MACOS materials had not been rated by the MPAA, and parents feared they had little control over the implicit moral lessons their children might learn at school through watching the films.[55]

Conservative pundits supported Conlan's arguments in newspapers across the country: that MACOS was the product of secular humanism and as such was turning children against their parents. According to James J. Kilpatrick, a well-known conservative columnist and interlocutor on *60 Minutes'* "Point-Counterpoint," progressive and liberal educators were misguided in their praise of MACOS.[56] They claimed the curriculum merely raised value issues and did not coerce the students into a particular answer, but this was precisely the problem: "The barely concealed purpose of MACOS is indeed to teach children how to think—to think, that is, as Dr. Bruner would like them to think."[57] Conlan argued, and Kilpatrick agreed, that the Netsilik people, as a moral model for learning about humanity, were too violent and too primitive and would ultimately break down the traditional American values families were struggling to instill in their children. Kilpatrick especially appreciated the scorn with which Conlan greeted MACOS's "ridicule" of "brave aggressive masculinity" and the curriculum's intention to "redefine the concepts of 'a real man' and 'a true woman.'"[58] On the basis of support provided by conservative activists, and with the energy of parents and teachers who had already mobilized in the fight over the evolutionary content of the Biological Sciences Curriculum Study a few years earlier, Conlan raised as many provocative questions as he could. Had MACOS received funding because the peer review system was broken? Were there financial irregularities with royalties owed to the federal government? Was the program teaching explicit values to the students? Should the books contain a disclaimer to the effect that although NSF funds had been used in developing the course, that did not constitute federal endorsement?[59]

Dow, meanwhile, wrote despairingly to Bruner, who in 1972 had joined the faculty at Oxford.[60] In his letter the evening before Conlan's scheduled appearance before the House, Dow wryly noted, "If you haven't already heard . . . Man: A Course of Study may become the best known and least used curriculum effort of the entire sixties."[61] The following day, before he could have received Dow's missive, Bruner alerted his friend that he was "very reluctant to get into a hoe-down with Kilpatrick and I am not sure how best to proceed by way of getting something into the Congressional Record." He also laughed at the irony of being "grouped in the final paragraph with Fred Skinner!"[62] Friends in England urged him to ignore the press and he was inclined to agree. Congress, however, presented a different scale of problem, and Bruner wrote that he stood "ready to enter the fray if anything can be served by it."[63] All of the other professors who had worked on MACOS in the 1960s had moved on to new projects, too. Back at Harvard, DeVore was caught up in events on campus—although he spoke out against the congressional attacks there, he did not travel to Washington.[64] As an explanation for his hesitancy to defend MACOS, DeVore later suggested that although he loved lecturing in front of classes, he disliked writing and published very little in a colloquial register.[65] Asen Balikci, once again based out of the Université de Montréal, was engaged in a new research project in Afghanistan. This left Dow as the sole public face of MACOS combatting, in Bruner's words, "the anti-intellectual hounds."[66]

Supporters of the National Science Foundation interpreted Conlan's criticisms of MACOS as an attempt to restrict their academic freedom and undercut the importance of basic research in the sciences—especially when he proposed that all NSF grants should be cleared by Congress before any monies were disbursed.[67] Senator Edward Kennedy spoke out against this plan, objecting that it would create an unreasonable administrative burden for both the NSF and for Congress. Even the staunch NSF critic William Proxmire blanched at the possibility of turning politicians into grant administrators.[68] For Conlan, the whole peer-review system seemed to rely on the questionable authority of scientific experts with little attention to the interests of taxpayers. These debates touched a nerve in the larger scientific community, and Kennedy submitted a number of letters he had received into the Congressional Record as a means of documenting their displeasure with the proposal.[69] David R. Mayhew, an associate professor and the director of graduate studies for political science at Yale wrote, "I can't think of a better way to destroy the NSF than to allow this Amendment to make it to the statute books. Academics are flaky enough in handing out money for scholarly projects, but, with all due respect, politicians would be a lot worse."[70] Several others compared this proposed direct government regulation of science to past events in the Soviet Union. Marshall M. Haith, a physician at the University of Denver, claimed that to make peer review a process of political judgment risked "creating a Lysenko system which can only impede free thought and reduce the likelihood of creative breakthroughs."[71] Representative James Symington, too, continued to respond to criticisms about MACOS, noting that "there is more criminal violence in one night of television than an Eskimo would encounter in a lifetime."[72]

Worried about these criticisms but convinced the program still had a future in American classrooms, the remaining MACOS team began to prepare a revised curriculum under the direction of Barbara Herzstein.[73] As a first step, they planned to expand the curriculum into a two-year course. This would allow them to add a historical dimension to the story of the Netsilik, preventing students from seeing the Netsilik as an unchanging culture. Biographies of individual members of the community and a new unit on contemporary Netsilik communities would be a good step in that direction, they thought. Across the animal units, new materials would clarify the importance of females within group structures. They considered, too, adding a booklet about an animal species in which males were predominantly responsible for raising the young, probably the phalarope, and an extensive unit on wild chimpanzees based on Jane Goodall's work in Gombe, Tanzania. Throughout, they sought to clarify the function of the existing examples—the *Observer's Handbook* allowed students to develop a "research-learning skill" rather than "collect data on unsuspecting subjects," and the *Many Lives of Kiviok* contained "tales, legends of a hero figure," not literal stories. None of these changes were ever realized. Due to the increasing bad publicity and fear of creating trouble for themselves, school boards hesitated to adopt the controversial curriculum. Sales of MACOS materials plummeted a full 70 percent between 1974 and 1975.[74]

Despite strong support in Congress for continuing the curriculum efforts of the NSF more broadly, Conlan succeeded in blocking the further disbursement of federal funds to support MACOS. At least rhetorically, even critics of MACOS agreed that the NSF had done an excellent job in the past on curriculum development but had walked onto shaky ground when they introduced implementation grants. As early as 1952, the NSF had made funds available for predoctoral training. They started supporting high-school curriculum development projects in 1956 and curricula for elementary school students a few years after that. Educators and officials at the NSF argued that no matter how many new materials they developed, these curricula would languish without a "push" because teachers would not know how to incorporate the new materials into their routine and the new programs would simply "gather dust."[75]

Responsive to this concern, in the late 1960s NSF started to provide funds for curriculum implementation to ensure materials and training could reach schools that otherwise would not be able to afford them. As the National Council for Social Studies argued in June of 1975, curriculum research, development, and implementation required substantial investment, making it a prohibitively expensive enterprise for everyone except the federal government and a very small number of private foundations. Not only should federal support for curriculum development and implementation be continued, the NCSS argued, funding should be significantly increased.[76] Yet the controversy was soon over and teachers' hopes for continued educational support dashed.

On February 10, 1976, Conlan once again raised the question of NSF's support of science curriculum development and implementation. This time, he received little sympathy from his fellow representatives.[77] The following month, the NSF nevertheless decided to dramatically curtail its investment in precollege science education.[78] The NSF's annual budget for curriculum development dropped to $5.5 million, enough to allow several ongoing projects to finish, but no monies were available for precollege teacher training. This was a far cry from the $40–50 million a year the agency had devoted to educational initiatives during the 1960s.[79] Peer-review policies at the NSF changed to rely less on the judgment of the program officer and more on soliciting increased feedback from external reviewers.[80] By August, Stever stepped down from his position as director of the National Science Foundation to become the first director of the Office of Science and Technology Policy under President Gerald R. Ford.[81] Richard Atkinson, who had served as NSF's deputy director throughout the MACOS fiasco, became acting director and then director of the NSF—the first social scientist to hold the post. One journalist found it "ironic that one of the effects of the curriculum movement is likely to be its own termination."[82]

The education experts Onalee McGraw and Dorothy Nelkin—one sympathetic to MACOS' critics, the other to its developers—each reflected on the lessons to be learned by the episode. McGraw published *Secular Humanism and the Schools: The Issue Whose Time Has Come* with the Heritage Foundation in 1976, who distributed copies for free to anyone who asked. She argued that humanistic education had replaced "basic education."[83] Humanism had caused "the precipitous deterioration of learning achievement in our schools."[84] MACOS provided her key example of an educational program that eroded students' beliefs in "eternal truths," which she explicitly related to the Ten Commandments. McGraw again suggested that "any persuasion of humanism that promotes a religious or irreligious belief is in violation of the constitutional separation of church and state." In support of these positions, she cited the logic of Conlan's arguments before Congress.[85] Her continued association with the Heritage Foundation allowed her to stay in touch with "hundreds of state and local groups" around the country.[86]

Nelkin, on the other hand, started with an article in *Scientific American* in which she distilled recent textbook controversies into three main themes for the magazine's scientific readers, also using MACOS as her prime example.[87] First, Nelkin suggested that a "non-negligible fraction" of Americans believed that science threatened their religious and moral values, sign of a larger disillusionment with science itself. Second, she suspected that this hostility emerged from the nexus of authority and professionalism behind the creation and initial enthusiastic reception of the new science curricula. Third, Nelkin posited that critics feared the meritocratic processes of science would erase a more egalitarian, pluralistic vision of society. All of these themes explained the conflict over MACOS in terms of textbook critics' anxieties with changes in postwar society. In Nelkin's slim volume of the same name, published by MIT Press in 1977, she noted that the fundamentalism of conservatives was matched by the literalism of scientists. Responsibility lay with them, too: "Scientists are convinced of the rationality and merit of their methods," she wrote, "and constantly dismayed by the popularity of nonscientific approaches to nature."[88] Both scientists and religious activists purported to provide a coherent vision of reality, but scientists had failed to recognize that their version of human nature was insufficient for many Americans.[89]

Balikci and DeVore remembered their experiences working with MACOS with remarkably different emotions. When Balikci had completed his PhD, he had been gripped by "the desert areas of Africa or southwest Asia, the Bedouin or the various Saharan tribes" but was asked to become an "Eskimologist" when working at the Canadian Museum, which in turn brought his research to the attention of Douglas Oliver and MACOS.[90] When Bruner took Oliver's place as director, Balikci never quite warmed to him, claiming that Bruner lacked an anthropological instinct and twisted his materials to fit a curriculum that ignored the rich ethnographic value of the films themselves. Rather than portraying the Netsilik on their own terms, he suggested that Bruner turned them into "a kind of caricature of a human being."[91] The Netsilik came to stand in for all humanity, with no cross-cultural comparisons with the !Kung, for example, as the first director had intended. "The man that really emerged out of Jerry Bruner's brochure is a bricoleur more or less of the American type," Balikci believed, "a kind of inventor, a jack of all trades who works in his basement a little bit like Edison, and who can invent practically anything, including a whole culture."[92] Worse, Balikci felt sideswiped by the furor over the curriculum. When he returned from his fieldwork in Afghanistan, finally exploring the desert, he found his films (and, he feared, his reputation as an ethnographer) ruined by the reaction against MACOS. "That was it," he said. "That was the end of my film ambitions."[93] Like Balikci, DeVore started his involvement with MACOS under Oliver's direction but found Bruner's entrance to be a welcome relief. DeVore had intended to travel to Kenya no matter what.[94] He embraced the switch from what he saw as Oliver's pseudo-evolutionary developmental sequence to one that emphasized behavioral complexity and cross-species comparisons—an approach very much in line with his own research.

MACOS refused to disappear entirely, however. The curriculum enjoyed a long life in private middle and high schools around the United States, especially in Friends schools.[95] When Tim and Patsy Asch moved to Canberra in 1976, their children participated in a newly adapted MACOS curriculum for Australians, in which the Netsilik materials were replaced with films and booklets about the people of the Western Desert.[96] West German teachers kept the Netsilik materials but compressed the entire year-long sequence to make room for a third component to the course—"Everyday Life and Politics"—which included lessons on the national system of government.[97] In a sense, this added component embodied the unspoken assumptions of participatory democracy and citizenship that course designers strove to embed in the American program.[98]

Conlan, too, largely exited the national political stage at the end of 1976. In a bruising Republican primary for a seat in the Senate, Conlan ran against fellow representative Sam Steiger. Hailed as "the year's meanest Senate campaign," the two conservatives had remarkably different styles. Steiger had moved out west years earlier from his native New York City and dressed in "urban cowboy fashion—pointed-toed cowboy boots, silver belt-buckle and pearl-buttoned shirts." In describing their two styles, one of Conlan's aides remarked that while Conlan was a scalpel, Steiger was meat-ax. Steiger shot back that he preferred to think of Conlan as a Roto-Rooter. The press labeled Conlan "aggressively Christian" and Steiger a "non-practicing Jew." Questions then arose about Conlan's tactics when someone slipped a letter under the door of Steiger's campaign offices that said "quit working for that Jew—remember, you have been warned." Nothing linked the letter to Conlan. However, when he toured churches to meet potential voters and announced, "a vote for Conlan is a vote for Christianity," his campaign increasingly looked anti-Semitic. Arizona Senator Barry Goldwater subsequently endorsed Steiger in the primary; the first time he had endorsed another politician in twenty-four years, and a sign of his concern with an emerging fissure among conservative Republicans.[99] When four prominent representatives from four different faiths called on Jimmy Carter and Gerald Ford, candidates for the presidential race in 1976, to "repudiate appeals to religious bigotry," they listed Conlan as

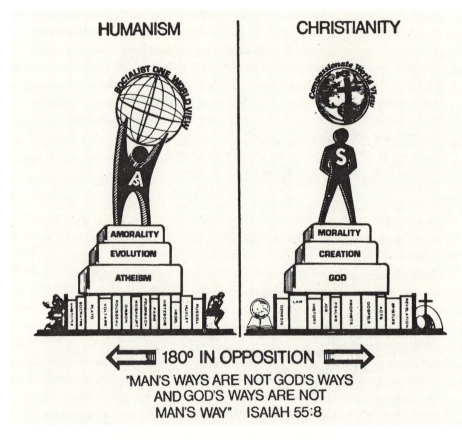

FIGURE 25. Resistance to secular humanism gathered steam in the following decades. Here, Tim LaHaye illustrates the three platforms of humanism—amorality, evolution, and atheism—resting on the work of (from left to right) Aristotle, Socrates, Plato, Voltaire, Rousseau, Gibbon, Weishaupt, Feuerbach, Paine, Nietzsche, Hegel, Huxley, and Russell. The supportive pillars of Christianity, however, came instead from reading Genesis, law, history, Job, Psalms, the prophets, Gospels, Acts, Epistles, and Revelation. Image: Tim LaHaye, *The Battle for the Mind* (Old Tappan, NJ: Fleming H. Revell, 1980), 80. Courtesy of the LaHaye Family.

an example of religious appeal gone too far. Steiger won the primary but lost the election.[100] Conlan had given up his seat in the House of Representatives to run for the Senate, so at the end of 1976 he returned to Phoenix.[101] The episode went down in political lore as a demonstration of the political dangers inherent to damaging primary campaigns.

In retrospect, the debates over MACOS amplified concerns about secular humanism within an increasingly visible evangelical community. Legal wrangling continued, too, and in 1979 Conlan coauthored an article in *Texas Tech*

Law Review with John W. Whitehead in which they expanded the logic of earlier reactions against MACOS into formal argumentation, this time with no reference whatsoever to the program.[102] During his 1980 campaign for president, the former governor of California Ronald Reagan deemed it expedient to repeat old arguments against MACOS, suggesting that the curriculum "indirectly taught grade school children relativism, as they decided which members of their family should be left to die for the survival of the remaining ones." Characterizing MACOS as a secular humanist curriculum, he added, "I don't recall the government ever granting $7 million to scholars for the writing of textbooks reflecting a religious view of man and his destiny."[103] After his election, Reagan's administration dismantled the NSF education directorate entirely, and for many scientists who had paid attention to the MACOS affair, the fate of the program "contributed materially" to that outcome.[104] In the coming decade, Conlan and Whitehead's use of "secular humanism" became a cornerstone of arguments for giving "equal time" to creationism and evolutionary thought in public schools.[105] Tim LaHaye, a San Diego preacher and founding member of the Moral Majority, promoted a crusade against "secular humanism" through writings like *Battle for the Mind* and the *Left Behind* novels (Figure 25).[106] Secular humanism again cropped up in many of the other classic evangelical texts of the period, including books by Jerry Falwell, Francis Schaeffer, and Phyllis Schlafly.[107]

Through the lens of these later heated debates, it now seems remarkable that anthropologists and psychologists in the mid-1970s were flummoxed by the storm brewing around them. By never referencing evolution in the narrow sense, MACOS designers believed they had protected themselves. Yet conservative textbook watchers worried over the express cultural relativism of scientific depictions of human nature. Bruner and DeVore seem to have believed this was a political, not scientific, battle and that the influence of the "anti-intellectual hounds" (as Bruner had termed them) would never impinge on their professional lives. This stance makes sense only if one considers the raucous campus radicalism of the late 1960s and early 1970s. Evolutionary anthropologists and biologists ignored the oratory flourishes of congressmen in order to concentrate on divisive activism flaring up within the academy, much closer to home.

PART V

Death of the Killer Ape

The animal in our nature cannot be regarded as a fit custodian for the values of civilized man.

—WILLIAM D. HAMILTON

IN APRIL 1974, five chimpanzees moved into the newly constructed Outdoor Primate Facility at Stanford University, situated at the intersection of the Linear Accelerator Center and the Jasper Ridge Biological Preserve. Students and faculty alike watched the chimpanzees cautiously explore the grounds.[1] According to the *Stanford Daily*, even the local cows followed their movements with solemn eyes.[2] David Hamburg had been working for years to make this happen. A professor of psychiatry and then human biology at Stanford, Hamburg imagined that chimpanzee research could help solve some of the most perplexing questions of human psychology—and he was especially fascinated by the problem of aggression. At Hamburg's invitation, in the fall of 1971 Jane Goodall spent the first of many semesters as a visiting professor at Stanford. Together they hailed the Stanford OPF as a unique facility that would allow them to study chimpanzees in a location that simulated their wild forest habitat and yet kept the animals constantly available for observation.[3] In the founding vision, the center balanced researchers' competing needs for naturalness and control.[4] Given the close alignment of the research goals at Stanford with Goodall's work in Tanzania, students started calling it Gombe West before the walls had been built or the chimpanzees arrived.

When she began her association with the university, Goodall had already been studying free-living chimpanzees in Africa for more than a decade. Thanks to this work and a new emphasis on genetic relatedness as the key factor in evaluating animal models of human behavior, over the course of the 1970s chimpanzees came to outshine baboons in colloquial science publications discussing the evolution of human nature.[5] Gombe Stream National

Park is located on the western edge of Tanzania, nestled against Lake Tanganyika. The area is accessible only by boat from the nearest town, Kigoma, seventeen miles to the south. Lake Tanganyika stretches north and south for almost 420 miles. This makes it the longest freshwater lake in the world, even though it averages only 50 miles in width, and it constitutes a natural boundary between nations in the region. Tanzania shared this watery border in the north with Burundi and Zaire (now the Democratic Republic of the Congo) and in the south with Zambia. From its position about 800 miles away in Tanzania's capital in Dar es Saalam, the US Embassy called Kigoma remote, even by Tanzanian standards, and Gombe even more so.[6] Together with field assistants, students, collaborators, and other visitors to the research station, Goodall had amassed a remarkable store of data on the daily habits, behaviors, and long-term social interactions of chimpanzees. More recently, researchers had also turned their attention to other primate inhabitants of Gombe, like baboons. After so many years, the accumulated notes, photographs, and film on this set of individuals was unparalleled for any wild animal population.[7] Yet the fate of Gombe had for years remained uncertain—attracted by the potential revenue, the Tanzanian government had contemplated opening the area to tourists. To guard against this possibility, in July of 1974—under the auspices of its director, Derek Bryceson, Goodall's new husband—Tanzania National Parks more fully integrated Gombe Stream Research Centre into its operations.[8]

Gombe buzzed with activity. Located at the nexus of several loosely intersecting institutional and financial networks, it seemed that everyone interested in primatology or the evolution of human behavior passed through at some point.[9] At Cambridge University, where Jane Goodall had earned her PhD under the guidance of Robert Hinde, primatological research formed part of a larger project to understand mammalian ecology and behavior. At Stanford, Hamburg's psychiatric training and undergraduate interest in human biology leant a far more anthropocentric bent to their perspectives.[10] Looking back on these years, Goodall recalled that both Hinde and Hamburg brought increased structure and internal consistency to the data she, Stanford undergraduates, and assorted graduate students and other researchers were gathering. A lively synergy emerged from this mix of researchers.[11] The primatologist William McGrew from Cambridge fondly remembered his encounters with Stanford undergraduates, describing them as "keen, fit, bright, California surfer-type kids," one of whom brought water skis to Gombe for entertainment. He also expressed enthusiasm for Goodall's prehensile toes and her sherry trifle, "heavy on the sherry."[12] Since the 1960s, field primatology everywhere had become more widespread, and Gombe provided a natural comparison for the new field sites popping up elsewhere in Africa and Asia.[13]

A relative newcomer to both chimpanzee research and life in Africa, Hamburg visited Gombe occasionally but never for a prolonged season. Tapping into the "extraordinary upsurge in research on the evolution of human behaviour in recent years," he dreamed of using insights gleaned from observing chimpanzees to understand the prevalence of violence in human societies.[14] He claimed it was easy to understand where his fascination with the causes of violence originated—as a Jewish kid growing up in Evansville, Indiana, he remembered the years leading up to the Second World War and assumed that the "capacity to justify and rationalize depreciation, scapegoating, hatred, and violence" were qualities limited to "psychotic individuals, degraded cultures, or aberrant political movements."[15] Only later had he come to appreciate the "ubiquity" of aggressive behavior among all people. It is difficult to find physical descriptions of Hamburg from the 1970s—he turned into the kind of scholar of whom people wrote down what he said rather than their impressions of the man himself. Photographs depict a slender man with a narrow chin and eyes that always looked directly at the camera. "Natural selection," Hamburg argued, "had shaped our ancestors in ways that suited earlier environments over millions of years." With rapid changes to human cultures in the last ten thousand years that far out-paced our ability to adapt physically, he questioned if humans remained "suited biologically" to the modern world. The key to understanding human nature, Hamburg surmised, relied on a scientific "understanding of the forces that moulded our species in the past"—especially the emotional ties that bind interpersonal relationships. Although humans love and protect each other, the history of humankind is also "replete with suspicion, hatred, and violence."[16]

In his publications from this time, Hamburg avoided commenting on politics except in vague generalities—urban crowding, pollution, aggression toward strangers, even the capacity of modern weapons to destroy. He never mentioned Richard Nixon or Watergate. Nor did Hamburg discuss the considerable international attention devoted to politically motivated violence that coalesced in the early 1970s around the term "terrorism." Assassinations, bombings, and hijackings perpetrated by bandits, rebels, guerrillas, and insurgents all came to be seen as part of this same grave problem.[17] Experts labeled small bands of men who used violence against innocents as a means of coercive intimidation "terrorists" and their work "a distinctive disorder of the modern world."[18] From an American perspective, the problem was both international—including the horrific kidnapping, attempted ransom, and killing of Israeli athletes and coaches at the Munich Olympics in October 1972—

as well as domestic. Governments around the world, including the United States, formulated new policies mandating that state officials could not negotiate with those they deemed terrorists, even in cases where diplomatic personnel were kidnapped, and devoted considerable funds to study terrorism as an international phenomenon.[19] Like many fellow scientists who had witnessed the highly politicized debates in which Desmond Morris or Paul Ehrlich became embroiled, Hamburg stayed away from political specifics like these, preferring to direct his research on primates at uncovering the root causes of violence in human lives (Figure 26).[20]

The efflorescence of field primatology and its potential for insight into human behavior inspired Hamburg and Goodall to co-organize a week-long conference, "The Behavior of Great Apes," in the summer of 1974, supported by the Wenner-Gren Foundation and hosted at the foundation's Burg Wartenstein Castle in the eastern foothills of the Austrian Alps.[21] Dian Fossey attended in order to share her recent research on mountain gorillas in Rwanda. Biruté Galdikas discussed her observations of orangutans in Borneo. Irven DeVore spoke, too—on aggressive male competition and social dominance in baboons and other primates. For several attendees, his talk (coauthored with Joseph Popp) constituted an introduction to what became known as sociobiological theory.[22] Other primatologists traveled from Japan, Scotland, the United States, the United Kingdom, and Holland, and their topics ranged widely. Their conference was only one of many that the Wenner-Gren Foundation sponsored, and most attendees were already well acquainted. Hamburg had met the physical anthropologist Sherwood Washburn in 1957, when they both held fellowships at Stanford's Center for Advanced Study in the Behavioral Sciences.[23] A couple of years later, Washburn organized the symposium "Social Life of Early Man," in which he, Hamburg, Michael Chance, and others spent almost two weeks talking shop at Burg Wartenstein.[24] Hamburg, together with Washburn and DeVore, later coordinated their efforts and arranged a yearlong study group at the Center for Advanced Study in the Behavioral Sciences to explore the evolution of primate behavior.[25] Hamburg and Goodall's conference "The Behavior of Great Apes" thus represented one in a series of symposia and meetings. In the 1970s, this dynamic, interdisciplinary group of scientists all knew each other quite well.

Of the three forms of evidence used to reconstruct universal visions of human nature, new data from field primatology—especially at Gombe—and paleontology proved particularly crucial in dispelling the killer ape notion that human uniqueness lay in our capacity to murder and make war on fellow members of our species. Among primatologists, the idea that humanity's capacity for murder had driven our evolutionary progress seemed far-fetched by 1975, largely due to a new appreciation of how aggressive other species of pri-

FIGURE 26. "Man's inhumanity to man continues unabated in the twentieth century." This anti-American poster appears in an essay David Hamburg wrote, "Ancient Man in the Twentieth Century," in which he celebrated the "extraordinary upsurge in research on the evolution of human behaviour in recent years." The text identified the poster as dating from the Vietnam War, although it was created by the Cuban artist Alfredo Rostgaard for OSPAAAL (the Organization of Solidarity of the People of Asia, Africa, and Latin America). In April of 1975 President Gerald Ford ordered the evacuation of all remaining US citizens and troops from South Vietnam, marking the end of American military involvement. David Hamburg, "Ancient Man in the Twentieth Century," in Vanne Goodall, ed., *Quest for Man* (New York: Praeger, 1975), 26–54, image 5, quote on 47. Poster depicting a US marine, poster collection, Hoover Institution Archives, https://digitalcollections.hoover.org/objects/11914.

mates could be. The introduction to *The Great Apes* volume that followed Goodall and Hamburg's conference noted that "species of apes are, to varying degrees, aggressive, dominance-seeking, and territorial. It is hard to reconcile the characterization, a decade ago, of the mild ape with the chimpanzee or gorilla now known occasionally to kill and eat the young of its own species."[26] Additionally, the primatologist Clifford Jolly advanced a hypothesis that might explain the lack of large canines in male humans based on his research with the primarily vegetarian Geladas in Ethiopia. Older theories of human evolution had posited that when our ancestors had learned to hunt with weapons, they no longer needed to tear things apart with their teeth. As our weapons grew, our canines shrank. Jolly proposed instead that the reduction of canines may have occurred as a function of adapting to a seed-eating lifestyle, like the grassland-foraging Geladas, long before humans learned to hunt.[27] Given the lack of paleontological evidence that could differentiate between these explanations, Jolly's theory provided a plausible alternative to Man the Hunter as a causal explanation for humanity's evolutionary history.[28]

Cultural anthropology provided a far more ambiguous signal. When Elizabeth Marshall Thomas and Colin Turnbull had chronicled the desert-dwelling !Kung and the Mbuti pygmies, who lived in the forests of the Congo, in widely read books, they described these cultures as peaceful hunter-gatherers.[29] In contrast, Robert Gardner and Karl Heider's depiction of the fearsome wars of the agricultural Dani "in the New Guinea stone age" had emphasized the potential violence of territorial disputes once cultures became agriculturally grounded in a particular region.[30] Turnbull's more recent experience with the mountain-dwelling Ik similarly led to his awakening horror at their apparent callousness and lack of mutual regard. After his experiences with the Man: A Course of Study program at Harvard, the ethnographic filmmaker Timothy Asch collaborated with Napoleon Chagnon in documenting the Yąnomamö culture on 16 mm.[31] They depicted Yąnomamö society, too, as characterized by raids and wars between neighboring villages. Their depictions of violence as inherent to Yąnomami social interactions would later erupt in controversy among anthropologists. In the 1970s, however, both the Ik and the Yąnomamö were cited as evidence that all human cultures, even when uncorrupted by Western influences, exhibited territorial violence.[32]

When Louis Leakey died in 1972, retrospectives of his life provided a perfect opportunity for paleoanthropologists, too, to distance their profession from the idea promulgated originally by Raymond Dart and rendered into colloquial science by Robert Ardrey: that human origins lay in our capacity to kill. Richard Leakey, Louis and Mary's second son, had started training in anthropology under the direction of his parents when he was just a boy. With the international fame of his family name and archeological finds of his own,

Leakey joined forces with the British anthropologist Roger Lewin and set about destroying the killer ape hypothesis in a series of books and articles published in the late 1970s and early '80s.[33] A review of these publications noted that "unlike many of his colleagues, Leakey does not believe that modern man is necessarily programmed for Armageddon."[34] Leakey and Lewin argued that proof of the inherently violent nature of humanity was severely lacking in the fossil record. The Second World War had clouded Americans' judgment, they reasoned, who, in looking for an explanation for the violence they had witnessed, had heartily embraced the concept of the killer ape and the macho male. The theory, Leakey and Lewin posited, implied such behavior was inevitable: that had made it extremely dangerous. They both believed strongly that violence and its consequences were purely cultural and as such could, in fact *should*, be eliminated.[35] Even when academic reviewers sympathized with this message, however, they lamented the errors in the work. One especially exasperated paleoanthropologist wrote that "the popularization of science ceases to be a praiseworthy activity when the science ceases to be science."[36]

In a more staid edited collection discussing the legacy of Louis Leakey's research, contributors took new hard looks at the anthropological, paleontological, and primatological evidence of human origins.[37] In eight "Hottentot villages" in Namibia, researchers had discovered a collection of skeletal elements that closely resembled those Dart had found associated with *Australopithecus* remains in South Africa in 1924. At the time, Dart had noted the absence of smaller bones and suggested that our ancestors had preserved the larger bones deliberately, using them whole as clubs and broken as daggers or knives—the first human tools. The cultural anthropologist C. K. Brain now provided a new explanation. In the community he studied, he wrote, people raised goats for their meat. After cooking and eating an animal, they would crack the bones for marrow and afterward toss them to their dogs. The dogs, in turn, gnawed what was left. As a result, only the most durable bones survived—the more fragile bones effectively disappeared. The physical sorting of remains Dart observed could therefore be achieved without villagers keeping only larger bones to turn them into weapons (or for any other reason). A reviewer of the volume put their success quite baldy: "Alas (for grisly romance, for Dart's 'predaceous armamentarium' and Robert Ardrey's 'man the killer'), . . . Occam's razor has cut away the club makers."[38]

When the world's living authority on the social behavior of ants, Edward O. Wilson, published *Sociobiology* in the summer of 1975, he too sounded the death knell of the killer ape, even though the idea that humans (at least the males) were by nature aggressive fit well with his characterization of human nature.[39] Rather than seeking to reconcile the violence of humans present with a more peaceful past, Wilson started with the opposite premise. If humans had started out being competitive, he reasoned, biologists needed to explain the origins of our capacity to cooperate to understand why we would sacrifice ourselves against our genetic best interests.[40] This question flipped understandings of human nature from earlier decades that assumed people are naturally cooperative and struggled to explain the recurrence of violence in human affairs.

In Wilson's focus on the biological conundrum of altruism, which he researched through an evolutionary focus on animal behavior, *Sociobiology* was both a turning point and a marker of how much had already changed in colloquial publications about human nature. Sociobiologists, as supporters of Wilson's research program came to call themselves, asserted their return to a Darwinian focus of natural selection acting on individuals (rather than groups) and as fundamentally concerned with reproduction (rather than mere survival). They believed they were placing professional evolutionary analyses of human behavior on a new, firm theoretical ground. Sociobiologists in the United States also argued that the comparative study of animal behavior was the crucial key that could unlock scientific studies of human nature. This argument would ultimately position them against interdisciplinary reconstructions of humanity's past that relied on the research of cultural anthropologists and paleontologists. Neither, sociobiologists contended, constituted a sufficient means of interpreting, much less addressing, the seemingly ubiquitous presence of violence in human cultures.

Sociobiologists were able to invest studies of animal behavior in the wild with such explanatory power because of the immense popularity of colloquial scientific publications from earlier decades. After the Second World War, scientists had eagerly cultivated public interest in progressive narratives of human evolution and in so doing produced a diverse array of biological and anthropological theories about what it meant to be human. In turn, politicized debates over humanity's aggressive tendencies in the 1960s—in response to books by Robert Ardrey, Konrad Lorenz, Desmond Morris, Lionel Tiger, and occasionally Robin Fox—made possible the sensationalized reception of sociobiology in the 1970s. Yet sociobiologists legitimized their own research program by demonizing such colloquial scientific publications as overly simplistic popularizations whose authors had misunderstood how evolution works. In essence, this community of self-styled sociobiologists argued that all recent theories of human social evolution, even Man the Hunter, had assumed that the social group to which a male belonged was the primary beneficiary of his territorial defense, aggression, and hunting (whether in terms of safety, family structure, or nutrition). By exploring how these behaviors benefited the male himself, sociobiologists argued they were doing something new.

For critics of the idea that humans have innate behavioral tendencies, however, the demise of the killer ape became the latest in a series of evidentiary defeats that disproved the biological basis of human behavior. These critics saw in sociobiology the same assumptions about sex and race that they had already been fighting against for a decade. Critics therefore linked sociobiological theories of masculine aggression and female sexual coyness with older naturalistic explanations of IQ differences between racially identified groups, claims that XYY men were more aggressive than their XY peers, and even eugenicists' beliefs that the future of humanity should be shaped by controlling the country's reproductive future.[41] By conjoining these debates (which previously had been rather disparate) under the larger umbrella of "biological determinism," feminists and other left-leaning scientists already skeptical of the killer ape easily transitioned into critics of sociobiology, seeing this purportedly new evolutionary framework as merely the latest in a series of disreputable theories that had already been disproven in previous guise. For them, the killer ape theory may have died, but only in its most caricatured form—sociobiology represented the same threat.

Arguments for and against sociobiology proved incredibly resilient, polarizing (on the one hand) essentialist explanations, grounded in comparisons with other animals, of why humans behave the way we do and (on the other) attempts to identify historical causes of persistent social injustice. This logic at first seems backward. After all, evolutionary thinking surely required acknowledging deep time. However, by comparing humans to other animal species, sociobiologists endorsed a universal human nature that had become fixed in humanity's evolutionary past. Males were territorially aggressive and females were sexually coy because these were both features of a transcendent "human nature." Detractors of sociobiologists denied these categories as fundamental and looked instead for biological, sociological, anthropological, and historical variation in human cultures. The ensuing debates internally divided research communities, pitting forward-looking biologist against biologist and well-intentioned anthropologist against anthropologist. In this highly variegated intellectual landscape of evolutionary theory in 1975, there was perhaps only one shared belief: that the killer ape theory of humanity had been nothing but simplistic puffery.

These final chapters, about aggression in chimpanzees and the contested reception of *Sociobiology*, thus highlight two increasingly divergent legacies of evolutionary theories of human behavior: one a synchronic vision of humanity as a remarkable species of animal, in the publications of E. O. Wilson and others, and the other diachronic, defined by humanity's complex history that remained current among advocates of interdisciplinary approaches to human biology, such as those advocated by Jane Goodall and David Hamburg. Vocal criticisms of sociobiological thinking ironically amplified the perceived currency of the human animal, and sociobiology permeated journalistic writing about human nature in the coming decades. Although the idea that humans, especially men, are innately aggressive has not gone away, what has faded is the belief that this very same aggression provided the secret ingredient to the unique natural history of humanity.

13

The New Synthesis

WHEN WILSON PUBLISHED *Sociobiology: The New Synthesis*, he had just turned forty-six. An interviewer described him as a "slender man" with "closely trimmed brown hair and just a trace of the soft accent of his native Alabama."[1] A zoology professor at Harvard, Wilson belonged to an intellectual community of like-minded scientists. The primatologist Irven DeVore had arrived on campus in 1963 thanks to the Man: A Course of Study program. Working with MACOS in turn had introduced the former history major Robert Trivers to the idea of natural selection; he decided to complete a PhD under DeVore's guidance. After Trivers's successful defense of his thesis in 1972, he too joined the faculty as an assistant professor of biology. By 1975, DeVore and Trivers together were training a cohort of graduate students and postdocs interested in applying evolutionary theory to animal and human behavior. A diverse group of students and faculty gathered on Thursday evenings for DeVore's "simian seminar" to talk for a couple of hours about evolution and behavior, often over a big pot of Texas chili and a few beers.[2] Conversation was never restricted to simians alone but drifted according to the interests of the participants, from Trivers's fascination with lizards to Alan Walker's paleoanthropological research.[3] The vociferous attacks on sociobiology galvanized this band of fellow travelers and provided a sense of shared identity. Mostly men, with a scant handful of women, the demographics of the group reflected the demographics of the institution. Inspired by the mathematical game theory of British theoretical biologists, especially William D. Hamilton, they conceived of natural and sexual selection as acting at the level of individuals, who in turn served as genetic-information processing units.[4] A trait could not spread in a population unless it conferred some advantage to the individuals who possessed it, allowing them to contribute more copies of their genes to the next generation of that population than other individuals.

Trivers and Hamilton first met at the "Man and Beast" conference sponsored by the Smithsonian and held in Washington, DC in May 1969.[5] Hamilton spoke more openly, and less mathematically, of his inspirations for tackling the evolution of cooperation than in his publications.[6] Hamilton had read Robert Ardrey's *African Genesis* shortly after it was published in 1961 and almost a decade later remained convinced of the book's basic point—that selfishness was more basic to animal nature than cooperation.[7] As humans became more social, Hamilton reasoned, we had also learned to exploit the pain of others. Our "vicious and warlike tendencies," therefore, made any system of so-called natural ethics problematic.[8] Given this, Hamilton puzzled over our equally obvious capacity to cooperate with strangers: how could our willingness to engage in altruistic behavior have evolved in a fiercely individualistic Darwinian framework? His preliminary answer was inclusive fitness. Hamilton posited that an animal typically shares half of her genes with a sister or daughter, a quarter with a niece or granddaughter, etc., so if she possessed a gene that caused her to behave altruistically to her brothers and sisters, she would selfishly give up her life if it meant saving more than two siblings, more than four nieces or nephews, etc.[9] Such apparent altruism was really genetic selfishness in disguise. Hamilton suggested that although kinship would moderate selfishness in humans as well, the complexity of our social relationships could never be explained by this cold principle alone—some cultures discouraged nepotism, and in other cases enmity between relatives had reached bitter and destructive conclusions. In general, though, he found it easy to imagine how, with increased intelligence and social awareness, "the hominid stock" would have naturally evolved toward greater cohesiveness within a community of loosely related individuals and increased warfare between less related communities. Hamilton concluded, therefore, that "the animal in our nature cannot be regarded as a fit custodian for the values of civilized man."[10] Trivers found his ideas invigorating, even if he occasionally had to lean in to catch exactly what Hamilton was saying.[11]

In his paper, Hamilton had worked with one of the classic tools of game theory—prisoner's dilemma matrices—and his reasoning reflected his deep intellectual connections to other British theorists such as John Maynard Smith, George R. Price, and George Williams.[12] In 1966 Williams had published *Adaptation and Natural Selection: A Critique of Some Current Evolutionary Thought*, which other evolutionists hailed for brilliantly illustrating how adaptation could take place without group selection (that is, without recourse to traits expressed at the level of populations rather than individuals).[13] Of particular concern to Williams were behaviors like alarm calls, in which the benefit appeared to accrue to the community at great cost to the individual making the call, who in so doing attracted the attention of the offending predators.[14] Williams argued that this apparent altruism could be explained through the survival of genetically related relatives in the area: even if the altruistic individual perished, her genes survived. Williams did not argue against the existence of group-level traits but rather that it was more parsimonious to explain evolution in terms of individual selection. Williams chose the ecologist Vero Copner Wynne-Edwards as a particular target.[15] In 1962 Wynne-Edwards had published a veritable tome, *Animal Dispersion in Relation to Social Behavior*, in which he spent 653 pages suggesting that animals and plants possessed the ability (through negative feedback mechanisms of various kinds) to self-regulate their populations.[16] Humans, he reckoned, had lost this ability with the cultural advent of agricultural practices—hence our population explosion in recent millennia. Despite Hamilton's withering critique of what amounted to decades of research, Wynne-Edwards never fought back, but neither did he change his mind. He instead sought to place his theories on more solid ground.

The attacks against Wynne-Edwards might have abated—after all, Williams sought merely to demonstrate that group selection was not necessary, not that it could not exist—except that Ardrey made the core of his third book on human nature, *The Social Contract*, published in 1970, about the brilliance and utility of Wynne-Edwards's ideas on group selection.[17] As a result, Wynne-Edwards's critics continued to attack him publicly at the same time as they proposed their own version of how natural selection operated and added his name to the same category of offenders as those 1960s popularizers from whom they wanted to distance their own theories. Group selection came to demarcate *Sociobiology* from other evolutionary theories of human behavior, a crucial component in arguments that it represented a new synthesis of evolutionary theory and the study of social behavior.

As sociobiology found its footing as a field, so did Trivers as one of its intellectual leaders. Based on insights from merging the American anthropological and zoological traditions in which he had been trained with British mathematical approaches to evolution, Trivers published a series of papers in the early 1970s that became crucial to sociobiological logic. Each paper wrestled with a different biological conundrum, and each made an indelible mark on the field. In the first of these papers, Trivers described his theory of reciprocal altruism as an intuitive concept—"you scratch my back and I'll scratch yours"—that could form the basis of social cooperation in human (and animal) societies.[18] Cooperation among relatives made sense to Trivers because any effort individuals expended on behalf of their relatives, even at great personal cost, would help spread the genes they shared in the next generation. More difficult was the question of why nonrelatives behaved altruistically toward one another when individuals that cheated the system could gain all of the benefit at no cost. By linking present cooperation to future benefit (a kind of anthropological gift-culture applied to evolutionary theory), Trivers noted that cheaters, if caught, would be excluded from such mutually beneficial networks of exchange.[19] He thus explained the evolution of cooperative social behavior as a balance between innate tendencies to behave kindly and to cheat—held in check with a healthy dose of communal suspicion—and all without recourse to a universal altruistic impulse.

Turning his attention to mate choice, Trivers reasoned that males and females invested different amounts of resources in their offspring: whereas a male peacock might father numerous chicks with several females, expending little energy taking care of his young, a female peacock devoted far more time, attention, and resources to ensuring the survival of the few hatchlings she could produce in a season.[20] This difference in parental investment meant that a male's reproductive capacity would far outstrip any female's potential reproduction. Thus, he expected females to be far pickier in their mate choices than males—Darwin had been right to emphasize female choice and male competition as the dominant mechanisms of sexual selection. Trivers put his finger on the pulse of the times: men and women, in his scenario, wanted different things out of sex and marriage. These underlying evolutionary logics provided a means of explaining the apparent conflict between the sexes in contemporary society.[21]

DeVore realized that fully embracing Trivers's ideas would mean turning his back "on everything I had understood up until that point in anthropology: my thesis, my professors, including Washburn."[22] He described his final acceptance of sociobiology as a "conversion experience." In the spring of 1972, DeVore and Wilson co-taught a graduate seminar to see if a science of sociobiology could be worked out. Sarah Blaffer Hrdy, a graduate student at the time, enrolled and found their enthusiasm infectious. Her first impressions of Trivers had been less than stellar—she found him "cocky" and began to think of him as "Hamilton's Harvard 'bull-dog'" for his role in reverently explicating Hamilton's mathematical notions for American scientists. She later came to appreciate what he was saying. "Like a shaman," she wrote, "he dove deep inside himself, resurfacing with extraordinary insights—often at great personal cost."[23] She recalled, too, that behind the scenes Wilson had strongly supported female students "in what is a very sexist field."[24] Over the course of that semester, she came to consider herself a sociobiologist, as did DeVore. Soon after, DeVore and Trivers left on a long trip to observe baboon and chimpanzee behavior in Kenya and Tanzania, as well the langurs in India, where Hrdy was conducting research for her dissertation.[25]

Conversations with DeVore and Trivers on that trip similarly transformed the way Richard Wrangham thought about animal behavior. Funded by David

Hamburg, Wrangham was conducting research at Gombe for his PhD at Cambridge. Wrangham began to devote his attention to sex differences in foraging and social behavior among the chimpanzees.[26] The ornithologist and evolutionary biologist Patricia Gowaty also remembers meeting DeVore and Trivers on their trip to Gombe. She shared with them that she was working at the New York Zoological Society, in the education department of the Bronx Zoo. As part of her duties, she had trained New York City teachers in the Man: A Course of Study program. Trivers told her he had written it.[27] (A couple of years later, she entered graduate school, eventually earning her PhD in zoology at Clemson.) Sociobiological reasoning thus traveled quickly through established networks of researchers already invested in explaining the evolution of social behavior in animals and people.

Trivers remembers the trip fondly, too, especially his fascination with weaning behavior in baboons and langurs, in which young animals would beg for milk long after their mothers tried to stop nursing—work that he would cite in another landmark paper on parent-offspring conflict.[28] He recalls meeting the Cambridge biologist Robert Hinde, who had advised Goodall's doctoral thesis and happened to be in Gombe visiting Wrangham. After Trivers gave an evening talk about his recent work, Hinde objected to the language with which he discussed the consequences of natural selection on the behavior of individuals. Trivers had argued that individuals "should" act a certain way according to evolutionary theory: "females ... should be able to guard against males who will only copulate and not invest subsequent parental effort," "parents should invest roughly equal energy in each sex," "selection should favor males producing ... an abundance of sperm," etc.[29] Hinde convinced him that this sounded like a series of moral imperatives. Trivers took his point. After that evening he instead explained how he "expected" animals to act. This both avoided Hinde's critique, he hoped, and also made clear that his theories would need to be tested in the field.[30]

Trivers's theories proved useful for Wilson's almost seven-hundred-page magnum opus in two significant ways.[31] First, he conceived of evolutionary theory primarily in terms of sex rather than survival. Second, Wilson adopted Trivers's easy translation of Hamilton's and Williams's mathematical precepts as powerful antidotes to group-selectionist thinking. Wilson invoked kin selection and sexual selection as novel theories undergirding a robust evolutionary approach to human nature. Despite the book's length (an interviewer for the *Boston Globe* noted it was "not light bedroom reading"), Wilson held out hope that "interested laypeople" would read it in addition to scientists and students.[32]

Wilson characterized the killer ape books of the 1960s as acts of advocacy rather than science, each propounding a different theory about the evolution of humanity in an equally nonfalsifiable fashion.[33] Into this category, he placed many of the authors we have encountered in previous chapters—Ardrey, Lorenz, Morris, Morgan, and even Fox and Tiger. Each of their books, Wilson suggested, might at first glance be seen as similar in orientation to *Sociobiology* because these authors proposed "comparing man with other primate species" in order to identify behaviors common to all humans. He noted, too, that they had been useful in calling scientific and popular attention "to man's status as a biological species adapted to particular environments," helpfully countering the all-too-pervasive view of human nature as infinitely malleable. However, and this was an intractable issue for Wilson, "their particular handling of the problem tended to be inefficient and misleading." In his characterization, each book advanced a single plausible hypothesis based on the authors' limited reading of scientific research and then extrapolated lessons from this single idea far past the limits of acceptable logic.[34]

Equally dismissive, the British zoologist Richard Dawkins connected his frustration with these books to their evolutionary arguments based on group selection. In 1976, less than a year after *Sociobiology* hit the bookstores, Dawkins published what would become the most successful colloquial rendition of the new evolutionary theory—*The Selfish Gene*.[35] A detail from one of Desmond Morris's surrealist paintings graced the cover (recall Figure 13). Robert Trivers penned a laudatory foreword and promised that Darwinian social theory, which Dawkins presented "for the first time . . . in a simple and popular form," should "give us a deeper understanding of the many roots of our suffering."[36] Dawkins opened the book by claiming as his purpose an examination of "the biology of selfishness and altruism."[37] Like Wilson, he acknowledged that earlier authors had made similar claims. He also baldly protested these authors' misrepresentation of basic evolutionary theory. "The trouble with these books is that their authors got it totally and utterly wrong," he wrote. "They made the erroneous assumption that the important thing in evolution is the good of the *species* (or the group) rather than the good of the individual (or the gene)."[38] According to Dawkins, even the anthropologist Ashley Montagu—a major critic of Ardrey, Lorenz, and Morris—had failed to put his finger on this fundamental difficulty because he too had rejected Alfred, Lord Tennyson's vision of nature as "red in tooth and claw."[39] Dawkins reserved particular scorn for Ardrey. Ardrey's popularization of Wynne-Edwards had brought group selection "out into the open," he wrote. Although group selection had long been "assumed to be true by biologists not familiar with the details of evolutionary theory," Dawkins claimed that in Ardrey's hands group selection had become a "theory to account for the whole of social order."[40] Hamilton penned a thoughtful and flattering review of *The Selfish Gene*, which delighted Dawkins.[41] Dawkins's dismissal of group selection as hopelessly ill-informed has remained one of the main attractions of the book, coupled with his lucid explanations of sexual selection, kin selection, and game theory.[42] *The Selfish Gene*'s engaging tone, gene's-eye-view of evolutionary theory, and embrace of gendered stereotypes ensured an enduring readership.

This intimate but quickly growing community of scientists needed funds to carry out their research, and funds were difficult to find. As the oil crisis continued, unemployment in the United States reached 9.2 percent, and inflation continued to rise. Economic downturns at home and abroad caused the Wenner-Gren Foundation to significantly tighten its belt.[43] The National Science Foundation was not likely to fund behavioral research, and the lion's share of excitement over recent trends in biology was directed at molecular and cellular research.[44] Funds from private philanthropy made up the difference, and a loose alliance formed between the sociobiologists, who needed money, and the first research directors of the Harry Frank Guggenheim Foundation: the anthropologists Robin Fox and Lionel Tiger, appointed in 1972, three years after Guggenheim's death. Under the guidance of Fox and Tiger, the HFGF funded research designed to unpack the origins of human aggression according to Guggenheim's original wishes, to improve "man's relation to man." Scientists could apply for small grants of approximately $10,000 to support research in a wide variety of fields—psychiatry, zoology, biological and cultural anthropology, psychology, endocrinology, and, not least, ethology—demonstrating the interdisciplinary nature of these issues at the time.[45] The directors' grantees included promising young researchers as well as scientists whom Fox and Tiger had read avidly in their early forays into ethology.[46]

After six years, the HFGF asked Fox and Tiger to reflect on their work so far as codirectors. In his self-evaluation, Fox maintained that under their guidance the foundation played a crucial role "in facilitating one important development in the thinking of social and biological scientists, namely the greatly increased recognition that evolutionary and genetic factors have an important influence on behavior."[47] Many small agencies and groups had contributed to this effort, he continued, but "the [Harry Frank Guggenheim] Foundation has played a role far out of proportion to the magnitude of its resources."[48] Fox's position was self-serving but not wrong. To highlight these areas of emphasis, Fox and Tiger reclassified grants awarded in previous years. In this new scheme, almost half of their grants fell into the categories of sociobiology and human and animal ethology, while the bulk of the remaining grantees devoted their research to brains, hormones, and behavior.[49] Most significantly, this reclassification illustrates how paleoanthropology and cross-cultural studies—two of the most crucial kinds of research exploring human nature in the 1960s—disappeared as significant research categories of interest to the HFGF.

For years, the Wenner-Gren Foundation had financed a remarkable quantity of primatological and anthropological research in the form of extended workshops, including the "Great Apes" conference organized by Goodall and Hamburg.[50] They also funded individual projects in this area, especially through a program called "The Origins of Man," established in 1965. Despite the program's success, the Wenner-Gren Foundation was forced to close "The Origins of Man" after only seven years.[51] Funding from the National Science Foundation could not have made up the difference. For most of the 1960s, the Division of Biological and Medical Sciences at the NSF had been divided into four sections that separately evaluated grants: molecular biology, genetic biology, systematic biology (including both zoology and botany), and ecology. Research on animal behavior fit within the jurisdiction of systematic biology (which also funded population biology and evolutionary biology), or within the independent laboratory-based Psychobiology Program—neither of which was likely to look kindly on using animal research to understand human behavior.[52] By 1975 the NSF was also deeply embroiled in Congressional debates over the Man: A Course of Study program and fending off Golden Fleece awards from skeptical congressmen. That year, the NSF reconstituted the BMS as part of a new Directorate of Biological, Behavioral, and Social Sciences, which included four divisions: cellular and molecular biology, environmental biology, behavioral and neural sciences, and the social sciences. Research on animal behavior and sociobiology fell cleanly into the third of these, but field scientists had to compete directly with laboratory-based research projects for funding—a hard sell. Moreover, social scientists were so frustrated with the lack of consideration and attention they received within the hierarchy of the NSF that they strongly considered establishing a separate National Social Science Foundation.[53] Trained as a psychiatrist, Hamburg enjoyed some success with grants from the National Institute of Mental Health, as had John C. Calhoun in seeking support for his research on overcrowding in rats.[54] As increasing numbers of students entered graduate school, the need for research funds could not have been more acute. Rather than compete for larger grants at federal agencies, then, many students of animal and human behavior found support for their research in small-scale philanthropic organizations.

As research directors of the HFGF until 1984, then, Fox and Tiger successfully created a space in which investigations of the human animal, especially as related to violence and dominance, could be supported and sustained. Without the funding Fox and Tiger provided to the burgeoning field of sociobiology, many of the scientists they supported might have focused on different questions and approaches for which they were able obtain funds from other sources. Even scientists like Wilson, DeVore, and Trivers could not transform the study of animal and human behavior in the United States without the help of insiders with financial resources at their disposal, insiders like Fox and Tiger. The HFGF thus funded American research at the intersection of ethology and biological anthropology at a crucial moment in the crystallization of sociobiology as a discipline. However grateful their grant recipients were, in their own research Fox and Tiger still appeared to be utilizing a version of natural selection out of sync with recent trends in evolutionary theory. Whether or not sociobiologists lumped Fox and Tiger in with the group selectionists, as did Wilson, or left them out entirely, like Dawkins, depended on preexisting fissures within the field.[55] Fox organized the conference "Kin Selection and Kinship Theory" in 1978, which put him back in the good graces of many neo-Darwinians, but the conference itself failed to generate enthusiasm among the social scientists he invited.[56]

Throughout his long book, Wilson's gambit was to move the study of behavior onto solid scientific ground by using the tools of ethology, population biology, and evolutionary theory. Difficulties arose when turning this analytical perspective on humanity. A great many critics of *Sociobiology* concentrated their objections on the final chapter—"Man: From Sociobiology to Sociology" (Figure 27). Wilson opened the chapter by asking his readers to "consider man . . . as though we were zoologists from another planet completing a catalog of social species on Earth."[57] He imagined this objectivity-through-alienation would allow readers to more easily appreciate the role evolution had played in the development of human bonding, the sexual division of labor, social stratification, our capacity for language, perhaps even the ease with which humans are indoctrinated, our esthetic impulses, and more. Wilson further suggested that by the end of the twenty-first century, sociological questions would be resolved by neurobiology and explanations at the cellular level. Psychology would no longer exist as a field. Biologists would be far closer to determining a "genetically accurate and hence completely fair code of ethics."[58] (The divisive politics of academic culture would be resolved by then, too, he hoped.) Wilson dreamt of a more unified, more complete knowledge of what it meant to be human and trusted that the study of animal behavior would substantially contribute to that understanding of our nature but without continued engagement with either cultural anthropology or paleoanthropology. Wilson sought to transform what had been a fundamentally interdisciplinary enterprise into a subset of biological inquiry.

Wilson's critics saw in his vision a reckless distortion of the present. At the American Association for the Advancement of Science meeting in 1977, Sherwood Washburn (DeVore's advisor at Chicago) vehemently spoke out against the entire sociobiological perspective as applied to humanity. The problem had started long before Wilson, he suggested. Popular interest in animal behavior had spilled over into attempts to use this knowledge to solve

FIGURE 27. "At the threshold of autocatalytic social evolution two million years ago." It is easy to read this image of *Homo habilis* defending their discovery of a fallen dinothere from rival predators from the critical perspective of *Sociobiology*'s detractors: vicious humans, wielding found materials as weapons (stones, sticks), males responsible for hunting and scavenging on the savannah. The females appear to be safely tucked out of sight, presumably gathering a few roots and vegetables while nursing their young. Anthropologists critiqued this vision as out of sync with recent research in the field. Illustrated by Sarah Landry, for *Sociobiology: The New Synthesis, Twenty-Fifth Anniversary Edition* by Edward O. Wilson, Cambridge, Mass.: The Belknap Press of Harvard University Press. Copyright © 1975, 2000 by the President and Fellows of Harvard University.

humanity's problems. But however fun readers might have found the writings of Ardrey, Morris, and Lorenz, Washburn noted, "they should be considered as good science fiction liberally laced with animal anecdotes" that could never form a useful "guide to human behavior in any way whatsoever." Wilson had built on the success of these books and made "precisely the same mistakes." Washburn contended that Wilson's "facile attribution of genetic causes to contemporary human social behaviors" repeated errors of the deeper past, too, echoing the genetic reductionism of social Darwinism and eugenics. In fact, he postulated, the practice was so common it needed a name: "gene-itis, the genetic disease." To avoid misunderstanding, Washburn clarified that he was not directing his comments at the interdisciplinary study of human biology, which he admired greatly, nor against the study of animal behavior or evolution. He was instead speaking out against the enthusiastic application of genetics in inappropriate situations, against ignoring history, against postulating genes for behaviors with no empirical knowledge of the underlying connections. (This is where his crescendo reached its climax.) "Gene-itis has been a part of our Western European culture for more than a century. It is time to cure that disease, whether it appears in the guise of eugenics, racism, or sociobiology" (Figure 28).[59]

FIGURE 28. E. O. Wilson had great confidence in the power of sociobiological theory to solve the riddle of human nature. "If man finally understands himself and finds there is no mystery," he told an interviewer mere weeks after publishing *Sociobiology*, "he may feel very lost indeed." Wilson divided modern biology into "two great enterprises"—one preoccupied with the "machinery of life" and the evolutionary perspective bent on understanding "the ultimate meaning of life." Wilson's optimism and Sherwood Washburn's cautionary tone regarding the "gene-itis" sweeping the nation are both captured in the illustration chosen by the *New York Times* to accompany Wilson's essay. Quotes: Nina McCain, "Sociobiology: New Theory on Man's Motivation," *Boston Globe*, 13 July 1975, A1, A4. Illustration by David G. Klein for Edward O. Wilson, "Evolutionary Biology Seeks the Meaning of Life Itself," *New York Times*, 27 November 1977, 188. Courtesy of David Klein, Pointmade Animation.

Washburn's criticisms of sociobiology were well practiced by 1977, and he joined a chorus of other disapproving voices. These tensions widened preexisting differences of opinion into deep intellectual rifts between scientists already invested in understanding human nature. This morass of opinion, whether framed as problems best solved by further investigations into ethology, kinship patterns, sexual selection, mathematical models, or paleoanthropology, left plenty of room for disagreement. Yet the differences between these men and women were nothing compared to the hostility generated by scientists outside the field who considered an evolutionary approach to human behavior to be inherently deterministic.

14

The Old Determinism

IN 1974, A FEW MONTHS more than a year before E. O. Wilson's *Sociobiology* appeared on American bookshelves, the paleontologist Stephen Jay Gould penned an essay for his popular column in the American Museum of Natural History's colloquial science magazine *Natural History*. In his early thirties, Gould embraced writing for colloquial audiences.[1] He summarized the objections to the killer ape theory, noting that "we have been deluged during the past decade by a resurgent biological determinism, ranging from 'pop ethology' to outright racism." He continued, "With Konrad Lorenz as godfather, Robert Ardrey as dramatist, and Desmond Morris as raconteur, we are presented with the behavior of man, 'the naked ape,' descended from an African carnivore, innately aggressive and inherently territorial." He did not stop there, adding Lionel Tiger, Robin Fox, and Carleton Coon to his list of offenders who together had crafted "a crude biological determinism" in their writings about human nature.[2]

Gould's objection to the depiction of humans as mere animals neatly captured the essence of later critiques of sociobiological reasoning. "How satisfying it is to fob off responsibility for war and violence upon our presumably carnivorous ancestors," he railed. "How convenient to blame the poor and the hungry for their own condition." This pop ethology story, he suggested, depended on "imperfect" analogies with the behavior of other animals and "scrappy" evidence from hominid fossils. Gould feared the popularity of these ideas as spread through the success of movies like Stanley Kubrick's *2001: A Space Odyssey* and *A Clockwork Orange*. Especially the latter, he posited, asked viewers to consider the false challenge to liberalism posed by these theories of innate human violence: "Shall we accept totalitarian controls for mass deprogramming or remain nasty and vicious within a democracy?" That debates over sociobiology erupted so quickly and so virulently spoke to critics' preexisting concerns with evolutionary depictions of human nature, concerns that had been forged and honed when coming to terms with the social implications of colloquial scientific publications in the previous decade. For Gould, the issue of biological determinism was far from "an abstract matter to be debated within academic cloisters."[3]

Debates over sociobiology within academic circles instantly polarized into arguments over nature versus nurture, biology versus culture, as the primary determinants of why humans behave the way we do.[4] Scientists on both sides of the issue accused the other of allowing politics to interfere with clearheaded scientific analysis. Sociobiology's critics mobilized out of a concern that sociobiologists were using their authority as scientists to advance ideas and concepts that at best lacked rigorous proof and at worst reframed social policy in the language of natural order.[5] That sociobiologists did not intend for their theories to be used as the basis for social policy was irrelevant. Physicists had created the atomic bomb; mathematicians were crucial to operations research; chemists not only perfected incendiary bombs and napalm, they kept the pesticide and herbicide industries running, too.[6] If the not-so-Cold War had taught scientists anything, sociobiologists' detractors argued, it should have been that they had a moral obligation to choose their research topics carefully. This precept extended to conflicts at home, where courts and politicians used biological and anthropological research to prop up discriminatory social policies, they suggested, as well as abroad, where the efforts of scientists in creating bombs and other weapons of war were deployed to devastating effect.[7] Wilson responded to these accusations by calling them politically motivated "academic vigilantism" and then by dismantling point-bypoint what he saw as his critics' fundamental (even willful) misreading of his book.[8] "Evidence that human nature is to some extent genetically influenced is in my opinion decisive," he wrote—to argue otherwise was less a question of science than one of "political censorship."[9]

One of the first broadsides came from the Boston-based Sociobiology Study Group, which published a lethal response to *Sociobiology* in the *New York Review of Books*.[10] The fifteen signatories included several professors who worked at Harvard University alongside Wilson, DeVore, and Trivers: Gould, the microbiologist Jon Beckwith, and the biologists Ruth Hubbard and Richard Lewontin. They saw Wilson's book—especially his final chapter—as promulgating myths about the biological basis of human behavior that would be deployed to bolster harmful stereotypes about gender and race, turning them into purported biological facts. From their perspective, the crude ethologizing about aggression that had become so popular in the 1960s had finally begun to fade when sociobiology appeared on the scene and stirred everything up again. They lamented that sociobiologists attracted a devoted following despite the continuing lack of evidence connecting genes and behavior; mathematical algorithms and sociobiologists' rejection of group selection only gave them the appearance of increased intellectual rigor. Underneath the rhetoric, the Sociobiology Study Group suggested, Wilson's theorizing was just as vulgar as Ardrey's speculations. For sociobiologists, their colleagues' willingness to object to the scientific validity of their work, in print, worried them far more than the simultaneous resistance to evolutionary conceptions of humanity from outside the academy.

As their ranks grew, the Sociobiology Study Group aligned with an organization called Science for the People.[11] Science for the People had started in the very late 1960s to protest the deep association of scientists with the war in Vietnam.[12] Some members hoped to find a way to conduct their own research outside the military-industrial-academic complex. Others joined because they firmly believed the world needed a watchdog organization to keep tabs on the potential misuses of scientific research. Science for the People soon began to pay attention to debates over human nature as fundamentally linked to issues of social justice. In a society where science carried great authority, they believed, it was up to professors with a conscience to speak out and defend the interests of people who might not have a public voice of their own and also, very much, to encourage "the people" to find their own voices.

When Wilson published *Sociobiology* in June of 1975, the "Genetic Engineering Group" reviewed the book for the *Science for the People* newsletter, in an attempt to counteract what they saw as the overwhelmingly positive attention Wilson was garnering in traditional presses.[13] They drew a direct parallel between Wilson's ideas and those of biological determinists like the psychologists Arthur Jensen or Richard Herrnstein. In 1969 Jensen had published a 123-page paper in the *Harvard Educational Review* in which he argued that continued differences in IQ scores between black and white children represented aggregate heritable differences in educability.[14] The divergence in average IQ scores between racially identified groups remained well documented and uncontroversial. At issue was Jensen's reliance on genetic rather than environmental factors in accounting for that difference.[15] Science for the People had devoted the entire March 1974 issue of their magazine to the IQ controversy, calling attention to the similarities of Jensen's views to those advanced more recently by Richard Herrnstein.[16] (Herrnstein would later coauthor with Charles Murray *The Bell-Curve: Intelligence and Class Structure in American Life*, once again entering public limelight by courting controversy.[17]) Editors worried, too, about the implications of biological-determinist reasoning regarding the XYY controversy.[18] The members most concerned about such trends and their "science fiction allure," especially among high school students, had banded together and called themselves the Genetic Engineering Group, also based in Boston.[19] Like the leadership of Science for the People more generally, this group despaired of being able to effect real change in dis-

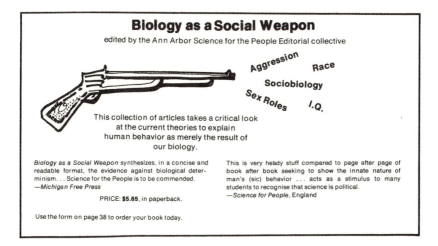

FIGURE 29. Illustration of the difficulties with theories of biological determinism and advertisement for a pamphlet published by the Ann Arbor Science for the People Editorial Collective that contained essays from the 1975 symposium, "Biological Determinism: A Critical Appraisal." Each of the words fired from the gun—race, IQ, sex roles, aggression, sociobiology, and aggression—represents a key theme of the essays in the volume. When Paul Ehrlich reviewed the volume he concluded, "In spite of some silliness and occasional bad science, this book should at least be sampled by anyone concerned with the political and social uses to which the biological sciences are put," ("Sociopolitical Aspects of Biology," *BioScience* 28, no. 6 [1978]: 403). Image from *Science for the People* 10, no. 4 (1978): 14.

cussions about the genetic basis of human behavior. By mid-1975 they decided that although grassroots organizing had accomplished smaller goals, such as bringing an end to XYY-screening of newborn children at the Boston Hospital for Women, the group needed a more public face in the form of publications in widely read journals and newspapers and interactions with local advocacy groups.[20]

According to the Genetic Engineering Group, believing arguments that science could ever be apolitical required blinding oneself to reality. As sociobiologists had chosen to ignore this reality, the group argued, they should then be held responsible for the uses and abuses of their ideas (Figure 29). "The theories put forth by the sociobiologists and their predecessors help to support maintenance of the *status quo* and to convince people that revolutionary changes in social relations (e.g. class structure and sex roles) are impossible."[21] The group identified three additional criticisms of the book that would echo through later critiques published by individual members: Wilson's too-easy slippage between animal and human behavior, his denial of cultural transmission in humans, and last, his reliance on "a speculative recon-struction of human prehistory" that relied on disproven assumptions about men as hunters and women as gatherers.[22] Despite these flaws, the Genetic Engineering Group said they took sociobiological theory seriously because it seemed likely to lend energy to a growing wave of interest in biological determinism.[23]

To alert new members to the dangers of sociobiological thinking and to galvanize the community of their existing membership, Science for the People began screening and discussing a film called *Sociobiology: Doing What Comes Naturally* at campuses around the country.[24] The twenty-minute film featured interviews with Irven DeVore, Robert Trivers, and E. O. Wilson—the three Harvard scientists particularly associated with the theory. In 1972 a camera crew had arrived in Cambridge and asked to speak with each of them about their research. DeVore later recalled being told that the interviews were to be used in a Canadian television series and that they would be sent a transcript of the show before it aired. Because he heard nothing, he forgot about it and assumed the film had never been made. Then, in 1976, he received a mass mail-ing announcing the stand-alone documentary. He wrote for a copy of the film but received no response. Earlier that year, DeVore had suffered a heart attack. Although he took a break from formal teaching, the simian seminars contin-ued in his kitchen, and the group saw themselves as inspired by two dual goals: understanding the evolutionary principles behind human behavior and defending the group against attacks by others on campus, including those mounted by Science for the People.[25] (Given all of these distractions, it is easy to appreciate why DeVore devoted his time to these more local controversies and not to the simultaneously evolving scandal over MACOS on the floor of Congress.) In December of 1976, DeVore attended a screening of *Sociobiology: Doing What Comes Naturally* to see what all the fuss was about.[26]

The film introduced each of its main characters sensationally. "Super-theorist, Harvard biologist, Robert Trivers," the voice-over intoned, as view-ers caught a glimpse of Trivers, with shoulder-length hair and a long brown wool coat, striding through a crowd of students toward the camera. "At Har-vard, a noted anthropologist has been studying baboon societies for more than a decade, a specialist in the origins of behavior, Irven DeVore." DeVore appeared relaxed, speaking from a couch in his office, with African statues gracing the window over his shoulders. "The dream? To connect the behavior, the biological evolution of lower life forms, and to project them to under-standing behavior in man, biologist Edward O. Wilson." Viewers met Wilson in his laboratory, surrounded by ants and microscopes. The film used Trivers to introduce the power of natural selection to modify human social behavior; DeVore to suggest that men and women, like male and female baboons, serve distinct biological functions in human societies; and Wilson to postulate that such biological roles are spread throughout the animal kingdom. In *Doing What Comes Naturally*, innate behavioral differences dividing the sexes accounted for why it is that men are intimately associated with warfare (and women are not). In Trivers's comments for the camera, he noted that warfare had a strong biological component to it. Soldiers would "either inseminate them [women] on the spot or take them back as concubines." Victors, through such biological spoils of war, acted to increase the number of their surviving offspring. This difference in fertility could have provided a selective edge to those men in the past who went to war, "and that's how you could begin to build in certain genetic predispositions towards warfare." Sociobiology, the movie suggests, provided a formidable set of tools that identified the mathematical rules governing the social behavior of all animals, including humans (Figure 30).

Although the film presented sociobiological thinking as a novel and exciting branch of research, at the screening critics of sociobiology emphasized the theory's continuities with earlier, "out of date" research traditions. It helped that the rules of "doing what comes naturally," as conceived by the film, reinforced the stereotypical gendered behaviors with which anthropologists had been visibly wrestling for the last decade—men became human when they learned to hunt, women when they stayed by the hearth to guard and nurture their young, and both when they shared the fruits of their efforts with the community. Yet the act of sharing had taken on a new significance for sociobiological theory. Rather than seeing cooperation as a basic human trait, the film chose to emphasize Trivers's statements about the inevitable conflicts that arise between all mothers and their children and between men and women. Each governed by their own evolutionary and cultural dictates, the desires of one individual necessarily led to disagreements with others.

DeVore was disappointed by what he saw and, we can imagine, harried by the conversation that followed. Worse, he had begun to receive a variety of complaints and letters from colleagues about the film and its message. To head off further criticisms, DeVore wrote a brief note to the American Anthropological Association's *Anthropology Newsletter* seeking to distance himself, Wilson, and Trivers, from the film.[27] He sought to combat what he saw as a gross misrepresentation of their ideas by arguing that they had nothing to do with the production of the film. The final product, he complained, was "a tasteless, sensationalized production that caricatured the field of sociobiology," all set to a "hard-rock musical background." (Indeed, the soundtrack and Trivers's flowing locks do more to convey the mid-1970s character of this moment than any other aspect of the film.) The filmmakers had also reproduced footage of baboon behavior that DeVore had created for MACOS. DeVore urged the

FIGURE 30. The central image of an advertising flyer for *Sociobiology: Doing What Comes Naturally*, distributed by Document Associates, 880 Third Avenue, New York, NY 10022. The historian Gar Allen handed me the flyer one day, which he had collected years earlier and kept safe in his office. The original lede read, "A new film on a new area of scientific inquiry with revolutionary implications for the disciplines of BIOLOGY, ANTHROPOLOGY, SOCIOLOGY, and PSYCHOLOGY." Flyer from the personal collection of the author. © The Cinema Guild, Inc.

EDC, who owned the footage, to seek an injunction against further distribution of the film and request the recall of all existing copies. He concluded his appeal in the *Anthropology Newsletter* by apologizing for "whatever costs, inconvenience, or embarrassment my colleagues may have suffered by the rental or purchase of this film," and he asked his fellow anthropologists to join him in discouraging the use and sale of the film. His statement did little to dissuade Science for the People. The Sociobiology Study Group of the Ann Arbor branch replied in the pages of the *Newsletter* that nowhere in DeVore's statement did he, Wilson, or Trivers disavow the content of their arguments as presented in the film. Despite the schlocky packaging, then, they deemed *Doing What Comes Naturally* a fair-enough representation of sociobiology's arguments and noted that Science for the People intended to continue showing the film, although they also promised to read DeVore's letter to the audience in advance of each screening.[28]

DeVore had further suggested that if anthropologists wanted to see a film in which sociobiological views were presented in a more mature form, they should instead watch *Nova*'s "The Human Animal"—an invitation Science for the People never accepted.[29] The Nova episode aired in March of 1977 and located Wilson, rather than Trivers, as the "vanguard of the sociobiologists." The voice-over at the beginning of "The Human Animal" explained that in the two years since the publication of his book, Wilson's claim that studies of animal behavior could uncover "universal features of social life" among humans had become increasingly controversial. DeVore featured in the film, too. Kinship relations, he said, are shaped by biological relatedness and are therefore bonds of both blood and cooperation. The episode noted that humans additionally share our resources with people in our communities to whom we are not related, thereby overcoming "our mammalian heritage of selfishness." (The sexual division of labor, though, remained.)

When the episode turned to the perspectives of sociobiology's critics, the charge against them was led by colleagues at Harvard, notably Richard Lewontin in zoology and Stephen Jay Gould in the Museum of Comparative Zoology, both of whom were members of the original Sociobiology Study Group that objected to the *New York Review*'s favorable review of *Sociobiology* in 1975.[30] *Nova* interviewed Lewontin first, highlighting the significant variation in definitions of hereditary "traits."[31] According to Lewontin, sociobiologists wanted to explain how natural selection could be responsible for spreading a trait like "cooperative hunting" through the male members of a single interbreeding population. In Lewontin's reconstruction of sociobiological logic, a trait could only spread in a population if it conferred an advantage to the individuals possessing it, but once established through selection acting over time, it became a property of the entire interbreeding community. In other words, if an entire population *today* shared the same hereditary trait, T, sociobiologists took this as evidence that *in the past* individuals possessing trait T had survived and reproduced at a greater rate than individuals without trait T—trait T was the result of natural selection. If biologists then observed trait T in both a human culture and in our closest living ape relatives (chimpanzees, say), sociobiologists further believed that this meant our shared common ancestor probably also exhibited trait T (otherwise the same trait would have had to evolve twice in two different populations living in different environments, and that seemed remarkably unlikely) and therefore that trait T was shared by all human cultures. Lewontin remained suspicious, however, that any complex behavior, even "cooperative hunting," could really be considered a single "trait."

Biologists were not yet sure of how genes influence human behavior, Lewontin insisted, and therefore could not define what they meant by a single trait, much less "divide up the complexity of human behavior into the bits and pieces that evolution is acting on." He suspected that behavior instead derived from the complexity of the human nervous system. Either way, he continued, biologists did not yet know—and that was the point. The filmmakers cut in a rejoinder from Wilson in which he described genetics as programming "an array of potentials" that individuals could either act upon or not, according to their culture. Wilson's claim resonated with Tiger and Fox's assertions that they could not be genetic determinists because they believed that hereditary traits in humans delimited the possible range of behaviors any individual might express—therefore any human behavior necessarily depended on both nature and culture, both genes and experience. In fact, we hear exactly this from one of Tiger's former graduate students later in the episode. Joseph Shepher remarked, "Man can change, can modify his own fate but not without any limits." Even powerful ideologies, Shepher suggested, "could only for a while change so basic a system in the human animal." Shepher further identified the maternal bond as exactly the kind of basic system inherent to female nature.[32] Despite the practice on many kibbutzim of communally raising children, Shepher posited, mothers yearned for more time with their own kids and in some cases even sought to eliminate the system.[33]

In addition to illustrating the differences among biologists, the *Nova* episode pitted Wilson's ideas against those of anthropologists. Certainly many anthropologists were critical of sociobiology and had tried to get the Council of the American Anthropological Association to issue a statement condemning it at the November 1976 annual meeting in Washington, DC. The council ultimately refused to pass the resolution, but the discussion lasted for the better part of an hour.[34] Yet the meeting also contained five sessions devoted to sociobiological themes. So the story of sociobiology's frosty reception among anthropologists was far more complicated than the narrator's implication that only a select few among them were willing to take Wilson's ideas seriously.

Questions of aggression played a key role in the *Nova* documentary, too. In the episode, Wilson described aggression as a "pattern or set of potentials" that vary according to species and to circumstances. Like all other animal species, humans possess the potential for aggression. Wilson carefully noted, however, that aggression should never been considered a singular drive in any species. (In effect, Lorenz's training in medicine and psychology had misled him.) Lewontin, in response, criticized sociobiologists for borrowing words typically associated with human social experience to describe animal behavior, words like "ritual," "courtship," and of course "aggression." In doing so, he suggested, these words became divorced from their historical and political context, not only through their metaphorical application in animals but then when applied once again to human behavior. People rarely participated in wars out of individualized feelings of aggression, Lewontin posited; they did so because of ideological conviction, or because they were drafted, or for economic reasons. "Wars are economic, social, and political phenomena," he argued, "which really have nothing to do with the meaning of aggression in an individual sense." At issue was a fundamental disagreement over how to conceptualize behavior in terms of biology, especially genetics. For Lewontin, genes were traits of individuals (for which sociobiologists had failed to provide sufficient data), while group phenomena like war fall within the purview of culture, not biology (where sociobiological analyses were irrelevant). As all humans are social creatures, necessarily human behavior can only be expressed as a function of culture, rather than mere biology. In fact, for Lewontin, that was the very quality that makes us human.

"The Human Animal" ended by asking viewers to imagine replaying the tape of human evolution again from the beginning: "According to sociobiology it would repeat itself almost exactly. Sociobiologists believe that our genes prescribe the ground rules of our lives; and that our future lies in our knowledge of them." Gould and Lewontin vehemently disagreed.[35] For them, the three cornerstones of sociobiology—determinism, adaptation, and noncontingency—were related but distinct problems. Could human behavior be reduced to genes or neurons? Had human social interactions been honed by natural selection? Were humans destined, by genes or design, to look and behave in the way we do? Gould and Lewontin answered no in all cases, and Gould continued to devote a portion of his essays in *Natural History* to correcting public misunderstanding of these concepts.[36] If the tape of human evolution were to be rewound and replayed, they insisted, it would turn out very, very differently.[37]

The primatologist Jeffrey Kurland later reviewed both films for the *American Journal of Physical Anthropology*. He had participated in DeVore's simian seminars as a graduate student at Harvard and worked on male aggression in Barbary macaques, among other projects, for which he also received funding from the Harry Frank Guggenheim Foundation. Kurland's reviews echoed DeVore's earlier encapsulation of their differences—one laughably horrific, the other fair and balanced. Six years after the films' releases, however, Kurland sounded a cautionary note. Too often, it seemed, unsuspecting audiences might think that biology and culture were antithetical. He lauded the *Nova* special for replacing such dualisms "with a more accurate, developmental perspective on the ultimate and proximate causes of behavior."[38] Perhaps, he continued, the controversial nature of the issues raised by sociobiology bespoke their social importance and philosophical intractability. Not only should biologists attempt to answer these questions, so should "anthropologists of all specializations and political persuasions" who could provide "facts; data; critical tests; iconoclasm; and most of all, the broad zoological, histological, and cross-cultural perspectives from which to view this curious primate, the human animal."[39] Kurland regretted the professional bifurcation of anthropology itself along the lines of nature-nurture but magnanimously avoided blaming either physical or cultural anthropologists alone—biologists, it seemed to him, were in another intellectual camp altogether. In contrast, his flippant review of *Sociobiology: Doing What Comes Naturally* suggested that no reader of the journal should take it seriously.[40] He claimed he had successfully avoided seeing the film for six years, primarily as a way of short-circuiting cocktail party conversations: "Haven't seen it, you know. Too busy just trying to keep up with the technical literature!"[41] He then critiqued the "upbeat, really 'with it' rock sound track featuring xylophones," the narrator who "sounds like the Marlboro Man," and the intercalation of the interviews with seamy shots of undergraduate men and women. He concluded by noting, "A sucker for adaptational explanations, I succumb to temptation. *S:DWCN* was actually put together by Science for the People in order to step up campus unrest and thus hasten the coming of the Revolution. In return for a share of the royalties, DeVore, Trivers, and Wilson were actually in on this cinematic Piltdown. No. I doubt it. Just . . . awful."[42]

From the perspective of female behavior, however, the two films looked remarkably similar. In fact, sociobiologists' embrace of "the male's natural physical freedom and the female's more vulnerable childbearing nature"—to the exclusion of both variation and changes in women's status in recorded human history—formed an independent basis of critique for many female scientists.[43] Multiple essays in two edited collections entitled *Genes and Gender*, published in 1978 and 1979, noted the apparent legitimacy conferred on "hereditarianism" by the 1973 awarding of the Nobel Prize in Physiology or Medicine to Konrad Lorenz, Niko Tinbergen, and Karl von Frisch and feared the transfer of this authority to sociobiology.[44]

The comparative psychologist Ethel Tobach and the endocrinologist Betty Rosoff organized the first "Genes and Gender" conference in 1977 and expected about fifty attendees. To cover expenses, each participant contributed three dollars, but as more and more women entered the room, they stopped collecting. At the end of the conference, over three hundred and fifty attendees resulted in a nest egg of three hundred dollars. When asked whether they would like their money back, participants voted instead to use the money to publish the proceedings.[45] This first conference and volume generated such enthusiasm that the organizers also decided to run a second conference the following year. Contributors to *Genes and Gender*—all of whom were scientists—resonated with the criticisms mounted by Science for the People (indeed, there was some overlap) that there was nothing "new" about sociobiology's conclusions. Like Science for the People, contributors to the *Genes and Gender* volumes worried that defenses of sexism and racism in the name of evolutionary theory were being used to support attacks on antiabortion legislation, to defeat the ratification of the Equal Rights Amendment, and to counter affirmative action programs. They argued that these efforts attempted to "pit women against Blacks, Hispanics, and other minorities in a period of increasing unemployment" at the precise moment when solidarity was needed most. By exposing the "myth of genetic destiny," attendees at the New York "Genes and Gender" conferences planned to combat sociobiology's apparent disregard for all forms of contemporary social injustice.[46]

Authors in the first *Genes and Gender* volume wrote colloquially and directed their critiques of sociobiology at nonscientific readers, warning them not to be taken in by the theory's glossy veneer of scientific authority. One contributor even highlighted Elaine Morgan's *Descent of Woman* as a "delightful" antidote to androcentric speculation in theories of human evolution.[47] Calling sociobiology "pseudoscience," the editors argued that theories of genetic determinism based on gender or race were effectively antihuman activities. By seeing human behavior as attempts by individuals to increase their genetic contribution to the next generation, sociobiologists oversimplified evolution, by "looking at the world through a telescope as if it were a microscope."[48] If the point of the first volume had been to counteract the "magazines, movies, TV shows, and radio [that] repeat this propaganda over and over," the second volume would show that the sociobiological "proof" on which such media claims were based was itself "unscientific."[49]

Contributors to the second volume decided that to have their feminist message taken seriously by scientists, they needed to speak in the professional language of science. In her essay "Social and Political Bias in Science," the neurophysiologist Ruth Bleier called attention to the important work of female primatologists in moving away from baboons as the standard model for theorizing human evolutionary trajectories.[50] She argued that the research of primatologists like Jane Lancaster and Thelma Rowell, for example, replaced the moral lessons of the masculine savannah with a more hopeful message of flexible, more egalitarian social structures, more peaceful interactions, and greater recognition of female contributions to the present and past survival of the species.[51]

Looking back on this period, Sarah Blaffer Hrdy—then a graduate student of DeVore and Trivers—described the "gening of America" as on a collision course with the "'gendering' of the social sciences."[52] As both a feminist and a sociobiologist, Hrdy felt caught in the middle. (She had resigned her membership with the AAAS in frustration that the association even considered a "ban" on sociobiology, but she later rejoined.[53]) According to Hrdy, one of the historical ironies of sociobiology was that through their emphasis on individuality, evolutionists paved the way for their own appreciation of female evolutionary strategies as distinct from those of males and therefore worthy of study in their own right. Attitudes among mainstream sociobiologists toward the contributions of females began to change slowly in the 1980s, with the next generation of feminist biologists.[54] Today, being a feminist evolutionist is no longer perceived as an oxymoron. That it ever was speaks to the underlying perception in the 1970s that primatology was friendly to women but evolutionary biology was hostile, despite considerable overlap between the two communities.[55]

As the controversy over sociobiology continued to grow, even Robin Fox and Lionel Tiger distanced themselves from *Sociobiology* as a text, although they continued to support the ideas it contained. Wilson had tried to vet his big book with appropriate readers. He sent a draft of the contentious final chapter on the evolution of human behavior to the members of the simian seminar at Harvard as well as to Fox.[56] According to Fox, he had unfortunately been traveling when Wilson was finishing the manuscript. He had agreed to read the draft, but it went to his Rutgers office, was then forwarded to his Princeton home, on to the wrong address at Oxford, and finally caught up with him in France, "far too late for his comments to be useful." Fox insisted that although Wilson was largely right in his conclusions, the chapter was too provocatively presented to be convincing at the end of an already-long book—he would have urged him to leave it out. DeVore similarly distanced himself after the publication of *Sociobiology*, claiming that Wilson owed to Trivers the bits of the book that were sound and persuasive; as for the bits that were not, he suggested, it was likely Wilson had misapplied some fact gleaned from elsewhere.[57] Trivers stood by Wilson, though, years after he was denied tenure at Harvard and left for sunny Santa Cruz.[58]

Debates about the legitimacy of sociobiological thinking thus amplified preexisting divides within both anthropology and biology over how best to conceptualize human nature and its evolutionary origins. The most dramatically concise moment in the intellectual battles over sociobiology took place near the conclusion of a two-day symposium at the 1978 annual meeting of the American Association for the Advancement of Science.[59] Co-organized by a biologist and an anthropologist, the symposium proposed a professional discussion between advocates and critics of sociobiology. Members of Science for the People celebrated the fact that the symposium presupposed the existence of a debate and mobilized their own events at the meeting, including a screening and discussion of *Sociobiology: Doing What Comes Naturally*. Wilson was scheduled to speak toward the end of the second day, but when he was about to begin, a group of ten people rushed to the front on the ballroom, shouting "Racist Wilson you can't hide, we charge you with genocide!" A couple of these agitators grabbed a pitcher of water sitting at the table for the convenience of the speakers and poured it over Wilson's head, adding: "Wilson, you are all wet!" The "ice-water squad" (as the historian Ullica Segerstråle dubbed them) then quickly dispersed. The organizers apologized to Wilson— the audience clapped and stood to acknowledge Wilson's perseverance. Gould, who had spoken earlier in the day, denounced such radical activism as inappropriate for an intellectual conversation. He, too, received a standing ovation. Wilson, still wet, calmly delivered his talk. (The ice-water squad, it became clear only later, had been organized by the International Committee against Racism—an independent activist group with loose ties to the Sociobiology Study Group in Boston.)

These events have been told and retold. I relate them here once more to demonstrate how Wilson and *Sociobiology* became a lightning rod directing colloquial and professional condemnation of biological determinism in human evolution to a central focus point. Relegating *Sociobiology* to "pop science"—and along with it the colloquial publications of Ardrey, Lorenz, and Morris, not to mention Lionel Tiger and Elaine Morgan—allowed Wilson's colleagues to continue the daily business of studying animal behavior, fossils, or human cultures. It also reflected a larger shift in what it meant for scientists to write in a colloquial voice about human nature. If scientists maintained a purely professional profile, they could expect certain rules of polite decorum.[60] If venturing into pop science, however, authors knew they would receive a different kind of scrutiny from general readers and colleagues alike.[61] In a 1977 interview, Gould noted that "anyone who generalizes and writes for the public, ipso facto, is going to be an object of great suspicion. People will say, even without reading you or knowing you, 'Oh Gould, he's just this waffler.'"[62] Gould spoke not only to defend his own writings, which ranged from dry articles in scientific journals to light-hearted explorations of scientific ideas in his *Natural History* columns, but also those of his "friend and 'alter-ego,'" the astronomer Carl Sagan.[63]

By the mid-1970s, then, scientists forcefully distinguished between professional writing for their colleagues and colloquial writing for "popular" audiences. Gould speculated that his colleagues assumed that writers could be good at only one. When Gould spoke about his desire to write for both professional and lay audiences, the role model that leapt to his mind was the Victorian geologist Charles Lyell. One of Darwin's inspirations and confidants, Gould noted, Lyell had developed "wonderful metaphors" for explaining the natural world to his readers. The days of intellectual paperbacks by Loren Eiseley, Theodosius Dobzhansky, or Margaret Mead, aimed at precisely this overlap, had already been forgotten.

Although the debates over sociobiology succeeded in generating phenomenal media attention—both negative and positive—Harvard occupied only one hill in the wide intellectual landscape of research on human evolution in the 1970s. At Stanford, three thousand miles away, the intellectual and moral stakes looked remarkably different.

15

Human Nature

BY THE MID-1970S, observations of great ape species living in natural conditions provided a wider range of lessons for human nature than had the strict social hierarchies observed in baboons a decade earlier.[1] When Jane Goodall and David Hamburg argued for the biological similarities shared by humans and chimpanzees, they also articulated a vision of human nature. They based this vision on biological relatedness rather than on ecological sympathy and implicitly questioned the gendered roles and social hierarchies that characterized baboon behavior as the most appropriate primate model for reconstructing the social and behavioral norms that might have characterized early human life on the savannah. Goodall's field studies of the great apes became a particular favorite of feminists—not only for the moral message they read in the apes' flexible social roles, but also because Goodall, like Dian Fossey and Biruté Galdikas, represented the success of young women transforming their chosen field. Goodall's early discoveries that chimpanzees manufactured tools, sticks with which to eat termites and masticated leaves with which to sponge up water, fit well with hypotheses that the origins of tool use lay in manufacturing aids for "gathering and processing food" rather than as weapons.[2] One of Hamburg's graduate students later recalled him warning her not to go overboard with sociobiology.[3] Baboons, upon further scrutiny, seemed neither as male-oriented nor as militarily structured as researchers had first thought.[4]

The rich observations at Gombe and other sites where scientists studied primate species in the wild suggested that chimpanzees formed long-lasting emotional relationships with other individuals, and intrigued Hamburg, as did the possibility of using the animals as experimental models for human psychological disorders. He had been especially taken by Goodall's description of the death of an older female, Flo, and three weeks later, the death of her adolescent son, Flint, who in his last moments returned to the spot at which she had died. Flint had been unusually attached to his mother for his age, and the emotional trauma of losing her proved his undoing.[5] Flint's

inability to adapt to his new circumstances fit into Hamburg's larger preoccupation with behavioral coping mechanisms, "especially in relation to the
extraordinary human capacities for hatred and violence." As he shared in a
1972 letter to his colleagues when he stepped down as chair of psychiatry to
concentrate more fully on his research, any such insights into human nature
required interdisciplinary cooperation—basic research in both biological
and social sciences, clinical medicine, and education—that reckoned with
"the evolutionary origins of human behavior and [looked] toward our contemporary circumstances."[6]

Goodall's aspiration to study the behavior of free-living chimpanzees
in their natural habitat required minimal intervention into their lives. For a
while she had supplied bananas to simian camp visitors but curtailed that
practice because the chimpanzees fought with each other and with the local
baboons over this precious resource.[7] Goodall's research provided evidence
of chimpanzees as "highly intelligent, intensely social" and "capable of close
and enduring attachments," "rich communication through gestures, postures,
facial expressions, and sound," using tools "effectively," which they make "with
some foresight." According to each of these criteria, however, chimpanzees fell
far short of human potential to love, use language, and manufacture complex
technologies. They also might cooperate to hunt, kill, eat other small mammals, and share food, but not as thoroughly as we do; they were "highly excit-
able and aggressive" yet not warlike.[8] Goodall and Hamburg invoked not only
behavioral parallels between the two species, but also chromosomal, immu-
nological, neuroanatomical, and molecular affinities as well.[9] Across many
dimensions, then, research on the behavior of great apes both confirmed our
animal heritage and maintained our unique status as human. Rather than subject the chimpanzees of Gombe to experiments that might alter their natural
behavior, the Stanford Outdoor Primate Facility allowed for a greater variety
of experiments on animals already in captivity and provided observational
training to Stanford students.

Undergraduate enthusiasm for primate research at Stanford had built syn-
ergistically upon rising excitement over the new Human Biology program.
The idea of an interdisciplinary major in human biology with an emphasis
on behavior had been floating through Hamburg's mind since the late 1960s,
when he and Joshua Lederberg taught an undergraduate special course,
"Man as Organism."[10] Washburn wrote to Hamburg, suggesting that Irven
DeVore would be a perfect person to join the program and provide a center
of intellectual gravity. (They ultimately chose to build on existing resources
on campus rather than seek external hires for the Program.[11]) Washburn also
suggested that any interdisciplinary program would need focus given how
many departments and disparate introductory courses would be involved—

human and animal behavior could be just the "unifying synthesis" they had discussed. The HumBio program became reality in 1969, when the Ford Foundation awarded Stanford almost two million dollars for a five-year trial and the university senate voted to approve HumBio as a new undergraduate interdepartmental program.[12] In the words of Norman Kretchmer, its first director, "Our program concerns man as an organism, his adaptation to other men and to nature, his ability to control and to live with his environment, and the mechanisms by which these factors relate to his biological and social evolution."[13] Similar hopes for behavioral synthesis had driven the MACOS program at Harvard—no wonder DeVore had leapt to Washburn's mind as a good fit.

When Hamburg resigned as chair of the Department of Psychiatry, he cited among other factors the startling success of the HumBio major.[14] Colin Pittendrigh offered the first official course in the HumBio program in the spring quarter of 1970: "Human Biology 1: Man and Nature." Hamburg and Goodall co-taught "Human Biology 2: The Evolution of Human Behavior." Select students could observe chimpanzee behavior at Stanford with Hamburg, and up to eight also received funding to travel to Tanzania for six months and participate in behavioral research under Goodall's tutelage at Gombe (Figure 31). Organizers expected around 50 students to register in the program and were astounded when 427 signed up. A year later, HumBio boasted 240 majors. By 1973 it was the third largest major on campus and offered students a variety of research experiences and possibilities of wider community engagement. The future was bright.

Then, late in the night on May 19, 1975, a heavily-armed rebel group from Zaire crossed Lake Tanganyika, beat up the local staff, and kidnapped four Western researchers from Gombe, including Steve Smith and Carrie Jane Hunter (Human Biology undergrads at Stanford), Emilie van Zinnicq Bergmann (a general administrator for the camp, from Holland), and Barbara Boardman Smuts (a graduate student working with David Hamburg and Jane Goodall).[15] The armed men attacked and beat several Tanzanian members of camp, demanding to know where the white people were. Mzee "Rashidi" Kikwale, one of the field assistants, could not hear out of one ear for months after he was struck on the side of his head with a gun. The park warden, Etha Lohay, and a student, Addie Lyaruu, two young Tanzanian women, ran from house to house warning others to hide.[16] When these local workers insisted that the four they had already captured were the only ones, the armed men eventually left, and in this lie they protected not only Goodall but other students as well. The remaining residents of camp waited until morning, when they could raise someone on the two-way radio to report the chaos of the night. Nobody knew the location of their missing friends or even if they were still alive.

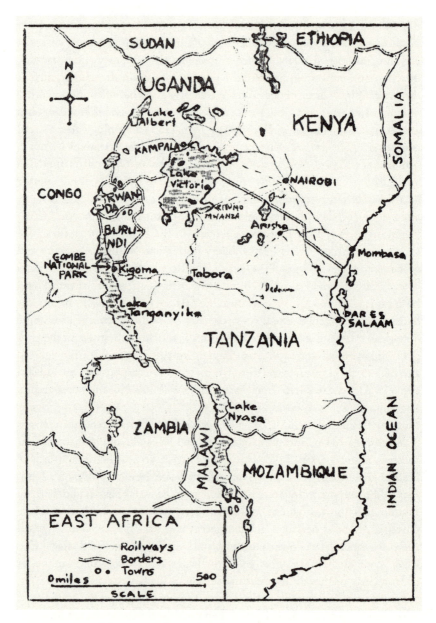

FIGURE 31. Hand-drawn map of East Africa distributed to Stanford students to help them plan for their trip, showing Gombe National Park's location on Lake Tanganyika. Students were told to bring swimwear because it would be impossible to buy locally, and the camp would supply snorkels, masks, and flippers for recreational swims. "Stanford Primate Research Newsletter," n.d., Box 452, Folder 5, Hamburg Papers, Columbia University Archives.

Six days later, on May 25, twenty-four-year-old Smuts walked in to Kigoma and asked to speak with US Ambassador W. Beverly Carter. She carried letters written by herself and the other captives to the US and Dutch ambassadors to Tanzania, and the president of Tanzania, demanding £200,000 (or $500,000), weapons, and the immediate release of members of the Parti de la Révolution Populaire (PRP) currently jailed in Tanzania.[17] If the list of conditions were not met within sixty days, the letter stated, the students believed their captors would "not hesitate to shoot us."[18] The Tanzanian government publicly refused to accede to the will of men they deemed terrorists.[19] Under the leadership of Laurent-Désiré Kabila, the PRP sought to oust the Zairian government, who in turn considered them enemies of the state. US foreign policy newly dictated that American officials could not negotiate with terrorists, out of concern that such cooperation would serve to encourage future attacks.[20] It seemed unlikely that any of these players would budge.

Hamburg swung into action. Given his role in sending undergraduates and graduate students to Gombe in the first place, Hamburg felt personally responsible for their well-being. Kenneth Stephen Smith was one of the HumBio students selected to travel to Gombe in 1975 and one of the students captured the night of May 19. He had written a letter to a friend describing his experiences with "pretty fatiguing" twelve-hour follows but also wrote that in early May, at the end of the rainy season, the weather was beautiful, the jungle lush, and Gombe itself was "peaceful and isolated." Accessible only by boat, he wrote, "we get very little news from the outside and when we do . . . it's about two weeks old." He closed, "I'll be waiting on the edge of my seat until I hear from you."[21] Emily Polis, another student present at Gombe that May, recalled that this apparent peace was a façade. Only a few miles north lay Burundi, locked in a long civil war. "When the wind was just right, we could hear gunfire and explosions echoing over the valleys that separated us. Escaping refugees would sometimes stop for food on their way to villages in Tanzania to the south, seeking safe haven."[22] Polis found safety in the forest around Gombe the night of the kidnapping and remembers the hours she spent next to Goodall, who in turn was seeking to comfort and keep quiet her eight-year-old son, Grub.[23]

Hamburg mobilized his scientific connections as the families of the kidnapped researchers mobilized their personal ones—together they raised and borrowed $460,000. It took a month to gather the funds. At the US Embassy in Dar es Salaam, US Ambassador W. Beverly Carter oversaw negotiations between Hamburg and representatives of the PRP over the details of how they would proceed. Carter proved crucial to the negotiations thanks to his good relations with Tanzania's president Julius Nyerere. His blackness did not hurt him, he believed, and he was known in Tanzania as someone who had been active in the US civil rights movement before joining the Foreign Service. At the State Department he had then specialized in African Affairs, and his wife developed an abiding interest in African art. When Carter accepted the position as ambassador, they began collecting local pieces, with a substantial number from Makonde sculptors and painters in southern Tanzania. The Art in Embassies Program at State also allowed them to exhibit a variety of pieces from well-known African American artists at the ambassador's residence. To see the collection, Nyerere had come to visit their home, a rare occurrence and a minor coup for Carter.[24]

As Hamburg contemplated the journey from Dar es Salaam to Kigoma, where the trade of money for people would take place, a close colleague at Stanford (the biologist Joshua Lederberg) even wrote to the game theorist and conflict strategist Thomas Schelling to ask if he could provide Hamburg with any guidance.[25] Schelling replied at length but feared he had no practical advice. Hamburg, Smith, and the money succeeded in garnering the release of Emilie Bergmann and Carrie Hunter on the night of June 27.[26] A month later, Smith remained a hostage.[27] Carter convinced President Nyerere to release several PRP members held in Tanzanian jails to assist in the negotiations, although Nyerere insisted on doing so quietly and without any publicity.[28] Then Smith was finally freed.[29] (Carter described Smith's release as "a very James Bond kind of episode ... which I really can't talk too much about."[30]) By July 29, Smith had been debriefed about his experiences, and all four kidnappees had left the country. Hamburg returned to Stanford, exhausted but relieved at the students' safe return.

Although Ambassador Carter had been in constant contact with the State Department throughout these events, he was recalled to Washington, where Secretary of State Henry Kissinger reprimanded him for failing to adhere to official US policy "that terrorists cannot negotiate with American officials."[31] His pending ambassadorship to Copenhagen was revoked and he returned to a desk job at State. Not only did Hamburg object vociferously to his demotion, so did members of the Black Congressional Caucus. Their continued attention eventually led to Carter being assigned a new ambassadorship, this time in Liberia.[32]

The kidnapped young people considered themselves lucky because throughout the entire experience, their captors had kept them together. There were very few amenities, but one student had demanded they be allowed to keep a copy of Niko Tinbergen's *The Herring Gull's World*.[33] Their only other reading material was supplied by the PRP and consisted of Mao Zedong's *Little Red Book* (in French) and small political treatises (in Swahili)—part of the PRP's efforts to "re-educate" the Westerners. They would read passages of

Tinbergen's book alone or aloud to each other in remembrance of the comforts of home. It sounded like poetry.[34]

The science writer Gale Peterson based her account of the kidnapping at Gombe and their aftermath on interviews with Goodall and others.[35] Peterson emphasized the cocoon of safety Derek Bryceson wove around his recent bride, seeking to protect Goodall physically and emotionally. As director of the National Parks in Tanzania, he felt he was doing everything he could to aid the process, with far more resources than she had at her disposal. Peterson suggests that Bryceson's strategy also isolated Goodall from the negotiations and the parents of the young researchers, who had travelled to Tanzania to work for the release of their children. Perhaps, Peterson implies, Bryceson was anti-American (certainly, the American families thought so) and questioned their presence at the camp in the first place. Few published accounts date from the 1970s, in part because the US government asked everyone to stay silent about the specifics of what had happened. Nobody wanted to create a precedent that could lead to further kidnappings nor did they want to legitimate the PRP or its actions. Only when Kabila became president of Zaire in 1997, renaming it the Democratic Republic of the Congo, did the young people he had kidnapped come forward with their stories.[36] They thought, perhaps naively, that public outrage over his past might help to oust Kabila from power. Yet in power he remained until he was assassinated by one of his bodyguards less than four years later.

The year 1975 was a difficult one for Gombe, not only in terms of human violence, but also for the chimpanzees in the park.[37] Starting in 1974, researchers had noticed what they thought might be attacks of raiding parties from one chimpanzee group within Gombe on another. By 1977, males of the Kasakela group succeeded in disbanding the Kahama group. All Kahama males had either been killed or had disappeared. In the pages of *National Geographic*, Goodall described this as "the first known extermination of one chimp community by another." Additionally, in August of 1975 (shortly after Smith's release), an adult chimpanzee (Pom) and her adolescent daughter (Passion) killed and ate Glika's three-week-old chimpanzee infant—a member of their own troop. Over the course of the next fourteen months, they would do so at least twice again. In the wake of the kidnapping, however, these events took time to process. Responsibility for recording data was transferred to local employees, who also learned to decide which animals should be followed, what data should be gathered, and how to continue research at Gombe under these new circumstances. Goodall checked in daily via two-way radio from Dar es Salaam and for the next couple of years was able to return in person occasionally, but rarely for prolonged visits. Tanzania banned other expatriates from extended research. Anthony Collins returned for a few short visits in the early 1980s, as did Richard Wrangham to conduct research on chimpanzee medicinal plants and to study the behavior of baboons in Gombe, but only in the 1990s did the site fully reopen to Western researchers.[38]

Without a PhD on site, funding for Gombe from the Grant Foundation evaporated. Acting on the advice of a friend in California, Goodall founded the nonprofit Jane Goodall Institute, so that private donors could support her research efforts from afar. Because of the kidnapping and its aftermath, publications surrounding these events took a number of years to come out. Smuts and Wrangham returned to Stanford to work on her dissertation and to bring order to the Gombe data, respectively. Wrangham had been conducting research in Ethiopia and first heard news of the kidnapping over the radio in Addis Ababa. Ten days later, after Smuts had returned to safety, he received a message that he was in danger of being kidnapped by members of the Tigrean People's Liberation Front. Jittery from the news at Gombe, he decided to leave. As he was doing so, he ran into a friend and colleague also conducting research in the area and passed along the warning. She decided to stay and several weeks later was captured with her husband and their two children. The Tigrean People's Liberation Front held the family for a year, whereupon they were released unharmed. When he heard the news, Wrangham called to welcome her home and was slightly horrified when she related that her family had luckily been captured with their binoculars—over the course of the year they had managed to conduct research in areas to which they would not have otherwise had access.[39] Thinking maybe he did not possess sufficient grit for fieldwork after all, Wrangham considered leaving primatology for good. Then he received a call offering him a job as a one-year replacement for Irven DeVore at Harvard. (DeVore had just survived his heart attack and they needed someone to cover his classes.) So from Stanford, Wrangham flew to Harvard, joined DeVore's weekly simian seminar, and found new intellectual inspiration from the camaraderie and intellectual energy of the group. Smuts, for her part, returned to East Africa in 1976 for a pilot study on baboon behavior in the Masai Mara Game Reserve and then conducted further research at the Gilgil Baboon Project in Kenya.[40]

Goodall's professional world had collapsed. After the successes she and Hamburg had enjoyed in previous years—garnering long-term support for Gombe, finally working out the details of Gombe West, and creating a vibrant community of undergraduate students and researchers in Tanzania—in the wake of the kidnapping, she could access chimpanzee behavior at Gombe only through the eyes of others or under occasional armed guard. The Tanzanian government had banned Western researchers from Gombe. Stanford refused to allow students to return. When Goodall ventured to Stanford several months later, she found that the flat she was supposed to rent had been let to someone else. Friends were less friendly. A continued professional relationship with HumBio proved impossible, and she received notice in December that her contract would not be renewed.[41] Hamburg, too, created private distance between them, although he continued to support her publicly and remained interested in primatology and evolution throughout his career.[42]

When Goodall did publish on the violence of the chimpanzees at Gombe, she hesitated to assign causality to the events. In the case of the infanticides, she suggested they would "be of great interest to students of behaviour, but cannot be meaningfully discussed until more incidents ... have been observed."[43] She had every reason to be cautious—infanticide was already a hot-button issue.

Sarah Blaffer Hrdy had published an article in early 1974, based on her research with Hanuman langurs on Mt. Abu in western India, suggesting competition between males might explain the evolution of infanticide in primate species.[44] Hrdy drew inspiration from Trivers, Wilson, and the community of sociobiologists at Harvard, but she also remembered reading Loren Eiseley's *The Immense Journey* as a girl.[45] She suggested that infanticide might evolve if stiff competition existed between rival males, so that any one male was unlikely to have access to females for an extended period of time, and if females could conceive nonseasonally, so that when a female lost an infant it would be both possible and advantageous for her to ovulate quickly. Hrdy wanted to explain how and why infanticide might come to play a role in the reproductive success of individual langurs, especially the males who seemed to be responsible for most of the killings. The researchers at Gombe had found her arguments intriguing. Ann Pusey later recalled Goodall recommending Hrdy's paper to her and their excited discussions about Hrdy's framing of infanticide as an evolutionary conundrum.[46]

The biological anthropologist Phyllis Dolhinow, who had trained with Sherwood Washburn and now taught at UC-Berkeley, disagreed with Hrdy's basic premise. Dolhinow had also studied langur behavior, albeit in northern India. "In normal troops langur males do not kill infants," she argued.[47] Dolhinow suggested that infanticide might occur in populations under huge environmental strains, like the residents of John Calhoun's overcrowded rodent cities, or if humans had inexorably transformed natural conditions—none of which she considered normal. In letters to Washburn, Dolhinow wrote that she took disproving the "killer ape" theory of human evolution as a prime motivation for her research.[48] From her perspective, Hrdy seemed to be espousing the same disastrous logic of innate aggression as had Robert Ardrey and Konrad Lorenz. Hrdy, of course, saw it differently. Much like other members of the Harvard community, she emphasized the importance of individual selection rather than group advantage—to her, Dolhinow's position relied on the assumption that killing offspring could never be in the best interest of the species.

As early as 1965, Goodall had noted the psychological difficulty some female chimpanzees had with pregnancy and birth. Mothers often carried premature stillborn infants in their arms for a few days before ultimately letting them go. So the idea of deliberate infanticide came as a shock. A decade later, researchers observed seven infants within the group killed by adult chimpanzees. Goodall divided these cases of infanticide into two biologically distinct categories. In some cases, "stranger" females had entered the area and were attacked by males who killed (and often ate) their young offspring (who were typically 1.5–2 years old). This type of infanticide could make sense as an evolutionary adaptation, she reasoned. Her logic echoed both Dolhinow and Hrdy—when males killed offspring to whom they were not related, they ensured that no group resources would be expended on unrelated individuals and, perhaps, the mothers would also become reproductively available to the males of her new group more quickly.[49] In three (perhaps four) other instances, however, an adult female within the community killed and ate the infants of females with whom she was well acquainted. Explaining this proved far more difficult.

The main culprit was Passion, an adult female chimpanzee. Passion was one of the original mothers of the troop when Goodall first started her observations in the 1960s and in the meantime had borne two offspring, Pom and Prof ("Prof" had been shortened from "Professor Hamburg"). With Pom and Prof in close proximity, Passion attacked and then the small group ate Glika's three-week-old infant. Pom would later "murder" (in Goodall's words) two additional infants, sharing the meat with Pom and Passion as if it came from "normal prey." One chimp baby was found dead with deep bite marks in its skull. Several others disappeared. Without direct observation of the circumstances of these cases, Goodall speculated that perhaps Pom and Passion were involved here, too. Between 1974 and 1976, five infants were born but only one survived (Frodo, born to Fifi, by then an experienced mother). Writing about the incidents in *National Geographic*, Goodall suggested Passion had badly mothered Pom. Perhaps Passion's "callous and indifferent" attitude might explain why Pom and Prof had so readily adopted her cannibalistic and murderous behavior.

Goodall worried that such behaviors might take place regularly but so rarely that researchers had simply never before observed them. Then a small glimmer of hope appeared. In 1978, first Passion, then Pom had infants of their own. Goodall named Passion's son Pax, anticipating that his birth might bring an end to the infanticides. After Pom gave birth, she hid from her mother for two weeks. Goodall wondered whether Pom hid out of fear that Passion would try to kill her vulnerable infant. Baby chimps stopped disappearing. Perhaps Passion's horrific behavior *had* been abnormal, born out of psychological trauma. Goodall refused to speculate—only time and more data would prove the issue.[50]

The ongoing debate between Hrdy and Dolhinow may well have stayed Goodall's willingness to interpret her results according to either rubric.[51] If infanticide occurred regularly but rarely in a population, then it would make sense, sociobiologically, to ask what evolutionary advantage such a behavior might confer on females such that it spread through a species. If infanticide was due to individual deviance or abnormal environmental conditions, then even asking the question made no sense. In subsequent years, researchers have observed relatively few instances of female infanticide (or attempted infanticide), but even these rare occasions led Goodall and Pusey to conclude that "these observations suggest that female infanticide may be a significant, if sporadic, threat, rather than the pathological behavior of one female."[52] (Pusey had been present at Gombe in May of 1975, but the evening of the kidnapping she had been away in Nairobi, receiving medical treatment for hookworm.[53])

Although Goodall implied an adaptive reason for the infant killings by stranger males might exist, she simultaneously held back from promoting sexual selection, through male-male competition for access to fertilization opportunities, as the likely cause. In *The Chimpanzees of Gombe*, she wrote that "from an evolutionary point of view, males can afford to be more aggressive, and more violently aggressive, than females because, at least among the higher primates, they are not directly involved in caring for the young." Yet the "main reason" for this sex difference she ascribed to "neurological and endocrinological mechanisms involved in the expression of the behavior."[54] Against this background, and lacking further data, the oddity of the female-female violence of Pom and Passion's infanticidal attacks could be explained only by appealing to something outside the normal order of nature—psychological disturbance.

Aggression as a result of territoriality made clearer sense to Goodall.[55] Her original study group had fractured into two communities, but the newly formed Kahama group to the south lasted for only five years before the Kasakela group brutally enforced their control of the entire region. The males of the Kahama group—Godi, Dé, Goliath, Charlie, and Sniff—and one of the females—Madame Bee—were each in turn attacked and beaten, either disappearing or dying afterward (Figure 32). The attack on Sniff at the end of 1977 "marked the final stage in the annihilation of the Kahama community."[56] However, even this basic association of aggression and territoriality was marred by doubt.

FIGURE 32. As a result of the chimpanzee violence at Gombe, *National Geographic* depicted the denizens of the park as "warmongering apes." This drawing depicts five males pummeling and biting Goliath, whom they then "abandoned to die." Illustration by David Bygott, for Jane Goodall, "Life and Death at Gombe," *National Geographic*, May 1979, 592–621. Copyright © 1979 by National Geographic Creative.

Margaret Power, for example, attacked Goodall's interpretation of male aggressive behavior as normal, claiming that Goodall herself had induced unnatural stress in the population when provisioning the chimpanzees with a highly nutritious and artificial food source—bananas.[57] Chimpanzees really were the peaceful species earlier researchers had assumed, Power argued, and by analogy so were humans. In his foreword to Power's *The Egalitarians*, Ashley Montagu (ever eager to enter the fray against biological determinism) hailed the book as "a gold mine of new ways of looking at old problems," a correction to "many endemic errors," and an attempt to "set the record straight" regarding normal chimpanzee behavior.[58] Power reasoned that the feeding created frustration in the chimpanzees and baboons who found it impossible to fulfill their desire for bananas, and competition increased both between members of the same species and also between chimpanzees and baboons. When violence around the bananas had become especially intense, Goodall had changed the feeding structure to dramatically curb the numbers of bananas available.[59] According to Power this only increased the frustration of the chimpanzees and the result was the violence of 1974–1977. The violence had not been territorial in origin but instead arose from individual anxiety, Power concluded.[60] Montagu had suggested as much immediately following Goodall's original publications.[61] Given the late date of Power's book, it seems designed to provide an alternative picture to that described by Goodall's massively comprehensive *The Chimpanzees of Gombe* published a few years earlier.[62]

Goodall later reflected that the kidnapping, including the "shock and misery" that came with it, had not changed her view of "human nature." Nor did it change anyone else's. The events of the kidnapping were traumatic and exceptional for the people involved but garnered media attention in the United States only because three of the kidnappees were white Americans—little attention was paid to the underlying political and social unrest that led to their capture. Yet after the "intercommunity violence and cannibalism" that Goodall saw for the first time among the chimpanzees, her view of "chimpanzee nature" was never the same. "Suddenly I found out that under certain circumstances they could be just as brutal, that they also had a dark side to their nature. And it hurt."[63]

The kidnapping and subsequent events profoundly affected Hamburg as well. According to one of his students, Hamburg had never really enjoyed spending time in Africa, and the negotiations after the kidnapping weighed heavily on him.[64] Three months after returning to California, he decided to accept a position as president of the Institute of Medicine at the National Academy of Sciences, its branch specializing in health policy. The psychologist Betty Hamburg, his wife, simultaneously accepted a new appointment at the National Institute of Mental Health. In a letter to friends and colleagues, they wrote that they both yearned for "a rare opportunity for public service" and were "very hopeful of doing something of unusual social value."[65] Leaving Stanford and basic research behind was not a step Hamburg took lightly—he initially planned for a leave of absence and officially resigned a few years later. Hamburg believed that from his new position he could make an immediate difference in ending violence by focusing on the immediate health and economic concerns of developing countries. After the autumn of 1975, Hamburg and Goodall never again taught Evolution of Human Behavior, Stanford students did not return to Gombe, and in 1977 Seymour Levine (already a faculty member at Stanford) replaced Hamburg as director of the Stanford Outdoor Primate Facility.[66] Hamburg remains reluctant to publicly discuss the kidnapping in Tanzania.[67]

Gombe changed, too, as Tanzanians continued research in the absence of Western scientists and took on greater responsibility for the intellectual mission of the center. Indeed, after the kidnapping, Gombe missed only a single day of behavioral data on the chimpanzees.[68] Tanzanian researchers also began to receive credit in print for their work. For example, in Goodall's 1977 article on infanticide, she notes all observers by name and in the acknowledgments wrote, "This paper could not have been written without the observations of many students and I am especially indebted to the Tanzanian field assistants who observed most of the incidents reported in this paper."[69] In her 1979 article in National Geographic, Goodall wrote that after the kidnapping, it no longer seemed safe to send students or any Western researchers to work in Gombe. "Gradually, as the seasons passed, the Tanzanian staff developed new self-confidence and with it a new enthusiasm and reliability. Today Gombe is as flourishing a research center as ever it has been."[70] She especially thanked Hilali Matama, who became the head researcher at Gombe, and Mzee Rashidi Kikwale, who had accompanied her on her first forays through the Tanzanian mountains.[71] In the wake of decolonization, former "field assistants" in many disciplines, including archeology, primatology, and zoology, began to be seen by Western scientists as research collaborators—several traveled to receive university degrees abroad and then returned to continue their research, becoming coauthors as well—but that would take time.[72] As researchers dispersed from Gombe and made other plans, the field sites where one could conduct behavioral research on primates continued to multiply.

Reactions to the violence at Gombe also rendered visible the polarization of nature and culture as the primary causal frameworks through which scientists in the coming years would interpret human, even chimpanzee, behavior. The kidnapping was set against a growing realization that the community of chimpanzees Goodall had described as gentle creatures when *National Geographic* first brought her research to national and international attention was itself erupting into conflict. For Goodall, the murder of fellow chimpanzees forever altered her picture of chimpanzee behavior, making it far more terrible and far more human than she had previously supposed. Before coming to a full understanding of the rare events, she concluded, researchers needed more observations, more data on which to base their theories. Hamburg reacted differently. After helping negotiate the release of the kidnapped students, he left basic research behind, devoting his professional attentions to supporting access to better health care and economic aid in developing countries, thereby hoping to relieve political tension and ultimately abate violence around the world. Their contrasting solutions to a personal encounter with violence and its aftermath illustrate the dichotomous lessons scientists could take from the world around them.

Sociobiology-inclined researchers saw male aggression as biologically normal, building on a robust literature about the human animal. The violence of the male chimpanzees reinforced what researchers already knew, but with a twist. In an earlier decade, humans' capacity to kill each other was one feature distinguishing us from other species. Now that same violence bespoke a common, aggressive legacy (at least for roving bands of young males, if not everyone else).[73] Yet female violence remained problematic. Were females as bestial as males? That struck many evolutionists as unlikely. Perhaps instead, females and women were driven to violence out of psychological grief or depression. This question would not be resolved quickly, and evolutionary explanations of violence in females have remained controversial, especially when that violence is directed at their young. Taking aggression for granted, sociobiologists sought to understand why animals ever cooperated, and in sexual selection and kin selection they thought they had answers. Sex, parenting, and animal families might look like cooperation, but when males and females sexually unite, each unconsciously follows a competitive strategy evolved over many generations to give birth to the next generation and in so doing perpetuates his or her individual genetic lineage. Predator warning systems, heroism in battle, and other seeming sacrifices to the perpetuation of the group were also genetic selfishness in disguise. In sacrificing his own life, a valorous male might protect his genetic legacy already embedded in the chromosomes of his offspring, siblings, and cousins.

For critics of sociobiology, violence remained a key issue, too. In their reconstruction, sociobiological explanations were not only wrong, they drew attention away from the possibility of identifying the real causes of social violence—whether cultural or environmental. Hamburg, for his part, joined the ranks of scientists who were increasingly divided over such issues, not because sociobiology was a new theory, but because it was old.[74] In the eyes of its many critics, sociobiology was merely the latest articulation of a research program into the animal basis of human nature that had already failed. When violence and prejudice became personal, biological theories of aggression and human nature became inadequate.

Coda

What were the secrets of the animal's likeness with, and unlikeness from man?

—JOHN BERGER

THE UNITED NATIONS pronounced 1986 the International Year of Peace. Seizing the opportunity, a group of twenty scientists and scholars convening for the sixth International Colloquium on Brain and Aggression drafted a collaborative statement they called the Seville Declaration on Violence. They asserted that "it is scientifically incorrect to say that we have inherited a tendency to make war from our animal ancestors." Four other precepts followed: that violence had never been "genetically programmed" in human nature, that selection for aggressive behavior in humanity's past had not taken place, that humans do not possess a "violent brain," and that a phenomenon as complex as war could never be reduced to a single "instinct." Violence was a projection of culture, the authors insisted, not nature. They sent the statement to a variety of scientific organizations, seeking endorsements for their words.[1]

The American Anthropological Association submitted the issue to their eighty-five hundred members via mail ballot. Of the twenty-two hundred ballots returned, almost 80 percent voted in favor of endorsing the Seville Declaration.[2] Robin Fox—a professor of anthropology at Rutgers and with Lionel Tiger codirector of the Harry Frank Guggenheim Foundation from 1972 to 1984—wrote to the AAA's *Anthropology Newsletter* with what he called "a murmur of dissent," deriding the statement as an "exercise in self-righteous piety."[3] Of course, aggression could never be the *cause* of war, he countered, but it could (and did) serve as a tool for motivating people already convinced by fanatical purpose. For Fox, the phrasing of the declaration provided evidence that the authors had misunderstood the science they sought to discredit. Aghast that leaders in his field thought any scientific question could be settled by majority vote, he equated the position of scientists studying the biological basis of human behavior in the United States with that of geneticists in the Soviet Union who had been under the political thumb of Trofim Lysenko in the 1950s.[4] Fox reported that upon reading his letter, colleagues who disagreed with the Seville statement had confessed to him that they had not spoken out themselves because they feared the condemnation of their peers who agreed with the "dogmatic, assertive tone of the declaration."[5] As an alternative to (in his opinion) UNESCO's doomed quest to articulate a shared "universalist ideology," Fox counseled pragmatism; the relationship between predation and murder in chimpanzees needed further study.[6] From Fox's perspective, the authors of the Seville Declaration sought to shut down research on the human animal at a moment when the evolutionary study of primate and human behavior was finally blooming.

Jonathan Benthall, the director of the Royal Anthropological Institute and founding editor of *Anthropology Today*, wrote a brief editorial in his newsletter supporting Fox's "gladiatorial" "polemic" (the language with which he imagined feminists would condemn Fox).[7] Benthall surmised the authors of the Seville Declaration had in mind the popular success of books by Robert Ardrey, Konrad Lorenz, and Desmond Morris when they expressed their doubts about aggression as a legitimate scientific concept. (In his estimation, these authors had thoughtfully contributed to the scientific interpretation of human nature, whether or not his colleagues now sullied their names.) Benthall's silence on the question of sociobiology reenacted the distance E. O. Wilson and others sought to place between their own writings and this earlier generation of colloquial scientific publications. Blame for hostility toward evolution landed conveniently on the shoulders of Ardrey, Lorenz, and Morris, rather than on the more professionalized authors who followed. Precise analysis, Benthall continued, would do more to counter the outrageous claims of racism or sexism than any vague statement. He trusted that evolutionists would continue to dismantle the tenets of creationism too. By equating all attempts to employ scientific knowledge for political purposes as inherently antiscientific in spirit (including feminism, anti-Semitism, and creationism), he cast Fox, the evolutionists, and himself as scientific truth-seekers in a wilderness of political confusion.[8] The Seville Declaration's short renunciation of violence as intrinsic to human nature provided him with a convenient target.

In the decades that now separate us from the battles over human nature in the 1960s and 1970s, readers have not lost their passion for epic evolutionary dramas in which the entirety of human history unfolds before their eyes.[9] Yet when students today respond to the question "What makes us human?" they are far more likely to invoke neurological facts than paleontological ones. The public battlefield over violence and cooperation has shifted to new ground in the mind and brain sciences—and the Seville Declaration emerged, after all, from a conference on brain research. Despite the apparent polarization of scientists writing about human nature into culture- and biology-oriented positions, the intellectual landscape defined by scientists working on the interaction between culture and biology has continued to flourish. Consider the six thousand members of the American Anthropological Association who never responded to the ballot and the position of scientists who identified, despite the divisive politics of the era, as both feminist and evolutionary.[10] That anthropologists, zoologists, and paleontologists were divided over aggression as an innate or learned component of human nature will strike no one who has read this book as surprising.

Beginning with postwar commitments to racial equality and ending in the battles over sociobiology in the 1970s, *Creatures of Cain* has shown that both male and female animals were endowed with new agency in postwar evolutionary theories. Although male hunting behavior was crucial to evolutionary theories based in familial masculinity and cooperation, only in the mid-1960s did colloquial evolutionary depictions of men acquire a competitive, aggressive edge that complemented images of women as sexually coy. When the idyllic dream of the nuclear family fractured, evolutionary roles for aggressive males and coy females took their place, each exemplifying why the sexes could never seem to agree. Along the way, the book's narrative ranged from educational policy to Hollywood movies, all the while unraveling and reweaving the tangled history of evolutionary theories of humanity's essential nature.[11]

Fundamental questions about the nature of humanity have proven crucial to recruiting and inspiring generations of students to pursue careers in the natural and social sciences, yet are difficult to find in professional journals. When evolutionists today look back to Charles Darwin, they commonly invoke the last sentence of his magisterial *On the Origin of Species*. It starts, "There is grandeur in this view of life." Colloquial scientific publications proved crucial to communicating this sense of awe at how dramatically humanity's physical and social constitution has changed over our long evolutionary history. Scientists and lay readers alike have long recognized the profound ways in which humans differ from all other animals. We learn from the past to more successfully confront the future—together. Yet thanks to our tight emotional connections, humans also wound each other both physically and psychologically. The difficulty of reconciling the best and the worst of our natures has constituted a powerful conundrum, reconfigured time and again by new theories of an essential, unitary human nature. Each author—Eiseley, Mead, Ardrey, Morgan, Tiger, Wilson, Goodall, and so many more—attempted to capture the ineffable, with verve, with conviction, and inevitably with blinders. My point was not to resolve the tensions between nature and nurture, antagonism and empathy, but to illustrate the generative tension produced by holding open such unanswerable, necessarily speculative questions within a thriving scientific community.

As evolutionary frameworks for generalizing from animal to human behavior have changed in the intervening years, so too has scientists' appreciation of what it means to be *animal*.[12] Rather than seeing animals as fiercely individualistic quasi-automata, intent on self-preservation at all costs, researchers now study food sharing and cultural traditions in chimpanzees, macaques, dolphins, and elephants, to name a few of the most well-known mammalian cases.[13] As a result, recent colloquial accounts of human nature have once again stressed the importance of altruism, cooperation, and group cohesion for success in our highly social lives.[14] Far from a return to Loren Eiseley's cinematic progressivist sweep through the ages, some of these evolutionary theories of human nature are couched in the language of rational choice theory.[15] Others continue to be written in the mode of evolutionary anthropology, drawing more heavily on the interdisciplinary commitments embraced by Sherwood Washburn.[16] In fact, evolutionary perspectives in the twentieth century have proven fruitful across the political spectrum.

These new tales of humanity's remarkable capacity to cooperate differ from the colloquial scientific discussions that have occupied this book in a number of key respects. Complementing assumptions about the aggressive nature of men, women in the Cold War were often characterized as cooperative, deeply attuned to the needs and desires of their companions, and empathetically bonded to their children and families.[17] Yet recent authors have not veered into feminist utopianism. No mainstream scientist has claimed that our female ancestors drove the cognitive and intellectual development of humanity through teaching others how to share. Instead, recent colloquial scientific publications have characterized these traits as representative of all humanity and testify to a self-conscious embrace of those primatologists who challenged gendered stereotypes in their research on primates' social behavior.[18] Women are now considered just as capable of competition as men, and men just as capable of empathy as women.[19] One of the most promising features of the rapidly proliferating literature on the evolution of cooperation is that it has brought experts on animals and experts on humans back into dynamic conversation.[20]

Ongoing synergistic frictions between colloquial and professional conversations about human nature are a continuing legacy of the Cold War decades in which Americans turned to experts on animal behavior to answer fundamental questions about what it means to be human. Today, our long, contingent journey from ape to human no longer provokes fear; it provides hope.

List of Archives

American Philosophical Society, Philadelphia, PA
    Theodosius Dobzhansky Papers, Mss.B.D65
    Henry Allen Moe Papers, Mss.B.M722
    Ashley Montagu Papers, Mss.Ms.Coll.109
    George Gaylord Simpson Papers, Mss.Ms.Coll.31
Bancroft Library, University of California, Berkeley, CA
    Sherwood Larned Washburn Papers, 1932–1996, BANC MSS 98/132c
Rare Book and Manuscript Library, Columbia University, New York, NY
    David Hamburg Papers, 1949–2005, CA#0005
    Stein and Day Publisher Records, 1963–1988, MS#1197
Monroe C. Gutman Library, Special Collections, Graduate School of
    Education, Harvard University, Cambridge, MA
    Peter B. Dow - Man: A Course of Study Records
Harvard University Archives, Cambridge, MA
    Papers of Jerome Seymour Bruner, Unprocessed Accessions 10823,
        11380, 14728
    Papers of Irven DeVore, Accession 14898
Harry Frank Guggenheim Foundation, New York, NY
    Institutional History, assembled and kept by Karen Colvard
Library of Congress, Manuscript Division, Washington, DC
    Margaret Mead Papers and the South Pacific Ethnographic Archives,
        1838–1996, MSS32441
NARA, Archives II, College Park, MD
    Record Group 307, Records of the National Science Foundation,
        Accession Number 307-98 166, UDUP/Entry 1-Agency
        Organizational History and Miscellaneous Background, Box 20,
        Folder: CLO-MACOS
National Anthropological Archives, Smithsonian Museum of Natural
    History, Department of Anthropology, Suitland MD

Papers of Timothy Asch

John Marshall Ju/'hoan Bushman Film and Video Collection,
1950–2000

Special Collections and University Archives, Rutgers University
Libraries, New Brunswick, NJ

MC 190, Robert Ardrey Papers, 1955–1980

Schlesinger Library, Radcliffe Institute for Advanced Study, Harvard
University

Berkshire Conferences on the History of Women, MC 606

Non-Print Media Services Library, University of Maryland, College Park,
MD

University of Pennsylvania Archives, Philadelphia, PA

Loren Corey Eiseley, 1907–1977, Papers, 1913–1987, UPT50 E036

List of Interviews

Patsy Asch	30 November 2011	by phone
Jerome Bruner	4 October 2011	New York, NY
Irven DeVore	4 November 2009	Cambridge, MA
	6 August 2011	
Phyllis Jay Dolhinow	25 October 2013	Berkeley, CA
Peter Dow	25 November 2011	New York, NY
Robin Fox	8 November 2011	Rocky Hill, NJ
David Hamburg	13 January 2017	Washington, DC
Karl Heider	1 September 2011	by phone
Sarah Blaffer Hrdy	27 October 2011	Citrona Walnut Farms, Davis, CA
Elaine Morgan	9 November 2011	by phone
Anne Pusey	23 May 2012	Duke University, Durham, NC
Lionel Tiger	10 November 2011	New York, NY
Robert Trivers	25 January 2014	New Brunswick, NJ
Richard Wrangham	7 May 2012	Harvard University, Cambridge, MA
Adrienne Zihlman	24 October 2011	Santa Cruz, CA

NOTES

Acknowledgments

1. Loren Eiseley, *All the Strange Hours: The Excavation of a Life* (1975; Lincoln: University of Nebraska Press, 2000), 186.

Introduction

1. Robert Ardrey, *The Territorial Imperative: A Personal Inquiry into the Animal Origins of Property and Nations* (New York: Atheneum, 1966); Konrad Lorenz, *On Aggression*, trans. Marjorie Kerr (New York: Harcourt, Brace & World, 1966); Desmond Morris, *The Naked Ape: A Zoologist's Study of the Human Animal* (New York: McGraw-Hill, 1967).

2. S. Dillon Ripley, "Preface," in *Man and Beast: Comparative Social Behavior* (Washington, DC: Smithsonian Institution Press, 1971), 6, quoting David G. Mandelbaum, "Violence in America: An Anthropological Perspective," *Economic and Political Weekly* 3, no. 43 (1968): 1649–1653, quote on 1651.

3. Ardrey invited this biblical association and gave the title "Cain's Children" to the final chapter of his earlier book, *African Genesis: A Personal Investigation into the Animal Origins and Nature of Man* (New York: Atheneum, 1961); see also Loren Eiseley, "The Intellectual Antecedents of *The Descent of Man*," in *Sexual Selection and the Descent of Man*, ed. Bernard Campbell (London: Heinemann, 1972), 1–16; and M. F. Ashley Montagu, "The New Litany of 'Innate Depravity,' or Original Sin Revisited," in *Man and Aggression*, ed. M. F. Ashley Montagu (New York: Oxford University Press, 1968), 3–17.

4. See Robert Wokler, "Perfectible Apes in Decadent Cultures: Rousseau's Anthropology Revisited," *Daedalus* 107, no. 3 (1978): 107–134. As Wokler points out, Ardrey's opposition to Rousseau softened in his third book, *The Social Contract: A Personal Inquiry into the Evolutionary Sources of Order and Disorder* (New York: Atheneum, 1970).

5. Charles Osgood, *An Alternative to War or Surrender* (Urbana: University of Illinois Press, 1962), 19.

6. Benjamin Spock, *The Common Sense Book of Baby and Child Care* (New York: Duell, Sloan and Pearce, 1946); Alfred C. Kinsey, Wardell B. Pomeroy, and Clyde E. Martin, *Sexual Behavior in the Human Male* (Philadelphia: W. B. Saunders, 1948); Alfred C. Kinsey, Wardell B. Pomeroy, Clyde E. Martin, and Paul H. Gebhard, *Sexual Behavior in the Human Female* (Philadelphia: W. B. Saunders, 1953); Elisabeth Kübler-Ross, *On Death and Dying* (New York: Macmillan, 1969).

7. Mark Greif explored the contemporaneous crisis in literary depictions of human nature, but he dismissed the relevance of scientific ideas in contributing to these discussions: Mark Greif, *The Age of the Crisis of Man: Thought and Fiction in America, 1933–1973* (Princeton, NJ: Princeton University Press, 2015), 11.

8. Most (but not all) of the contributors to the colloquial science literature during these decades were men, an aspect of the genre that came under fire in the later 1960s, as we will see in Part 3. Notable exceptions included Margaret Mead, *Coming of Age in Samoa: A Psychological Study of Primitive Youth for Western Civilization* (New York: W. Morrow, 1928); Ruth Benedict, *The Chrysanthemum and the Sword: Patterns of Japanese Culture* (Boston: Houghton Mifflin, 1946); and Rachel Carson, *Silent Spring* (Boston: Houghton Mifflin, 1962).

9. Celebrity culture itself changed in the postwar US thanks to this same diversification of media forms, which contributed to the increasing celebrity of scientists; see Declan Fahy and Bruce Lewenstein, "Scientists in Popular Culture: The Making of Celebrities," in *Routledge Handbook of Public Communication of Science and Technology*, ed. Brian Trench and Massimiliano Bucchi (New York: Routledge, 2014), 83–96. On scientists' participation in television, movies, and museums, see Gregg Mitman, *Reel Nature: America's Romance with Wildlife on Film* (Cambridge, MA: Harvard University Press, 1999); David Kirby, *Lab Coats in Hollywood: Science, Scientists, and Cinema* (Cambridge, MA: MIT Press, 2011); Marcel Chotkowski LaFollette, *Science on American Television: A History* (Chicago: University of Chicago Press, 2013); and Karen Rader and Victoria Cain, *Life on Display: Revolutionizing U. S. Museums of Science and Natural History in the Twentieth Century* (Chicago: University of Chicago Press, 2014).

10. As recent scholarship has shown, the opposition of "popular" and "professional" breaks down when historians take into account the multifarious publics participating in conversations about science: Bernard Lightman, *Victorian Popularizers of Science: Designing Nature for New Audiences* (Chicago: University of Chicago Press, 2007); Katherine Pandora and Karen A. Rader, "Science in the Everyday World: Why Perspectives from the History of Science Matter," *Isis* 99, no. 2 (2008): 350–364; Jonathan Topham, ed., "Focus: Historicizing 'Popular Science,'" *Isis* 100, no. 2 (2009): 310–368; Bruce Lewenstein, "Experimenting with Engagement," *Engineering Ethics* 17, no. 4 (2011): 817–821; James Secord, *Visions of Science: Books and Readers at the Dawn of the Victorian Age* (Chicago: University of Chicago Press, 2015).

11. Taking seriously Jim Secord's contention that science is a form of communication, one can helpfully think of professional and colloquial registers as reflecting the diglossic character of science as a form of communication in which "high" and "low" forms of a language easily coexist; see James A. Secord, "Knowledge in Transit," *Isis* 95, no. 4 (2004): 654–672; and Charles A. Ferguson, "Diglossia," *Word* 15 (1959): 325–340. I could have used the term "vernacular" to contrast with "professional," but that term has gained currency among historians of science designating a shared intellectual terrain between elite science and local knowledge: Katherine Pandora, "Knowledge Held in Common: Tales of Luther Burbank and Science in the American Vernacular," *Isis* 92, no. 3 (2001): 484–516; Pamela Smith, *Body of the Artisan: Art and Experience in the Scientific Revolution* (Chicago: University of Chicago Press, 2006); Helen Tilley, "Global Histories, Vernacular Science, and African Genealogies; Or, Is the History of Science Ready for the World?" *Isis* 101, no. 1 (2010): 110–119.

12. For example, Hayward Cirker, "The Scientific Paperback Revolution: A Traditional Medium Assumes a New Role in Science and Education," *Science* 140, no. 3567 (1963): 591–594. As

a result, postwar scientists who wrote in a colloquial mode would not have thought of themselves as participating in a broader "citizen science" that welcomed nonscientists as contributors to the process of data gathering, processing, and knowledge construction; e.g., Gowan Dawson, Chris Lintott, and Sally Shuttleworth, "Constructing Scientific Communities: Citizen Science in the Nineteenth and Twenty-First Centuries," *Journal of Victorian Culture* 20, no. 2 (2015): 246–254.

13. For example, Stephen Hilgartner, "The Dominant View of Popularization: Conceptual Problems, Political Uses," *Social Studies of Science* 20, no. 3 (1990): 519–539.

14. On the interdisciplinary nature of evolutionary science in the 1950s, see, e.g., Vassiliki Betty Smocovitis, *Unifying Biology: The Evolutionary Synthesis and Evolutionary Biology* (Princeton, NJ: Princeton University Press, 2006); and Joe Cain and Michael Ruse, eds., *Descended from Darwin: Insights into the History of Evolutionary Studies, 1900–1970* (Philadelphia: American Philosophical Society, 2009).

15. For example, Donna Haraway, *Primate Visions: Gender, Race, and Nature in the World of Modern Science* (New York: Routledge, 1989); Lorraine Daston and Gregg Mitman, eds., *Thinking with Animals: New Perspectives on Anthropomorphism* (New York: Columbia University Press, 2004); Gregory Radick, *The Simian Tongue: The Long Debate about Animal Language* (Chicago: University of Chicago Press, 2007); Henrika Kuklick, ed., *A New History of Anthropology* (Oxford: Blackwell, 2008). Non-Western cultures conceived of nature, primatology, and humanity in rather different terms; e.g., Kinji Imanishi, *A Japanese View of Nature: The World of Living Things*, trans. Pamela Asquith (New York: RoutledgeCurzon, 2002).

16. Just think, for example, of the dramatic rise in futuristic science-fiction literature of the time, for example, the twinned projects of Arthur C. Clarke's novel *2001: A Space Odyssey* (New York: New American Library, 1968) and Stanley Kubrick's film *2001: A Space Odyssey* (Metro-Goldwyn-Mayer, 1968), 161 min.

17. For example, Bruce Lewenstein, "The Meaning of 'Public Understanding of Science' in the United States after World War II," *Public Understanding of Science* 1 (1992): 45–68; Mitman, *Reel Nature*; Kirby, *Lab Coats in Hollywood*.

18. John Napier, *The Roots of Mankind* (Washington, DC: Smithsonian Institution Press, 1970), 2.

19. On the especially fluid nature of authority within the age of the counterculture, see David Kaiser, *How the Hippies Saved Physics: Science, Counterculture, and the Quantum Revival* (New York: W. W. Norton, 2011); Michael Gordin, *The Pseudoscience Wars: Immanuel Velikovsky and the Birth of the Modern Fringe* (Chicago: University of Chicago Press, 2012); Patrick McCray, *The Visioneers: How a Group of Elite Scientists Pursued Space Colonies, Nanotechnologies, and a Limitless Future* (Princeton, NJ: Princeton University Press, 2012); David Kaiser and W. Patrick McCray, eds., *Groovy Science: Knowledge, Innovation, and American Counterculture* (Chicago: University of Chicago Press, 2016).

20. For one attempt to wrest this authority back onto historical grounds for nonhuman species, see David Sepkoski, *Rereading the Fossil Record: The Growth of Paleobiology as an Evolutionary Discipline* (Chicago: University of Chicago Press, 2012).

21. Rae Goodell, *The Visible Scientists* (Boston: Little, Brown, 1977), which is based on her 1975 PhD dissertation from Stanford University.

22. Goodell, *Visible Scientists*, 3.

23. Historians of Victorian science have carefully attended to questions of authority and expertise, e.g., Melinda Baldwin, "The Victorian Bookshelf," *Historical Studies in the Natural Sciences* 45, no. 4 (2015): 610–620; Gowan Dawson and Bernard Lightman, eds., *Victorian Science and Literature*, 8 vols. (London: Pickering & Chatto, 2011–2012); James A. Secord, *Victorian Sensation: The Extraordinary Publication, Reception, and Secret Authorship of* Vestiges of the Natural History of Creation (Chicago: University of Chicago Press, 2000).

24. Raymond A. Sokolov, "Talk with Stephen Jay Gould," *New York Times*, 20 November 1977, BR4; Myrna Perez Sheldon, "Stephen Jay Gould, An Evolutionary Heretic" (unpublished manuscript, 2017).

25. Zuoyue Wang, *In Sputnik's Shadow: The President's Science Advisory Committee and Cold War America* (New Brunswick, NJ: Rutgers University Press, 2009).

26. Nathaniel Comfort, *The Science of Human Perfection: How Genes Became the Heart of American Medicine* (New Haven, CT: Yale University Press, 2012); Daniel Kevles, *In the Name of Eugenics: Genetics and the Uses of Human Heredity* (New York: Knopf, 1985).

27. The genetic code and computing provided fertile grounds for imagining both human reasoning and the very building blocks of life as forms of information: Lily Kay, *The Molecular Vision of Life* (Oxford: Oxford University Press, 1993); Lily Kay, *Who Wrote the Book of Life? A History of the Genetic Code* (Stanford, CA: Stanford University Press, 2000); Soraya de Chadarevian, *Designs for Life: Molecular Biology after World War II* (Cambridge: Cambridge University Press, 2002); Dorothy Nelkin and M. Susan Lindee, *The DNA Mystique: The Gene as a Cultural Icon* (Ann Arbor: University of Michigan Press, 2004).

28. Erika Lorraine Milam, "The Equally Wonderful Field: Ernst Mayr and Organismic Biology," *Historical Studies in the Natural Sciences* 40, no. 3 (2010): 279–317.

29. See, most recently, Marga Vicedo, *The Nature and Nurture of Love: From Imprinting to Attachment in Cold War America* (Chicago: University of Chicago Press, 2013); Jamie Cohen-Cole, *The Open Mind: Cold War Politics and the Sciences of Human Nature* (Chicago: University of Chicago Press, 2014); Tania Munz, *The Dancing Bees: Karl von Frisch and the Discovery of the Honeybee Dance Language* (Chicago: University of Chicago Press, 2016).

30. Roy Richard Grinker, *In the Arms of Africa: The Life of Colin M. Turnbull* (New York: St. Martin's Press, 2000); Adrianna Link, "Documenting Human Nature: E. Richard Sorenson and the National Anthropological Film Center, 1965–1980," *Journal of the History of the Behavioral Sciences* 52, no. 4 (2016): 371–391; Nancy C. Lutkehaus, *Margaret Mead: The Making of an American Icon* (Princeton, NJ: Princeton University Press, 2008); Tracy Teslow, *Constructing Race: The Science of Bodies and Cultures in American Anthropology* (New York: Cambridge University Press, 2014).

31. Haraway, *Primate Visions*; Susan Sperling, "The Troop Trope: Baboon Behavior as a Model System in the Postwar Period," in *Science without Laws: Model Systems, Cases, Exemplary Narratives*, ed. Angela N. H. Creager, Elizabeth Lunbeck, and M. Norton Wise (Durham, NC: Duke University Press, 2007), 73–89.

32. During the Cold War, scientists used the term "hominid" to refer to humans and all extinct relatives that lived more recently than the common ancestor we share with other extant great apes. Today paleoanthropologists use "hominin" for this purpose, and "hominid" has come to designate a much larger taxonomic group that includes all great apes species, extant and extinct. Switching back and forth between the two terms when quoting and discussing

would cause confusion, so throughout the book I follow the convention of paleoanthropologists in this period and use the term "hominid" in the sense they meant.

33. Molecular biologists, paleontologists, and organismal biologists in these years also hotly debated how to reconcile insights from protein assays and (later) genetic divergence centered on both the pattern of evolution (how extant species were related to one another) and estimates of when the last common ancestor of two species had lived. Although contemporaneous with the events that unfold in this book, these debates did not venture into the fraught terrain of reconstructing what our potential ancestors had looked like or how they had behaved. On this history, see Marianne Sommer's *History Within: The Science, Culture, and Politics of Bones, Organisms, and Molecules* (Chicago: University of Chicago Press, 2016), especially the last third, which explores the career of Luigi Luca Cavalli-Sforza, whose work in molecular anthropology began in the early 1960s.

34. On the intellectual battles that have characterized the history of human paleontology, see Roger Lewin, *Bones of Contention: Controversies in the Search for Human Origins* (Chicago: University of Chicago Press, 1997).

35. John Marshall, dir., *The Hunters* (Somerville, MA: Documentary Educational Resources, 1957), 72 min. and *Bitter Melons* (Somerville, MA: Documentary Educational Resources, 1971), 30 min.; Colin Turnbull, *The Forest People* (New York: Simon and Schuster, 1962). Much of the existing scholarly literature on postwar cultural anthropology is biographical: Nancy C. Lutkehaus, *Margaret Mead: The Making of an American Icon* (Princeton, NJ: Princeton University Press, 2008); Grinker, *In the Arms of Africa*; Patrick Wilcken, *Claude Lévi-Strauss: The Poet in the Laboratory* (New York: Penguin Press, 2010).

36. Robert Gardner and Karl G. Heider, *Gardens of War: Life and Death in the New Guinea Stone Age*, introduction by Margaret Mead (New York: Random House, 1968).

37. On the mid-century professionalization of ethology, or the scientific study of animal behavior, see Richard W. Burkhardt Jr., *Patterns of Behavior: Konrad Lorenz, Niko Tinbergen, and the Founding of Ethology* (Chicago: University of Chicago Press, 2005). On field studies in primates, see Haraway, *Primate Visions*, Radick, *Simian Tongue*, and Georgina Montgomery, *Primates in the Real World: Escaping Primate Folklore and Creating Primate Science* (Charlottesville: University of Virginia Press, 2015).

38. Eric Sandeen, *Picturing an Exhibition: The Family of Man and 1950s America* (Albuquerque: University of New Mexico Press, 1995); Fred Turner, *The Democratic Surround: Multimedia and American Liberalism from World War II to the Psychedelic Sixties* (Chicago: University of Chicago Press, 2013), esp. 181–212.

39. Mark Borrello, *Evolutionary Restraints: The Contentious History of Group Selection* (Chicago: University of Chicago Press, 2010); Oren Harman, *The Price of Altruism: George Price and the Search for the Origins of Kindness* (London: Bodley Head, 2010); Ullica Segerstråle, *Nature's Oracle: The Life and Work of W. D. Hamilton* (Oxford: Oxford University Press, 2013); Paul Erickson, *The World Game Theorists Made* (Chicago: University of Chicago Press, 2015).

40. Richard Dawkins, *The Selfish Gene* (New York: Oxford University Press, 1976).

41. Bruno Latour, "Drawing Things Together," in *Representation in Scientific Practice*, ed. Michael Lynch and Steven Woolgar (Cambridge, MA: MIT Press, 1990), 19–68. On evolutionary images, see Marianne Sommer and Veronika Lipphardt, eds., "Visibility Matters: Diagrammatic Renderings of Human Evolution and Diversity in Physical, Serological and

Molecular Anthropology," special issue, *History of the Human Sciences* 28, no. 5 (2015); Theodore W. Pietsch, *Trees of Life: A Visual History of Evolution* (Baltimore: Johns Hopkins University Press, 2012); Marianne Sommer, *Bones and Ochre: The Curious Afterlife of the Red Lady of Paviland* (Cambridge, MA: Harvard University Press, 2007); Greg Myers, "Every Picture Tells a Story: Illustrations in E. O. Wilson's *Sociobiology*," in *Representation in Scientific Practice*, 231–265.

42. Charles Percy Snow, *The Two Cultures and the Scientific Revolution* (New York: Cambridge University Press, 1959).

43. On the gendering of the X and Y chromosomes in these same decades, see Sarah Richardson, *Sex Itself: The Search for Male and Female in the Human Genome* (Chicago: University of Chicago Press, 2013).

44. Kubrick, dir., *2001: A Space Odyssey*; Sam Peckinpah, dir., *Straw Dogs* (Cinerama, 1971), 117 min.

45. E. O. Wilson, *Sociobiology: A New Synthesis* (Cambridge, MA: Belknap Press of Harvard University Press, 1975).

Part One. The Ascent of Man

Loren Eiseley, "An Evolutionist Looks at Modern Man," *Saturday Evening Post*, 26 April 1958, 28, quote on 120.

1. Loren Eiseley, *The Immense Journey* (New York: Random House, 1957). He described his hearing loss and the circumstances that led him to write *The Immense Journey* in Eiseley, *All the Strange Hours: Excavation of a Life* (1975; Lincoln: University of Nebraska Press, 2000): 107, 173–180.

2. Eiseley, *Immense Journey*, 76–77. I use "man" as an actor's category throughout the first half of the book in order to call attention to anthropologists' self-conscious shift to "human" later in the book.

3. John Krige and Jessica Wang, eds., "Nation, Knowledge, and Imagined Futures: Science, Technology, and Nation-Building, Post 1945," special issue, *History and Technology* 31, no. 3 (2015).

4. Matthew Wisnioski, *Engineers for Change: Competing Visions of Technology in 1960s America* (Cambridge, MA: MIT Press, 2012); Lizabeth Cohen, *A Consumer's Republic: The Politics of Mass Consumption in Postwar America* (New York: Vintage Books, 2008).

5. "Men of the Year," *Time*, 2 January 1961, 40.

6. Marcel Lafollette, *Science on the Air: Popularizers and Personalities on Radio and Early Television* (Chicago: University of Chicago Press, 2008); Gregg Mitman, *Reel Nature: America's Romance with Wildlife on Film* (Cambridge, MA: Harvard University Press, 1999).

7. Hayward Cirker, "The Scientific Paperback Revolution: A Traditional Medium Assumes a New Role in Science and Education," *Science* 140, no. 3567 (1963): 591–594, quote on 591. At the time, Cirker served as president of Dover Publications, a leading publisher of intellectual paperbacks.

8. National Geographic, *School Bulletin*, 6 December 1965.

9. Melinda Gormley, "Pulp Science: Education and Communication in the Paperback Book Revolution," *Endeavour* 40 (2016): 24–37.

10. Daniel J. Kevles, *The Physicists: The History of a Scientific Community in Modern America* (New York: Knopf, 1977).

11. Zuoyue Wang, *In Sputnik's Shadow: The President's Science Advisory Committee and Cold War America* (New Brunswick, NJ: Rutgers University Press, 2008).

12. Eiseley, *All the Strange Hours*, 141–143.

13. Loren Eiseley, *Darwin's Century: Evolution and the Men Who Discovered It* (Garden City, NY: Doubleday, 1958) and *The Firmament of Time* (New York: Atheneum, 1960).

14. Eiseley, *Immense Journey*, 56.

15. Eiseley, *Immense Journey*, 6.

16. Theodosius Dobzhansky, *The Biological Basis of Human Freedom* (New York: Columbia University Press, 1956); Jacob Bronowski, *The Ascent of Man* (Boston: Little, Brown, 1973).

17. On Dobzhansky, see Mark Adams, ed., *The Evolution of Theodosius Dobzhansky: Essays on His Life and Thought in Russia and America* (Princeton, NJ: Princeton University Press, 1994). On Morgan's lab, see Robert Kohler, *Lords of the Fly:* Drosophila *Genetics and the Experimental Life* (Chicago: University of Chicago Press, 1994).

18. Francisco Ayala and Timothy Prout, "Theodosius Dobzhansky," *Social Biology* 23, no. 2 (1976): 101–107, quote on 106.

19. Ghillean T. Prance, "In Memoriam: Theodosius Dobzhansky," *Brittonia* 28, no. 1 (1976): 85; Bentley Glass, ed., *The Roving Naturalist: Travel Letters of Theodosius Dobzhansky* (Philadelphia: American Philosophical Society, 1980).

20. Theodosius Dobzhansky, "The Ascent of Man," *Social Biology* 19, no. 4 (1972): 367–378.

21. Theodosius Dobzhansky, "Species after Darwin," in *A Century of Darwin*, ed. S. A. Barnett (London: Heinemann, 1958), 19–55, on 27–28.

22. Although Ruth Benedict and Margaret Mead were among the most well-known anthropologists of the early 1950s, ironically most media attention devoted to anthropology as a discipline centered instead on recent finds in archeology and physical anthropology: Evon Z. Vogt, "Anthropology in the Public Consciousness," in *Yearbook of Anthropology* (Chicago: University of Chicago Press, 1955), 357–374.

23. Some of her colleagues criticized Mead's promulgation of cultural relativism, although the strongest objections found their way to print only after she had died. See Derek Freeman, *Margaret Mead and Samoa: The Making and Unmaking of an Anthropological Myth* (Cambridge, MA: Harvard University Press, 1983); cf. Paul Shankman, *The Trashing of Margaret Mead: Anatomy of an Anthropological Controversy* (Madison: University of Wisconsin Press, 2009); Nancy Lutkehaus, *Margaret Mead: The Making of an American Icon* (Princeton, NJ: Princeton University Press, 2008).

24. Betty Friedan, *The Feminine Mystique* (New York: Dell, 1963); Friedan took particular issue with Mead's *Male and Female: A Study of the Sexes in a Changing World* (New York: W. Morrow, 1949).

25. Amei Wallach, "Margaret Mead's Pacific," *Newsday*, 14 December 1984, C18; "The Hall That Mead Built," *Natural History*, February 1985, 72–73.

26. As a graduate student at Columbia in the 1920s, Mead studied with Franz Boas and Ruth Benedict; in the postwar era she drew strength from Boas' legacy; he had died in 1942; Tracy Teslow, *Constructing Race: The Science of Bodies and Cultures in American Anthropology* (New York: Cambridge University Press, 2014).

27. Margaret Mead, "What Is Human Nature?," *Look*, 19 April 1955, 56–62, quote on 57.

28. Mead, "What Is Human Nature?," 62.

29. Bernard Lightman, ed., *Global Spencerism* (Boston: Brill, 2016).

30. Edward Steichen, *The Family of Man: The Photographic Exhibition created by Edward Steichen for the Museum of Modern Art* (New York: Simon and Schuster, 1955). On the international legacy of the project, see Fred Turner, "The Family of Man and the Politics of Attention in Cold War America," *Public Culture* 24, no. 1 (2012): 55–84.

31. John P. Jackson Jr., *Social Scientists for Social Justice: Making the Case against Segregation* (New York: New York University Press, 2001). Conservatives also found their scientific experts: see John P. Jackson Jr., *Science for Segregation: Race, Law, and the Case against Brown v. Board of Education* (New York: New York University Press, 2005).

32. Marshall Flaum, dir., *Miss Goodall and the Wild Chimpanzees* (National Geographic Specials, 1965), 60 min.; Jacob Bronowski, *The Ascent of Man*, thirteen-episode television miniseries (BBC, Time-Life Television Productions, 1973)—the first episode, "Lower than the Angels," covered all human history before the advent of agriculture. See also Bronowski, *Ascent of Man*. On the stunning success of postwar educational television shows about nature and science see Mitman, *Reel Nature* and Marcel Chotkowski LaFollette, *Science on American Television* (Chicago: University of Chicago Press, 2013).

33. John Rudolph, *Scientists in the Classroom: The Cold War Reconstruction of American Science Education* (New York: Palgrave, 2002). By the time MACOS appeared in print, ESI had changed its name to the Education Development Center.

34. Popularized by George Peter Murdock, *Social Structure* (New York: The Free Press, 1949); cf. Joanne Meyerowitz, ed., *Not June Cleaver: Women and Gender in Postwar America* (Philadelphia: Temple University Press, 1994).

35. On armchair theorizing, see Harro Maas, "Sorting Things Out: The Economist as an Armchair Observer," in *Histories of Scientific Observation*, ed. Lorraine Daston and Elizabeth Lunbeck (Chicago: University of Chicago Press, 2011), 206–229.

36. Eiseley, *All the Strange Hours*, 182. Carefully restored, the library has since been designated a historic landmark and today is officially known as the Fisher Fine Arts Library on Penn's campus at 220 South 34th Street.

37. Eiseley, *All the Strange Hours*, 139.

38. Ayala and Prout, "Theodosius Dobzhansky," 106.

39. Theodosius Dobzhansky, *The Biology of Ultimate Concern* (New York: New American Library, 1967), 7.

40. Eiseley, *All the Strange Hours*, 90. On Darwin's views on religion throughout his life, see Janet Browne, *Charles Darwin: Voyaging* and *Charles Darwin: The Power of Place* (Princeton, NJ: Princeton University Press, 1996 and 2003).

Chapter One. Humanity in Hindsight

1. Evon Z. Vogt, "Anthropology in the Public Consciousness," *Yearbook of Anthropology* (1955): 357–374. On journalists' excitement over dramatic human fossil finds, see Lydia Pyne, *Seven Skeletons: The Evolution of the World's Most Famous Human Fossils* (New York: Viking, 2016).

2. Wilifrid Le Gros Clark, *The Foundations of Human Evolution* (Eugene: Oregon State System of Higher Education, 1959), 9.

3. Charles Darwin and Alfred Russel Wallace, "On the Tendency of Species to form Varieties; and on the Perpetuation of Varieties and Species by Natural Means of Selection, [Read July 1st, 1858]," *Journal of the Proceedings of the Linnean Society of London: Zoology* 3 (20 August 1858): 45–50, in John van Whye, ed., The Complete Work of Charles Darwin Online (2002–), http://darwin-online.org.uk/.

4. John Napier, *The Roots of Mankind* (Washington, DC: Smithsonian Institution, 1970; London: George Allen & Unwin, 1971), 17.

5. Margaret Mead, "Cultural Determinants of Behavior," in *Behavior and Evolution*, ed. Anne Roe and George Gaylord Simpson (New Haven, CT: Yale University Press, 1958), 480–503; Vassiliki Betty Smocovitis, "Humanizing Evolution: Anthropology, the Evolutionary Synthesis, and the Prehistory of Biological Anthropology, 1927–1962," *Current Anthropology* 53, no. S5 (2012): S108–S125, S114–S121.

6. This trend that started even while he was alive: see Janet Browne, "I Could Have Retched All Night: Charles Darwin and His Body," in *Science Incarnate: Historical Embodiments of Natural Knowledge*, ed. Christopher Lawrence and Steven Shapin (Chicago: University of Chicago Press, 1998), 240–287.

7. Adrian Desmond and James Moore, *Darwin's Sacred Cause: Race, Slavery and the Quest for Human Origins* (New York: Allen Lane, 2009); Richard Weikart, *From Darwin to Hitler: Evolutionary Ethics, Eugenics, and Racism in Germany* (New York: Palgrave Macmillan, 2004); cf. Robert J. Richards, *Was Hitler a Darwinian? Disputed Questions in the History of Evolutionary Theory* (Chicago: University of Chicago Press, 2013). If you are looking for an engaging biography, you cannot go wrong with Janet Browne's excellent *Charles Darwin: Voyaging* and *Charles Darwin: The Power of Place* (New York: Knopf, 1995 and 2002; Princeton, NJ: Princeton University Press, 1996 and 2003).

8. Julian Huxley, the grandson of Darwin's contemporary and defender Thomas Henry Huxley and a zoologist in his own right, even tried to coordinate a sea expedition retracing the young Charles' voyage in the *Beagle*. (Finances collapsed early on.) On the importance of the centennial conferences for mid-century evolutionists, see Vassiliki Betty Smocovitis, "The 1959 Darwin Centennial Celebration in America," *Osiris* 14: 274–323; and, for a more anthropological angle, Smocovitis, "Humanizing Evolution."

9. Loren Eiseley, *Darwin's Century: Evolution and the Men Who Discovered It* (Garden City, NY: Doubleday Anchor Books, 1958).

10. S. A. Barnett, ed., *A Century of Darwin* (Cambridge, MA: Harvard University Press, 1958), xii.

11. Charles Darwin, *On the Origin of Species*, introduction by Ernst Mayr (Cambridge, MA: Harvard University Press, 1964); Morse Peckham, ed., *The Origin of Species: A Variorum Text* (Philadelphia: University of Pennsylvania Press, 1959).

12. Martin Rudwick, *The Great Devonian Controversy: The Shaping of Scientific Knowledge among Gentlemanly Specialists* (Chicago: University of Chicago Press, 1985); and Rudwick, *Scenes from Deep Time: Early Pictorial Representations of the Prehistoric World* (Chicago: University of Chicago Press, 1992).

13. Browne, *Power of Place*, quotes on 316, 15, 62.

14. C. H. Waddington, "Theories of Evolution," in *Century of Darwin*, ed. Barnett, 1–18, quote on 5; see also Napier, *Roots of Mankind*, 18.

15. On the history of the synthesis between genetics and evolutionary theory, see Julian Huxley, *The Modern Synthesis* (London: G. Allen & Unwin, 1942); Ernst Mayr and William B. Provine, *The Evolutionary Synthesis: Perspectives on the Unification of Biology* (Cambridge, MA: Harvard University Press, 1980); Vassiliki Betty Smocovitis, *Unifying Biology: The Evolutionary Synthesis and Evolutionary Biology* (Princeton, NJ: Princeton University Press, 1996); William Provine, *The Origins of Theoretical Population Genetics* (Chicago: University of Chicago Press, 2001); and Joe Cain and Michael Ruse, eds., *Descended from Darwin: Insights into the History of Evolutionary Studies, 1900–1970* (Philadelphia: American Philosophical Society, 2009).

16. Theodosius Dobzhansky, "Living with Biological Evolution," in *Man and the Biological Revolution* (Toronto: York University Press, 1976), 21–45, quote on 21.

17. Theodosius Dobzhansky, "Species after Darwin," in *Century of Darwin*, ed. Barnett, 19–55, quote on 19.

18. For a similar sentiment, see Julian H. Steward, "Evolutionary Principles and Social Types," in *Evolution after Darwin: II. The Evolution of Man: Man, Culture, Society*, ed. Sol Tax (Chicago: University of Chicago Press, 1960), 169–186, on 169.

19. Wilfrid Le Gros Clark, "The Study of Man's Descent," in *Century of Darwin*, ed. Barnett, 173–205, quote on 173.

20. Le Gros Clark, "Study of Man's Descent," quote on 174.

21. Bernard Campbell, "Man for All Seasons," in *Sexual Selection and Descent of Man* (London: Heinemann, 1972), 40–58, quote on 41.

22. The paleoanthropologist John Napier preferred to think of Darwin's theory of natural selection as kind of metaphorical "missing link," providing subsequent scientists a platform from which to move forward: Napier, *Roots of Mankind*, 17.

23. Barnett, *Century of Darwin*, xv. Barnett could not resist adding, "The autographed copy of Marx's *Capital* remains uncut at Downe House."

24. Alfred Russel Wallace, "The Origin of Human Races and the Antiquity of Man Deduced from the Theory of 'Natural Selection,'" *Journal of the Anthropological Society of London* 2 (1864); Wallace, "The Limits of Natural Selection as Applied to Man," in *Contributions to the Theory of Natural Selection: A Series of Essays* (London: Macmillan, 1870).

25. Eiseley, *Immense Journey*, 84.

26. In Darwin's letter to Wallace dated 14 April 1869, the original paragraph reads: "If you had not told me I sh^d have thought that they had been added by some one else. As you expected I differ grievously from you, & I am very sorry for it. I can see no necessity for calling in an additional & proximate cause in regard to Man. But the subject is too long for a letter. I have been particularly glad to read y^r discussion because I am now writing & thinking much about man." Darwin Correspondence Project, Letter no. 6706, accessed 7 August 2016, http://www.darwin project.ac.uk/DCP-LETT-6706.

27. Eiseley, *Immense Journey*, 89. The secretary of the Smithsonian Institution Leonard Carmichael echoed this comparison in "Absolutism, Relativism, and the Scientific Psychology of Human Nature," in *Relativism and the Study of Man*, ed. Helmut Schoek and James W. Wiggins (Princeton, NJ: D. van Nostrand, 1961), 6.

28. Eiseley, *Immense Journey*, 85.

29. Eiseley, *All the Strange Hours*, 91.

30. On the changing interpretations of Neanderthal remains from the late nineteenth century to the late twentieth, see Marianne Sommer, *Bones and Ochre: The Curious Afterlife of the Red Lady of Paviland* (Cambridge, MA: Harvard University Press, 2007).

31. J. S. Weiner, K. P. Oakley, and W. E. Le Gros Clark, "The Solution of the Piltdown Problem," *Bulletin of the British Museum (Natural History): Geology* 2, no. 3 (1953): 139–146; J. S. Weiner and K. P. Oakley, "The Piltdown Fraud: Available Evidence Reviewed," *American Journal of Physical Anthropology* 12, no. 1 (1954): 1–7; J. S. Weiner, "The Evolutionary Taxonomy of the Hominidae in the Light of the Piltdown Investigation," in *Selected Papers of the 5th International Congress of Anthropological and Ethnological Sciences, Philadelphia 1956* (Philadelphia: University of Pennsylvania Press, 1960), 741–752.

32. Weiner, Oakley, and Le Gros Clark, "Solution of the Piltdown Problem," 145. On the longer history of fluorine absorption as a means of dating fossils, see Matt Goodrum and Cora Olson, "The Quest for an Absolute Chronology in Human Prehistory: Anthropologists, Chemists and the Fluorine Dating Method in Paleoanthropology," *British Journal for the History of Science* 42, no. 1 (2009): 95–114.

33. Eiseley, *Immense Journey*, 95.

34. Weiner, "Evolutionary Taxonomy of the Hominidae," 751.

35. On the history of South African paleoanthropology and intertwined issues of race and human identity, see Christa Kuljian, *Darwin's Hunch: Science, Race, and the Search for Human Origins* (Auckland Park, South Africa: Jacana, 2016).

36. When Franz Weidenreich took over the analysis of these bones, he controversially claimed that the Peking Man fossils represented the ancient progenitor of all Chinese peoples. On paleoanthropological research in China, see Sigrid Schmalzer, *People's Peking Man: Popular Science and Human Identity in Twentieth-Century China* (Chicago: University of Chicago Press, 2008).

37. Quote and discussion in Theodosius Dobzhansky, *Evolution, Genetics, and Man* (New York: John Wiley, 1955), 328; Raymond A. Dart, "A Note on the Taungs Skull," *South African Journal of Science* 26 (December 1929), 648–658, quote on 651.

38. In de Beer's estimation, the teeth of modern humans resembled the milk teeth of *Australopithecus* fossils, whereas the adult teeth of *Australopithecus* more closely resembled gorilla teeth—further evidence for de Beer of humanity's paedomorphic development; De Beer, "Darwin and Embryology," in *Century of Darwin*, ed. Barnett, 155–172, on 166.

39. The span of the gap both lengthened with new means of dating fossils and shortened with the discovery of new specimens of known species.

40. A. S. Romer, "Darwin and the Fossil Record," in *Century of Darwin*, ed. Barnett, 130–152, on 150. See also S. L. Washburn and F. Clark Howell, "Human Evolution and Culture," in *Evolution after Darwin: II*, ed. Tax, 33–56, on 41.

41. Sherwood Washburn, "Behavior and Human Evolution," in *Classification and Human Evolution*, Viking Fund Publications in Anthropology 37, ed. Sherwood Washburn (New York: Wenner-Gren Foundation for Anthropological Research, 1963), 190–203, on 203.

42. Paul Farber, *Mixing Races: From Scientific Racism to Modern Evolutionary Ideas* (Baltimore: Johns Hopkins University Press, 2010).

43. Dobzhansky, "Species after Darwin," 38, 51.

44. Dobzhansky, "Species after Darwin," 28.

45. This topic formed a central focus of discussion at Washburn's conference "Classification and Human Evolution." See, e.g., Theodosius Dobzhansky, "Genetic Entities in Hominid Evolution," 347–362; George Gaylord Simpson, "The Meaning of Taxonomic Statements," 32–49; Ernst Mayr, "The Taxonomic Evaluation of Fossil Hominids," 332–346, and others in the volume from the conference, *Classification and Human Evolution*, ed. Washburn.

46. Ernst Mayr, *Systematics and the Origin of Species from the Viewpoint of a Zoologist* (New York: Columbia University Press, 1942).

47. Ernst Mayr, "Taxonomic Categories in Fossil Hominids," *Cold Spring Harbor Symposia on Quantitative Biology* 15 (1950): 109–118.

48. Theodosius Dobzhansky, *Evolution, Genetics, and Man* (New York: Wiley, 1955), *The Biological Basis of Human Freedom* (New York: Columbia University Press, 1956), *Mankind Evolving* (New Haven, CT: Yale University Press, 1962), and others. Dobzhansky, Simpson, and Mayr also contributed to the evolutionary picture provided in *The World We Live In* (New York: Time-Life Books, 1955), which emphasized the importance of evolution through isolation and adaptation, even if the book discussed humans only obliquely. See also Ruth Moore, *Evolution* (New York: Time-Life Books, 1962), for which Dobzhansky wrote the introduction, and F. Clark Howell's *Early Man* (New York: Time-Life Books, 1965).

49. Elwyn L. Simons, "Some Fallacies in the Study of Hominid Phylogeny," *Science* 141, no. 3584 (1963): 879–889.

50. Ian Tattersall, *The Strange Case of the Rickety Cossack and Other Cautionary Tales from Human Evolution* (New York: Palgrave Macmillan, 2015), 213; for a more intellectually substantial account of his concerns with Mayr's classification scheme, see "Species Concepts and Species Identification in Human Evolution," *Journal of Human Evolution* 22 (1992): 341–349.

51. For example, Theodosius Dobzhansky and L. C. Dunn, *Heredity, Race, and Society* (New York: Penguin Books, 1946); Edward Steichen, *The Family of Man*, prologue by Carl Sandburg (New York: Simon and Schuster for the Museum of Modern Art, 1955); Fred Turner, *The Democratic Surround: Multimedia and American Liberalism from World War II to the Psychedelic Sixties* (Chicago: University of Chicago Press, 2013), 181–212.

52. Eiseley, *Immense Journey*; J.B.S. Haldane, *The Unity and Diversity of Life* (Delhi: Ministry of Information & Broadcasting, 1958).

53. Theodosius Dobzhansky, *The Biology of Ultimate Concern* (New York: New American Library, 1967), 3.

54. Irven DeVore, "The Evolution of Social Life," in *Horizons of Anthropology*, ed. Sol Tax (Chicago: Aldine, 1964), 25–36, quote on 35.

Chapter Two. Battle for the Stone Age

1. Adrianna Link, "Documenting Human Nature: E. Richard Sorenson and the National Anthropological Film Center, 1965–1980," *Journal of the History of the Behavioral Sciences* 52, no. 4 (2016): 371–391.

2. Alfred J. Kroeber articulated the idea without calling it a "critical point" in "The Superorganic," *American Anthropologist* 19 (1917): 163–213; R. L. Lee and M. J. O'Brien, "The Concept of Evolution in Early Twentieth-Century Americanist Archeology," *Archeological Papers of the*

American Anthropological Association 7 (1997): 21–48. It seems to be Clifford Geertz who, in retrospect, used the phrase to encapsulate Kroeber's position, e.g., Clifford Geertz, "The Transition to Humanity," in *Horizons of Anthropology*, ed. Sol Tax (Chicago: Aldine, 1964), 38.

3. Geertz, "Transition to Humanity," 39.

4. On Jane Goodall's research and primatology as a means of reconstructing human nature, see Donna Haraway, *Primate Visions: Gender, Race, and Nation in the World of Modern Science* (New York: Routledge, 1989), plus the many books by Goodall herself in which she discusses her early research.

5. For example, Claude Lévi-Strauss, "Concept of Primitiveness," in *Man the Hunter*, ed. Richard B. Lee and Irven DeVore (Chicago: Aldine, 1968), 349–352; Ricardo Ventura Santos, Susan Lindee, and Vanderlei Sabastião de Souza, "Varieties of the Primitive: Human Biological Diversity Studies in Cold War Brazil (1962–1970)," *American Anthropologist* 116, no. 4 (2011): 723–735.

6. On social scientists' quest to establish themselves as experts on questions of racialized prejudice and discrimination, see Ellen Herman, *The Romance of American Psychology: Political Culture in the Age of Experts* (Berkeley: University of California Press, 1995); Daryl Scott, *Contempt and Pity: Social Policy and the Image of the Damaged Black Psyche, 1880–1996* (Chapel Hill: University of North Carolina Press, 1997); and John P. Jackson Jr., *Social Scientists for Social Justice: Making the Case against Segregation* (New York: New York University Press, 2001).

7. Clyde Kluckhohn, *Mirror for Man: The Relation of Anthropology to Modern Life* (New York: McGraw-Hill, 1949), 11.

8. "$10,000 Book Award Won by Harvard Man," *New York Times*, 15 March 1947, 11; Bernard Mishkin, "Science on the March," *New York Times*, 30 January 1949, BR15.

9. Kluckhohn, *Mirror for Man*, 41.

10. Kluckhohn, *Mirror for Man*, 102–103. On the history of race in Victorian England and its entanglement with theories of heredity, see Sadiah Qureshi, *Peoples on Parade: Exhibitions, Empire, and Anthropology in Nineteenth-Century Britain* (Chicago: University of Chicago Press, 2011) and George W. Stocking Jr., *Victorian Anthropology* (New York: Free Press, 1987).

11. Kluckhohn, *Mirror for Man*, 105.

12. Kluckhohn, *Mirror for Man*, 277.

13. M. F. Ashley Montagu, *Man's Most Dangerous Myth: The Fallacy of Race* (1942; New York: Columbia University Press, 1945), 93.

14. Ashley Montagu, "Foreword," in Kluckhohn, *Mirror for Man: The Relation of Anthropology to Modern Life* (Tucson: University of Arizona Press, 1985), xvi.

15. Jenny Bangham, "What Is Race? UNESCO, Mass Communication and Human Genetics in the Early 1950s," *History of the Human Sciences* 28, no. 5 (2015): 80–107; Michelle Brattain, "Race, Racism, and Antiracism: UNESCO and the Politics of Presenting Science to the Postwar Public," *American Historical Review* 112, no. 5 (2007): 1386–1413; UNESCO, *Four Statements on the Race Question* (Paris: UNESCO, 1969); Nadine Weidman, "An Anthropologist on TV: Ashley Montagu and the Biological Basis of Human Nature, 1945–1960," in *Cold War Social Science: Knowledge Production, Liberal Democracy, and Human Nature*, ed. Mark Solovey and Hamilton Cravens (New York: Palgrave Macmillan, 2012), 215–232.

16. Susan Sperling, "Ashley's Ghost . . . ," in *Cold War Social Science*, ed. Solovey and Cravens, 17–36, on 18.

17. Sherwood L. Washburn, "Thinking about Race," *Annual Report of the Board of Regents of the Smithsonian Institution* (1945): 363–378.

18. See Earnest Hooton, *Apes, Men, and Morons* (New York: G. P. Putnam's Sons, 1937).

19. Sherwood Washburn, "The New Physical Anthropology," *Transactions of the New York Academy of Sciences*, 2nd ser., 13, no. 7 (1951): 298–304. On the significance of Washburn's work to the evolutionary synthesis, see Vassiliki Betty Smocovitis, "Humanizing Evolution: Anthropology, the Evolutionary Synthesis, and the Prehistory of Biological Anthropology, 1927–1962," *Current Anthropology* 53, no. S5 (2012), 114–S116; Donna Haraway, "Remodelling the Human Way of Life: Sherwood Washburn and the New Physical Anthropology, 1950–1980," in *Bones, Bodies, Behavior: Essays on Biological Anthropology*, ed. George W. Stocking Jr. (Madison: University of Wisconsin Press, 1988), 206–259; and Shirley C. Strum, Donald G. Lindburg, and David Hamburg, eds., *The New Physical Anthropology* (Upper Saddle River, NJ: Prentice Hall, 1999).

20. Washburn, "Thinking about Race," 370.

21. Washburn, "Thinking about Race," 377.

22. Carleton S. Coon, *The Origin of Races* (1962; New York: Alfred A. Knopf, 1969), 658.

23. John P. Jackson Jr., *Science for Segregation: Race, Law, and the Case against Brown v. Board of Education* (New York: New York University Press, 2005), 99–178.

24. On the reception of Coon's book within the context of mid-twentieth century evolutionary theory and genetics, see Peter Collopy, "Race Relationships: Collegiality and Demarcation in Physical Anthropology," *Journal of the History of the Behavioral Sciences* 51, no. 3 (2015): 237–260.

25. John P. Jackson Jr., " 'In Ways Unacademical': The Reception of Carleton S. Coon's 'The Origin of Races,' " *Journal of the History of Biology* 34, no. 2 (2001): 247–285; Theodosius Dobzhansky, "Possibility That Homo Sapiens Evolved Independently 5 times Is Vanishingly Small," *Current Anthropology* 4, no. 4 (1963): 364–366; see also his "addendum" in Theodosius Dobzhansky, "Genetic Entities in Hominid Evolution," in *Classification and Human Evolution*, Viking Fund Publications in Anthropology 37, ed. Sherwood Washburn (New York: Wenner-Gren Foundation for Anthropological Research, 1963), 362.

26. A combination of radioactive dating and new fossil finds in Africa would eventually prove Coon wrong in detail as well as concept; see the last third of Marianne Sommer's *History Within: The Science, Culture, and Politics of Bones, Organisms, and Molecules* (Chicago: University of Chicago Press, 2016).

27. Sherwood Washburn, "The Study of Race," *American Anthropologist* 65, no. 3 (1963): 521–531.

28. Sherwood Washburn, "Evolution of a Teacher," *Annual Review of Anthropology* 12 (1983): 1–24.

29. Theodosius Dobzhansky, *Mankind Evolving: The Evolution of the Human Species* (New Haven, CT: Yale University Press, 1962).

30. See also Theodosius Dobzhansky, "Genetics and the Races of Man," in *Sexual Selection and the Descent of Man, 1871–1971*, ed. Bernard Campbell (Chicago: Aldine Publishing Company, 1972): 59–86, on 66.

31. Washburn, "Study of Race," 531; see also Lévi-Strauss, *Race and History* (Paris: UNESCO, 1952), 49: "We can see the diversity of human cultures behind us, around us, and before us. The

only demand that we can justly make . . . is that all the forms this diversity may take may be so many contributions to the fullness of all the others."

32. Robert Heine-Geldern, "Vanishing Cultures," *Scientific American*, May 1957, 39–45; Adrianna Link, "For the Benefit of Humankind: Urgent Anthropology and the Smithsonian's Center for the Study of Man, 1965–1968," in *Global Transformations in the Life Sciences, 1945–1980*, ed. Patrick Manning and Mat Savelli (Pittsburgh: University of Pittsburgh Press, forthcoming).

33. Colin Turnbull, *The Forest People* (New York: Simon & Schuster, 1961), vii. He followed this with a more scholarly monograph, *Wayward Servants* (Garden City, NY: Natural History Press, 1965).

34. Roy Richard Grinker, *In the Arms of Africa: The Life of Colin M. Turnbull* (New York: St. Martin's Press, 2000), 99.

35. Turnbull, *Forest People*, 115.

36. Turnbull, *Forest People*, quotes on 11, 12.

37. On long-standing romantic visions of forests in Western literature, see Robert Pogue Harrison, *Forests: The Shadow of Civilization* (Chicago: University of Chicago Press, 1993). These conceptions of the forest as a primeval human Eden would persist through Americans' experience with guerilla tactics in the Vietnam War, when the armed forces attempted to clear forest with Agent Orange.

38. Homer Bigart, "Harvard Expedition Discovers a Warrior Tribe in New Guinea," *New York Times*, 5 April 1961, 7.

39. Peter Kihss, "Governor's Son Is Missing off Coast of New Guinea," *New York Times*, 20 November 1969, 1.

40. "Gov. Rockefeller and His Wife Part; Will Get Divorce," *New York Times*, 18 November 1961, 1.

41. "Youth Was Warned on Trading Tactics," *New York Times*, 21 November 1961, 31.

42. Reuters, "Companion Is Rescued: Governor's Son Last Seen in Sea," *New York Times*, 21 November 1961, 1.

43. "Dutch Rejoin Navy Hunt," *New York Times*, 24 November 1961, 32.

44. "Rockefeller Due in City Tonight: Flying Back Convinced Only 'Miracle' Can Save Son," *New York Times*, 29 November 1961, 33.

45. "War Called on Account of Rain," display ad, *New York Times*, 25 September 1962, 76; Editors, "The Ancient World of a War-Torn Tribe," *Life*, 28 September 1962, 73–91.

46. Editors, "Ancient World," 78.

47. Robert Gardner and Karl G. Heider, *Gardens of War: Life and Death in the New Guinea Stone Age*, introduction by Margaret Mead (New York: Random House, 1968).

48. Gardner, foreword to *Gardens of War*, xi–xvi.

49. *Star Trek*, which ran on television from 1966 to 1969, appropriated this widespread anthropological principle as the Prime Directive, which Starfleet officers then violated on a regular basis.

50. Gardner and Heider, *Gardens of War*, 24.

51. Gardner and Heider, *Gardens of War*, 135.

52. Margaret Mead, introduction to *Gardens of War*, ix.

53. Conway Zirkle, "Human Evolution and Relativism," in *Relativism and the Study of Man*, ed. Helmut Schoek and James W. Wiggins (Princeton, NJ: J van Nostrand, 1961), 24.

54. Julian Steward, review of *An Anthropologist at Work: Writings of Ruth Benedict*, ed. Margaret Mead (Boston: Houghton Mifflin, 1959), *Science* 129, no. 3345 (1959): 322–323, quoted in Zirkle, "Human Evolution and Relativism," 43.

55. Claude Lévi-Strauss, *Les structures élémentaires de la parenté* (Paris: Presses universitaires de France, 1949); the revised second edition (1967) was translated into English in 1969.

56. Richard Lee and Irven DeVore, eds., *Man the Hunter* (Chicago: Aldine, 1968).

57. Theodosius Dobzhansky, *The Biological Basis of Human Freedom* (New York: Columbia University Press, 1956), 6. See also Alfred L. Kroeber, "Evolution, History, and Culture," in *Evolution after Darwin: II. The Evolution of Man: Man, Culture, Society*, ed. Sol Tax (Chicago: University of Chicago Press, 1960), 1–16.

58. On the importance of cultural relativism to anthropological theory after the Second World War and the lionization of Franz Boas, see Tracy Teslow, *Constructing Race: The Science of Bodies and Cultures in American Anthropology* (New York: Cambridge University Press, 2014).

59. George Gaylord Simpson, *The Meaning of Evolution* (New Haven, CT: Yale University Press, 1949), 337.

60. George Gaylord Simpson, *Tempo and Mode in Evolution* (New York: Columbia University Press, 1944).

61. Dobzhansky, *Biological Basis of Human Freedom*, 107–108; see also N. C. Tappan, "Primate Evolution and Human Behavior," in *Selected Papers of the 5th International Congress of Anthropological and Ethnological Sciences, Philadelphia 1956* (Philadelphia: University of Pennsylvania Press, 1960), 725–731.

62. Dobzhansky, *Biological Basis of Human Freedom*, 121.

63. Simpson, *Meaning of Evolution*, 310, quoted in Dobzhansky, *Biological Basis of Human Freedom*, 134.

64. Margaret Mead, "Cultural Determinants of Behavior," in *Behavior and Evolution*, ed. Anne Roe and George Gaylord Simpson (New Haven, CT: Yale University Press, 1958), 480–503, quote on 486. See also Julian Steward, "Evolutionary Principles and Social Types," in *Evolution after Darwin: II*, ed. Tax, 169–186, 183.

65. Mead, "Cultural Determinants of Behavior," 487.

66. From an experimental angle, see Rebecca Lemov, *World as Laboratory: Experiments with Mice, Mazes, and Men* (New York: Hill and Wang, 2005); Alexandra Rutherford, *Beyond the Box: B. F. Skinner's Technology of Behavior from Laboratory to Life, 1950s–1970s* (Toronto: University of Toronto Press, 2009); and Marga Vicedo, *The Nature and Nurture of Love: From Imprinting to Attachment in Cold War America* (Chicago: University of Chicago Press, 2013). On animal observations in the wild, see Richard W. Burkhardt Jr., *Patterns of Behavior: Konrad Lorenz, Niko Tinbergen, and the Founding of Ethology* (Chicago: University of Chicago Press, 2005); Gregory Radick, *The Simian Tongue: The Long Debate About Animal Language* (Chicago: University of Chicago Press, 2008); Georgina Montgomery, *Primates in the Real World: Escaping Primate Folklore and Creating Primate Science* (Charlottesville: University of Virginia Press, 2015); and Tania Munz, *The Dancing Bees: Karl von Frisch and the Discovery of the Honeybee Language* (Chicago: University of Chicago Press, 2016).

67. Marshall Sahlins, "Culture and Environment," in *Horizons of Anthropology*, ed. Tax, 132–147.

68. S. L. Washburn and Virginia Avis, "Evolution of Human Behavior," in *Behavior and Evolution*, ed. Simpson and Roe, 421–436, quote on 421.

69. Clifford Geertz, "The Transition to Humanity," in *Horizons of Anthropology*, ed. Tax, 44.

70. Stuart Altmann, ed., *Social Communication among Primates* (Chicago: University of Chicago Press, 1967), xi.

71. Joseph P. Kahn, "The Professor of Carnal Knowledge: Harvard's Irven DeVore Knows What Turns Students on to Biology: Sex Talk," *Boston Globe*, 26 March 1997, D1.

72. By the late 1970s, he had trained over half of the field primatologists active in the United States. As of 2006, a full 60 percent of field primatologists were academic descendants of Washburn: Elizabeth A. Kelley and Robert W. Sussman, "An Academic Genealogy on the History of American Field Primatologists," *American Journal of Physical Anthropology* 132, no. 3 (2007): 406–425.

73. John Russell Napier, *The Roots of Mankind* (Washington, DC: Smithsonian Institution Press, 1970), 173; Sherwood L. Washburn and Irven DeVore, "The Social Life of Baboons," *Scientific American* (June 1961): 62–71.

74. Washburn and Devore, "Social Life of Baboons," 71.

75. Irven DeVore and Richard B. Lee, preface to *Man the Hunter*, ed. Richard B. Lee and Irven DeVore (Chicago: Aldine, 1968), vii–ix.

76. Sherwood L. Washburn and C. S. Lancaster, "The Evolution of Hunting," in *Man the Hunter*, ed. Lee and DeVore, 293–303, quote on 293.

77. Washburn and Lancaster, "Evolution of Hunting," 303.

78. On models in the history of science, see Angela Creager, Elizabeth Lunbeck, and M. Norton Wise, eds., *Science without Laws: Model Systems, Cases, Exemplary Narratives* (Durham, NC: Duke University Press, 2007).

79. Richard B. Lee, "What Hunters Do for a Living, or, How to Make Out on Scarce Resources," in *Man the Hunter*, ed. Lee and DeVore, 30–48.

80. Richard B. Lee and Irven DeVore, "Problems in the Study of Hunters and Gatherers," in *Man the Hunter*, ed. Lee and DeVore, 1–12, on 7. At the same conference, Marshall Sahlins similarly argued that so-called primitive ways of life were in fact cultures of affluence ("Notes on the Original Affluent Society," in *Man the Hunter*, ed. Lee and DeVore, 85–89). This thesis has had a remarkable afterlife; see, e.g., Marshall Sahlins, *Stone-Age Economics* (New York: Routledge, 1972) and Murray Bookchin, *The Ecology of Freedom: The Emergence and Dissolution of Hierarchy* (Palo Alto, CA: Cheshire Books, 1982).

81. Lévi-Strauss, "Concept of Primitiveness," 349.

82. Irven DeVore, "Quest for the Roots of Society," in *The Marvels of Animal Behavior*, ed. Thomas B. Allen (Washington, DC: National Geographic Society, 1972): 393–409. Research in subsequent decades by Shirley Strum, Jeanne and Stuart Altmann, Barbara Smuts, and others would demonstrate that the social cohesion of baboon troops comes primarily from female-female interactions. See Shirley C. Strum and Linda M. Fedigan, eds., *Primate Encounters: Models of Science, Gender, and Society* (Chicago: University of Chicago Press, 2000).

83. Napier, *Roots of Mankind*, 134.

84. Susan Sperling, "The Troop Trope: Baboon Behavior as a Model System in the Postwar Period," in *Science without Laws*, ed. Creager, Lunbeck, and Wise, 73–89.

85. Dale Peterson, *Jane Goodall: The Woman Who Redefined Man* (Boston: Houghton Mifflin Company, 2006). On the National Geographic Society's tendency to fund young female researchers who added glamor to their magazine and television specials, see Haraway, "Women's Place Is in the Jungle," *Primate Visions*, 279–303.

86. Napier, *Roots of Mankind*, 120.

87. Virginia Morell, *Ancestral Passions: The Leakey Family and the Quest for Humankind's Beginnings* (New York: Touchstone, 1995), 242.

88. The Wilkie Foundation also funded Louis Leakey, Biruté Galdikas, George Schaller, Raymond Dart, Robert Ardrey, and many others. Later funding for research at Gombe came from the National Geographic Society, the Leakey Foundation, the William T. Grant Foundation, and the Jane Goodall Institute, as well as through grants to individuals to develop specific research projects from agencies, in the US, such as the National Science Foundation and the National Institutes of Health. As a psychiatrist, Hamburg was successful at securing NIMH grants to support both Gombe and Gombe West.

89. "An Oral History of Primatology, Jane Goodall, at Cambridge," University of Cambridge, Personal Histories Project, 28 April 2011, accessed 8 February 2018, https://sms.cam.ac.uk /media/1332629.

90. Robert Hinde, speaking at "An Oral History of Primatology, Jane Goodall, at Cambridge." On William H. Thorpe's position in British ethology, see Burkhardt, *Patterns of Behavior*, 301–345; and Radick, *Simian Tongue*, 251–267.

91. See, e.g., Jane van Lawick-Goodall and David A. Hamburg, "Recent Developments in the Study of Primate behavior," *Bulletin of the American Academy of Arts and Sciences* 27, no. 7 (1974): 36–48.

92. Virginia Morell, "Called 'Trimates,' Three Bold Women Shaped Their Field," *Science* 260, no. 5106 (1993): 420–425. On the long-standing practice of referring to research subjects with personal names, see Etienne Benson, "Naming the Ethological Subject," *Science in Context* 29, no. 1 (2016): 107–128.

93. Morell, *Ancestral Passions*, 243.

94. Kenneth MacLeish, "In Deepest Africa: A Startling Link to Mankind," *Life*, 24 November 1961, 84–98, on 90.

95. Guy Blanchard, dir., *Dr. Leakey and the Dawn of Man* (Washington, DC: National Geographic Specials, 1966), 60 min.

96. For example, Des Bartlett and Louis S. B. Leakey, "Finding the World's Earliest Man," *National Geographic*, September 1960, 420–435; and Louis S. B. Leakey and Hugo van Lawick, "Adventures in the Search for Man," *National Geographic*, January 1963, 132–152.

97. Harry L. Shapiro, "Louis S. B. Leakey, 1903–1972," *Saturday Review*, October 1972, 70–73.

98. L.S.B. Leakey and M. D. Leakey, "Recent Discoveries of Fossil Hominids in Tanganyika: At Olduvai and Near Lake Natron," *Nature* 202, no. 4927 (1964): 5–7; and L.S.B. Leakey, P. V. Tobias, and J. R. Napier, "A New Species of the Genus *Homo* from Olduvai Gorge," *Nature* 202, no. 4927 (1964): 7–9.

99. Ernst Mayr, "The Taxonomic Evaluation of Fossil Hominids," in *Classification and Human Evolution*, ed. Washburn, 332–346, 339.

100. Shapiro, "Louis S. B. Leakey."

101. George Gaylord Simpson, "The Evolutionary Concept of Humanity," in *Sexual Selection and the Descent of Man*, ed. Campbell, 17–39, quote on 24; Napier, *Roots of Mankind*, 209; A. Irving Hallowell, "Self, Society, and Culture in Phylogenetic Perspective," in *Evolution After Darwin: II*, ed. Tax, 309–372.

102. "Oldest Murder Unearthed," *Science News Letter* 79, no. 10 (March 1961): 147.

Chapter Three. Building Citizens

1. Patsy Asch, interview by author, 30 November 2011.

2. Jerome Bruner, interview by author, 4 October 2011.

3. Jerome S. Bruner, *Toward a Theory of Instruction* (Cambridge, MA: Belknap Press of Harvard University), 24. On the context of MACOS within broader science educational reforms, see Christopher Phillips, *New Math: A Political History* (Chicago: University of Chicago Press, 2015); and John Rudolph, *Scientists in the Classroom: The Cold War Reconstruction of American Science* (New York: Palgrave, 2002).

4. Jamie Cohen-Cole, *The Open Mind: Cold War Politics and the Sciences of Human Nature* (Chicago: University of Chicago Press, 2014).

5. Jerome S. Bruner, "Man: A Course of Study," *ESI Quarterly Report* (1965): 3–13, reprinted in Jerome S. Bruner, *In Search of Pedagogy*, Vol. 1 (New York: Routledge, 2006), quote on 90. See also Peter Dow's insider history of MACOS, *Schoolhouse Politics: Lessons from the Sputnik Era* (Cambridge, MA: Harvard University Press, 1991).

6. On the importance of film in anthropology and science, and the impossibility of achieving an unmediated vision, see Anna Grimshaw, *The Ethnographer's Eye: Ways of Seeing in Anthropology* (Cambridge: Cambridge University Press, 2001); and Gregg Mitman and Kelley Wilder, eds., *Documenting the World: Film, Photography, and the Scientific Record* (Chicago: University of Chicago Press, 2016).

7. Elting Morison, "The Gut Assumptions (Shores Still Dimly Seen but Touched with Rosy Fingered Dawn), January 4, 1963," Box 1, Folder 4, Peter B. Dow–Man: A Course of Study Records, Monroe C. Gutman Library, Graduate School of Education, Harvard University, Cambridge, MA (hereafter cited as MACOS Records).

8. Janet Whitla, interview by Peter Dow, 23 March 1976, Box 43, Folder 12, MACOS Records; John P. Ivens, "One Kind of Human Being: MACOS, the Human Sciences, and Governmentality," *European Education* 45, no. 3 (2013): 16–43.

9. Steven White (Director, Special Projects, ESI) to Douglas Oliver, 5 March 1962, Box 1, Folder 3, MACOS Records. ESI began as the corporate distribution arm of the Physical Science Study Committee. EDC, the ultimate publishers of the MACOS curriculum, grew rapidly and continues today as a nonprofit developer and publisher of educational materials.

10. Douglas Oliver, *Invitation to Anthropology* (Garden City, NY: Natural History Press, 1964), xii; Douglas Oliver, "An Ethnographer's Method for Formulating Descriptions of 'Social Structure,'" *American Anthropologist* 60, no. 5 (1958): 801–826.

11. Dow, *Schoolhouse Politics*, 58; see also Harry F. Wolcott, "The Middlemen of MACOS," *Anthropology and Education Quarterly* 38 (2008): 202–203; and Everett Mendelsohn, interview by Peter Dow, 31 January 1975, Box 42, Folder 25, MACOS Records.

12. Peter Burton Dow, "Innovation's Perils: An Account of the Origins, Development,

Implementation, and Public Reaction to 'Man: A Course of Study'" (EdD diss., Harvard University, 1979), quote on 127; Wolcott, "Middlemen of MACOS," 203; Kevin Smith, interview by Peter Dow, 17 October 1974, Box 43, Folder 8, MACOS Records.

13. "A Short History of the Social Studies Program [Spring 1965]," Box 3, Folder 15, MACOS Records.

14. Oliver, *Invitation to Anthropology*, 60–73.

15. Oliver, *Invitation to Anthropology*, 91.

16. John Marshall directed and produced a variety of filmic materials on the Ju/'hoansi, including *The Hunters* (Somerville, MA: Documentary Educational Resources, 1957), 72 min. and *Bitter Melons* (Somerville, MA: Documentary Educational Resources, 1971), 30 min. For Karl Heider's subsequent work on the Dani, see Robert Gardner and Karl G. Heider, *Gardens of War: Life and Death in the New Guinea Stone Age* (New York: Random House, 1969); and Karl G. Heider, dir., *Dani Films: Dani Sweet Potatoes* and *Dani Houses* (Somerville, MA: Documentary Educational Resources, 1974), 19 and 16 min. respectively.

17. On film at Cambridge in the 1960s, see Scott MacDonald, *American Ethnographic Film and Personal Documentary: The Cambridge Turn* (Berkeley: University of California Press, 2013); Jay Ruby, ed., *The Cinema of John Marshall* (Philadelphia: Harwood Academic Publishers, 1993).

18. Asen Balikci, *The Netsilik Eskimo* (Garden City, NY: Natural History Press, 1970). See also Mark Turin, "Interview with Asen Balikci," 12 January 2003, video file, published 11 November 2004, University of Cambridge DSpace, accessed 2 January 2017, http://www.dspace.cam .ac.uk/handle/1810/1467. Parts 2 and 3 discuss Balikci's experiences with the Netsilik and MACOS.

19. Turin, "Interview with Asen Balikci."

20. "Elementary Sequence—Outline of an Eskimo Unit for Grade One, June 12, 1963," Box 2, Folder 3, MACOS Records.

21. Asen Balikci, interview by Peter Dow, 17–18 October 1974, Box 42, Folder 6, MACOS Records. See Julian Steward, *Theory of Culture Change: The Methodology of Multilinear Evolution* (1955; Champaign: University of Illinois Press, 1972); and Asen Balikci, "The Netsilik Film Series," *Visual Anthropology* 22, no. 5 (2009): 384–392, on 386.

22. Julian Steward, "Causal Factors and Processes in the Evolution of Prefarming Societies," in *Man the Hunter*, ed. Richard B. Lee and Irven DeVore (Chicago: Aldine, 1968), 321–334; Robert A. Manners, "Julian Haynes Steward, 1902–1972," *American Anthropologist* 75, no. 3 (1973): 886–903.

23. Turin, "Interview with Asen Balikci." The narrative that follows is reconstructed from Turin, "Interview with Asen Balikci" and Balikci, interview by Dow.

24. Kevin Smith, October 17, 1974, interview by Dow, Box 38, Folder 8, MACOS Papers.

25. Balikci, interview by Dow.

26. Asen Balikci to Margaret Mead, 27 March 1964, Box D3, Folder 15: "Asen Balikci, 1962–1971," Timothy Asch Papers, National Anthropological Archives, Suitland, MD (hereafter cited as Asch Papers).

27. Asen Balikci to Margaret Mead, 28 September 1964, Box D3, Folder 15: "Asen Balikci, 1962–1971," Asch Papers.

28. Jerome Bruner, *The Process of Education* (1960; Cambridge, MA: Harvard University Press, 1977).

29. Jerrold Zacharias, interview by Peter Dow, 11 October and 15 November 1974, Box 43,

Folder 13, MACOS Records; "Social Studies Program, Spring Report 1965," Box 3, Folder 14, MACOS Records.

30. Blythe Clinchy, interview by Peter Dow, 31 October 1974; and Joseph Featherstone, interview by Peter Dow, 29 October 1976, Box 42, Folders 10 and 13, respectively, MACOS Records.

31. Dow, *Schoolhouse Politics*, 49–51, 112; "Appendix VIIa: The Human Past" and "Appendix VIIb: Jones—Sat. 23 June—Following Clinchy's presentation," Box 1, Folder 3, MACOS Records; "Broad Aims of the Unit on Human Origins" and "Unit on Evolution," Box 2, Folder 5, MACOS Records; "Meeting Regarding the Unit on Man's Place in Nature, October 15, 1964," Box 2, Folder 7, MACOS Records; "ESI Unit on Human Origins, December 1964," Box 3, Folder 3, MACOS Records.

32. Robert Adams, interview by Peter Dow, 31 October 1975, Box 42, Folder 2, MACOS Records. Adams resigned his position as part of the MACOS team shortly after this decision was made (Dow, *Schoolhouse Politics*, 70). Although Bruner deemphasized the Cities Unit in MACOS ("Minutes of the Social Studies Planning Committee, November 10, 1964," Box 4, Folder 14, MACOS Records), McAdams did produce a simpler version for the Anthropology Curriculum Study Project, based at the University of Chicago: M. Collier, "Official Reports: American Anthropological Association Council Meeting, Saturday, November 17, 1962 Chicago, Illinois—Report on the Anthropology Curriculum Study Project," *American Anthropologist* 65, no. 3 (1963): 677–678.

33. Jerome S. Bruner, *Toward a Theory of Instruction* (Cambridge, MA: Harvard University Press, 1966), 72.

34. Balikci, interview by Dow. The transcript says "crevice" (meaning, a crack in rock), but it seems more likely he and his sled fell into a crack in the snow and ice, a crevasse.

35. Father Guy Mary-Rousselière's master's thesis on local string games was eventually published as "Les Jeux de Ficelle des Arviligjuarmiut," *Bulletin Musée Nationaux du Canada* 233, no. 88 (1969).

36. MacDonald, *American Ethnographic Film*, 66.

37. John Marshall, "Filming and Learning," in *The Cinema of John Marshall*, ed. Jay Ruby (1993; New York: Routledge, 2012), 1–134, quote on 39.

38. Timothy Asch to Margaret Mead, 25 March 1960, Box C45, Folder: 1960 And–Azy, Margaret Mead Papers and South Pacific Ethnographic Archives, 1838–1996, MSS32441, Library of Congress, Washington, DC (hereafter cited as Mead Papers).

39. Timothy Asch to Margaret Mead, 8 January 1965, Box C72, Folder: 1965, An–Ay, Mead Papers. Mead responded generously: Mead to Asch, 26 January 1965, Box D2, Folder 17: "Asch Timothy, 1959–1965, 1972–1978," Mead Papers.

40. Adrianna Link, "Documenting Human Nature: E. Richard Sorenson and the National Anthropological Film Center, 1970–1984," *Journal of the History of the Behavioral Science* 52 (2016): 371–391. On the entanglement of media and ethnography earlier in the century, see Brian Hochman, *Savage Preservation: The Ethnographic Origins of Modern Media Technology* (Minneapolis: University of Minnesota Press, 2014).

41. Elizabeth Marshall Thomas, *The Harmless People* (London: Secker & Warburg, 1959). She would later publish the immensely popular *The Hidden Life of Dogs* (Boston: Houghton Mifflin Company, 1993), which spent nearly ten months on the *New York Times* best-seller list.

42. Thomas, *Harmless People*, 6, 7.

43. Thomas, *Harmless People*, 9.

44. Timothy Asch, *Dodoth Morning* (Documentary Education Resources, 1961), 17 min. The film starts with still photographs overlaid with explanatory text, a style he would later abandon in favor of less didactic techniques. Elizabeth Marshall Thomas, *Warrior Herdsman*, with photographs by Timothy Asch (New York: Knopf, 1965).

45. Christine Ward Gailey, "Richard Lee: The Politics, Art and Science of Anthropology," *Anthropologica* 45, no. 1 (2003): 19–26.

46. Patsy Asch, "Proposed Classroom Trials for the Bushman Unit, Sept. 1967–June 1968," April 1967, Series 3: Teaching Materials, Box 6, Folder: Preliminary Lesson Plans !Kung Bushman Unit, Spring, 1967, Asch Papers.

47. Asen Balikci to Margaret Mead, 28 December 1965, Box D3, Folder 15 Asen Balikci 1962–1971, Mead Papers.

48. On contemporary conventions of wildlife filmography, see Gregg Mitman, *Reel Nature: America's Romance with Wildlife on Film* (Cambridge, MA: Harvard University Press, 1999).

49. Quentin Brown, dir., *Netsilik Eskimo Series* (Somerville, MA: Documentary Educational Resources, 1970); Nancy C. Lutkehaus, "Man: A Course of Study: Situating Tim Asch's Pedagogy and Ethnographic Films," in *Timothy Asch and Ethnographic Film*, ed. E. D. Lewis (New York: Routledge, 2003): 57–73.

50. MACOS, *Using MEDIA, Trial Teaching Edition* (Cambridge, MA: Education Development Center, 1967). ("MEDIA" is the abbreviation for Making Education Dramatic, Interesting and Accurate.) Series 3: Teaching Materials, Box 8, Folder: Using MEDIA, Asch Papers.

51. Most of the MACOS booklets, slide shows, and teaching guides are available online through www.macosonline.org/course/, accessed 16 May 2012.

52. Irven DeVore, interview by Peter Dow, 20 October 1975, Box 41, Folder 12, MACOS Records.

53. Irven Devore, interview by Dow.

54. Young would later work with John Marshall, too, who also attributes to Young an improved ability to film personal interactions: Marshall, "Filming and Learning," 43.

55. Turin, "Interview with Asen Balikci."

56. Asen Balikci and Robert M. Young, *The Eskimo: Fight for Life* (Washington, DC: Education Development Center, 1970), 60 min.

57. Turin, "Interview with Asen Balikci."

58. "Conference on the Eskimo Unit, January 20 and 21, 1964," Box 2, Folder 3, MACOS Records.

59. "Conference on the Eskimo Unit, January 20 and 21, 1964," Box 2, Folder 3, MACOS Records.

60. MACOS, *The Many Lives of Kiviok* (Cambridge, MA: Education Development Center, 1967); Edward Fields, *The Songs and Stories of the Netsilik Eskimos* (Washington, DC: Education Development Center, 1970); Carter Wilson, *On Firm Ice* (Cambridge, Mass: Education Development Center, 1967).

61. Jay Ruby, "Anthropology as a Subversive Art: A Review of *Through These Eyes*," *American Anthropologist* 107 (2005): 684–687. For Balicki's defense of using filmic recreations for educational purposes, see Asen Balikci, "Reconstructing Cultures on Film," in *Principles of Visual Anthropology*, ed. Paul Hockings (The Hague: Mouton Publishers, 1975), 191–200.

62. Evans Clinchy to Asen Balikci, 19 September 1963, Box 4, Folder 1, MACOS Records.

63. Turin, "Interview with Asen Balikci."

64. A film along these lines was later produced with Asen Balikci's guidance: Gilles Blais, *Yesterday, Today: The Netsilik Eskimo* (Newton, MA: Education Development Center, 1972), 57 min. See also DeVore, interview by Dow.

65. Charles Laird, dir., *Through These Eyes* (Watertown, MA: Documentary Educational Resources, 2003), 55 min.

66. Asch, "Proposed Classroom Trials for the Bushman Unit."

67. MACOS, Selections from Field Notes, 1959 March–August, Irven DeVore Anthropologist (Washington, DC: Curriculum Development Associates, 1970).

68. Patsy Asch, phone interview by author, 30 November 2011; see also Tim and Patsy Asch, interview by Peter Dow, 23 November 1975, Box 42, Folder 4, p. 27, MACOS Records. The Dani films were dropped very early in the planning process.

69. DeVore, interview by Dow.

70. MACOS, *Baboon Communication* (Washington, DC: Curriculum Development Associates, 1970); Chelsea House, *Hello Darwin!* (New York: Association-Sterling Films, 1969), 50 min.

71. This view resonated strongly with Noam Chomsky's *Language and Mind* (New York: Harcourt, Brace & World, 1968).

72. MACOS, *Chimpanzees* (Washington, DC: Curriculum Development Associates, 1970).

73. MACOS, *Herring Gulls* (Washington, DC: Curriculum Development Associates, 1970), 22–23.

74. Niko Tinbergen, *The Herring Gull's World: A Study in the Social Behaviour of Birds* (London: Collins, 1953); Richard W. Burkhardt Jr., "Ethology's Traveling Facts," in *How Well Do Facts Travel? The Dissemination of Reliable Knowledge*, ed. Peter Howett and Mary Morgan (Cambridge: Cambridge University Press, 2011), 195–222.

75. MACOS, *Four: Herring Gulls* (Cambridge, MA: Education Development Center, 1970), 5; Louis Darling, *The Gull's Way* (New York: Morrow, 1965).

76. Chelsea House, *What Makes Man Human* (New York: Association-Sterling Films, 1968), 16 min.

77. Tinbergen, *Herring Gull's World*, 73.

78. See, for example, Jerome Bruner's revised introduction to *The Process of Education* (1960; Cambridge, MA: Harvard University Press, 1977), vii–xvi.

79. Dow, *Schoolhouse Politics*, 122.

80. Tim Asch refused to give up on the Ju/'hoansi material, believing he could adapt it for use with older students; Asen Balikci to "Teachers," on MIT letterhead, Box 3, Folder 12: "Teaching Materials-Film and Teaching," Asch Papers.

81. MACOS, *Three: Introductory Lessons: Salmon* (Cambridge, MA: Education Development Center, 1970).

82. Mendelsohn, interview by Dow; Kathy Sylva, interview by Peter Dow, 11 November 1974, Box 43, Folder 10, MACOS Records. For a complete elaboration of MACOS personnel, see Dow, *Schoolhouse Politics*.

83. Chelsea House, *What Makes Man Human*.

84. Chelsea House, *Hello Darwin!*

85. Chelsea House, *Hello Darwin!* The students also asked whether they thought they would

learn more in a different school, and almost all the students said yes—because other schools had gymnasiums or working heat and no broken windows in the winter. For more on the busing controversy in Boston, see J. Anthony Lukas, *Common Ground: A Turbulent Decade in the Lives of Three American Families* (New York: Random House, 1985).

86. For example, Nancy Hicks, "10-Year-Olds Encouraged to Think Big," *New York Times*, 16 November 1969, 75; Ruth Moss, "New Course Teaches Young Chicagoans about Shaping Man's Humanity," *Chicago Tribune*, 4 January 1970, I6; Maya Pines, "Jerome Bruner Maintains Infants Are Smarter Than Anybody Thinks," *New York Times*, 29 November 1970, 243; "Course on 'Man' Found Outstanding Success," *Hartford Courant*, 30 May 1971, 5B1.

87. Ruby, "Anthropology as a Subversive Art," 685.

88. All quotes this paragraph from "Teaching Man to Children," *Time*, 19 January 1970, 54.

89. Jerome Bruner, interview by author, 4 October 2011.

Part Two. Naturalizing Violence

Robert Ardrey, *African Genesis: A Personal Investigation into the Animal Origins and Nature of Man* (New York: Atheneum, 1961), quote on 348.

1. Robert Ardrey, *The Territorial Imperative: A Personal Inquiry into the Animal Origins of Property and Nations* (New York: Atheneum, 1966); Harry Frank Guggenheim (HFG) to Henry Allen Moe, Folder: Guggenheim, Harry Frank #5, Mss.B.M722, Henry Allen Moe Papers, American Philosophical Society, Philadelphia (hereafter cited as Moe Papers).

2. HFG to Moe, 10 November 1967, Folder: Guggenheim, Harry Frank #5, Moe Papers. Guggenheim referred to Moe as his "old and valued friend" in a letter to Paul Fitts, 21 January 1964, Folder: Guggenheim, Harry Frank #3, Moe Papers.

3. HFG to Moe, 19 December 1967, Folder: Guggenheim, Harry Frank #5, Moe Papers.

4. HFG to "All Hands—Man's Relation to Man Project," 16 December 1968, Folder: Guggenheim, Harry Frank, Foundation #2, Moe Papers.

5. "30 March 1965, State of New York, Department of State, James E. Allen Jr., Commissioner of Education of the State of New York," Folder: Guggenheim, Harry Frank, Foundation #1, Moe Papers.

6. HFG to Moe, 8 January 1964, Folder: Guggenheim, Harry Frank #2, Moe Papers.

7. Nadine Weidman, "Popularizing the Ancestry of Man: Robert Ardrey and the Killer Instinct," *Isis* 102, no. 2 (2011): 269–299.

8. Konrad Lorenz, *King's Solomon's Ring: New Light on Animal Ways*, trans. Marjorie Kerr Wilson, foreword by Julian Huxley (New York: Crowell, 1952); Konrad Lorenz, *On Aggression*, trans. Marjorie Kerr Wilson (New York: Harcourt, Brace & World, 1966).

9. On the friendship between Lorenz and Tinbergen (despite considerable political and scientific differences), see Richard W. Burkhardt Jr.'s dual biography, *Patterns of Behavior: Konrad Lorenz, Niko Tinbergen, and the Founding of Ethology* (Chicago: University of Chicago Press, 2005). While being held a prisoner of war by the Russians, Lorenz wrote a manuscript culminating in an extended discussion of human evolution. This manuscript circulated among his colleagues for a number of years, who published arguments and snippets in different form. The original manuscript was subsequently lost until his daughter discovered it among his papers after his death: *The Natural Science of the Human Species: An Introduction to Comparative Behav-*

ioral Research: The "Russian Manuscript" (1944–1948), ed. Agnes von Cranach, trans. Robert D. Martin (1992; Cambridge, MA: MIT Press, 1996).

10. Desmond Morris, *The Naked Ape: A Zoologist's Study of the Human Animal* (New York: McGraw-Hill, 1967).

11. Under Patterson and Guggenheim, *Newsday* was a successful conservative suburban daily newspaper (Monday–Saturday) in a tabloid format; Lee Smith, "The Battle for Sunday," *New York Magazine*, 25 October 1971, 34–39.

12. HFG to Moe, 19 April 1965, Folder: Guggenheim, Harry Frank #2, Moe Papers.

13. For example, Henry Allen Moe, "The Shortage of Scientific Personnel," *Science* 105, no. 2721 (1947): 195–198; and Moe, "The Power of Freedom," *American Association of University Professors* 37 (1951): 462–475.

14. HFG to Moe, 19 April 1965, Folder: Guggenheim, Harry Frank #2, Moe Papers.

15. Pendray also wrote popular books (including several science fiction novels under the pseudonym Gawain Edwards) and received a John Simon Guggenheim Fellowship in 1964.

16. G. Edward Pendray, "Summary of Progress, Man's Relation to Man Project," 15 April 1961 to 1 August 1965, Folder: Guggenheim, Harry Frank Foundation: Lindbergh, Charles A., Moe Papers.

17. James Gilbert, *Redeeming Culture: American Religion in an Age of Science* (Chicago: University of Chicago Press, 1997); David Kaiser, *How the Hippies Saved Physics: Science, Counterculture, and the Quantum Revival* (New York: W. W. Norton & Company, 2011); Matthew Wisnioski, *Engineers for Change: Competing Visions of Technology in 1960s America* (Cambridge, MA: MIT Press, 2012); Patrick McCray, *Visioneers: How a Group of Elite Scientists Pursued Space Colonies, Nanotechnologies, and a Limitless Future* (Princeton, NJ: Princeton University Press, 2013); Michael Gordin, *Pseudoscience Wars: Immanuel Velikovsky and the Birth of the Modern Fringe* (Chicago: University of Chicago Press, 2012); Kelly Moore, *Disrupting Science: Social Movements, American Scientists, and the Politics of the Military, 1945–1975* (Princeton, NJ: Princeton University Press, 2009). David Kaiser and Patrick McCray, eds., *Groovy Science: Knowledge, Innovation, and American Counterculture* (Chicago: University of Chicago Press, 2016).

18. Beth Bailey, *Sex in the Heartland* (Cambridge, MA: Harvard University Press, 1999); Elizabeth Fraterrigo, Playboy *and the Making of the Good Life in Modern America* (New York: Oxford University Press, 2009).

19. Elaine Tyler May, *Homeward Bound: American Families in the Cold War Era* (New York, 1988); Stephanie Coontz, *The Way We Never Were: American Families and the Nostalgia Trap* (New York: Oxford University Press, 1992); Joanne Meyerowitz, ed., *Not June Cleaver: Women and Gender in Postwar America, 1945–1960* (Philadelphia: Temple University Press, 1994); Wendy Kline, *Building a Better Race: Gender Sexuality, and Eugenics from the Turn of the Century to the Baby Boom* (Berkeley: University of California Press, 2001); James Gilbert, *Men in the Middle: Searching for Masculinity in the 1950s* (Chicago: University of Chicago Press, 2005); Alexandra Stern, *Eugenic Nation: Faults and Frontiers of Better Breeding in Modern America* (Berkeley: University of California Press, 2005).

20. William H. Masters and Virginia E. Johnson, *Human Sexual Response* (Boston: Little, Brown, 1966).

21. Erika Lorraine Milam, "Men in Groups: Aggression and Anthropology, 1966–1984," *Scientific Masculinities*, ed. Erika Lorraine Milam and Robert A. Nye, *Osiris* 30 (2015): 66–88.

22. HFG to General Charles A. Lindbergh, 10 May 1966, Folder: Guggenheim, Harry Frank, Foundation: Lindbergh, Charles A., Moe Papers.

23. Lindbergh to HFG, 26 April 1966, Folder: Guggenheim, Harry Frank, Foundation: Lindbergh, Charles A., Moe Papers.

24. HFG to Lindbergh, 10 May 1966, Folder: Guggenheim, Harry Frank, Foundation: Lindbergh, Charles A., Moe Papers.

25. Lindbergh to HFG, 29 May 1966, Folder: Guggenheim, Harry Frank, Foundation #27, Moe Papers.

26. HFG to Lindbergh, 1 June 1966, Folder: Guggenheim, Harry Frank, Foundation #27, Moe Papers.

27. HFG to Doolittle, 6 July 1966, Folder: Guggenheim, Harry Frank #6; HFG to Moe, 10 November 1967, Folder: Guggenheim, Harry Frank #5; HFG to Moe, 19 December 1967, Folder: Guggenheim, Harry Frank #5, all in Moe Papers. Based on this correspondence, it appears likely that Guggenheim had not yet finished *Territorial Imperative* at the time he penned his editorial.

28. [Harry F. Guggenheim], "The Mark of Cain," *Newsday*, 25 September 1967, 33.

29. Ardrey to HFG, 18 October 1967, Folder: Guggenheim, Harry Frank #5, Moe Papers.

30. HFG to Ardrey, 30 and 31 October 1967, Folder: Guggenheim, Harry Frank #5, Moe Papers.

31. Ardrey to HFG, 4 November 1967, Folder: Guggenheim, Harry Frank #5, Moe Papers.

32. HFG to Ardrey, 7 November 1967, Folder: Guggenheim, Harry Frank #5, Moe Papers. Ardrey had received a fellowship from the John Simon Guggenheim Memorial Foundation in 1937 for the project Creative Arts—Drama & Performance Art.

33. HFG to Moe, 10 November 1967, Folder: Guggenheim, Harry Frank #5, Moe Papers; see also Charles R. Lindbergh to HFG, 24 April 1968, Folder: Guggenheim, Harry Frank, Foundation #13, Moe Papers.

34. G. Edward Pendray to HFG, 10 January 1968, Folder: Guggenheim, Harry Frank #5, Moe Papers.

35. HFG to Pendray, 8 January 1968, Folder: Guggenheim, Harry Frank #5, Moe Papers. Lindbergh's comments should be understood in the context of his deep sympathy with the American eugenics movement. See, e.g., Andrés Horacio Reggiani, "Charles Lindbergh and the Institute of Man," in *God's Eugenicist: Alexis Carrel and the Sociobiology of Decline* (New York: Berghahn Books, 2007), 85–102.

36. For example, M. F. Ashley Montagu, ed., *Culture: Man's Adaptive Dimension* (Oxford, 1968).

37. As a cultural anthropologist, Montagu publicly dismissed Ardrey's arguments but until the later 1960s remained sympathetic to the man. See M. F. Ashley Montagu, ed., *Man and Aggression* (New York: Oxford University Press, 1968); Montagu, *The Nature of Human Aggression* (New York: Oxford University Press, 1976).

38. Ardrey to HFG, 17 January 1968, Folder: Guggenheim, Harry Frank, Foundation: Ardrey, Robert, Moe Papers.

39. Memo to Dr. Moe, 13 March 1968, Folder: Guggenheim, Harry Frank, Foundation #15, Moe Papers.

40. Edward Pendray, "Summary of Meeting: The Man's Relation to Man Project," 19 March 1968, Folder: Guggenheim, Harry Frank, Foundation, 1968 #41, Moe Papers.

41. Pendray, "Summary of Meeting." No one considered Ardrey: he lived happily in Rome, possessed no administrative experience, and lacked a PhD.

42. General Charles R. Lindbergh to HFG, 24 April 1968 (copied/sent to Moe: 13 May 13 1968), Folder: Guggenheim, Harry Frank, Foundation #13, Moe Papers.

43. Doolittle to HFG, 24 October 1968, Folder: Guggenheim, Harry Frank, Foundation #14, Moe Papers.

44. Ardrey to Moe, 24 April 1968, Folder: Guggenheim, Harry Frank, Foundation: Ardrey, Robert, Moe Papers.

45. HFG to Ardrey, 17 January 1969, Folder: Guggenheim, Harry Frank, Foundation #2, Moe Papers.

46. Ardrey to Morris, 31 January 1969, Box 1, Folder 26, Robert Ardrey Papers, 1955–1980, Special Collections and University Archives, Rutgers University Libraries, New Brunswick, NJ (hereafter cited as Ardrey Papers).

47. Truman Capote's *In Cold Blood* (New York: Random House) was also published in 1966, having been serialized in the *New Yorker* starting in September 1965.

48. Author interviews with Adrienne Zihlman (24 October 2011), Richard Wrangham (7 May 2012), Anne Pusey (23 May 2012), and Sarah Blaffer Hrdy (27 October 2011). Lionel Tiger (10 November 2011) and Robin Fox (8 November 2011) were quite influenced by these books, as detailed in Chapter 8. Irven DeVore, too, remembered the kerfuffle these books caused in professional circles: Devore, interview by author, 6 August 2011. See also Tiger's enthusiasm for *African Genesis* in his letter to Ardrey, 21 September 1962, Box 1, Folder 8, Ardrey Papers. Interviewees differed as to which book(s) they read and discussed with their peers, but all remembered at least one.

49. I take the concept of "public science" here from Bruce Lewenstein, "The Arrogance of 'Pop Science,'" *Scientist*, 13 July 1987, 12; Katherine Pandora, "Popular Science in National and Transnational Perspective: Suggestions from the American Context," *Isis* 100 (2009): 346–358; and James Secord, "Knowledge in Transit," *Isis* 95, no. 4 (2004): 654–672.

Chapter Four. Cain's Children

1. Robert Ardrey, *African Genesis: A Personal Investigation into the Animal Origins and Nature of Man* (New York: Atheneum, 1961): 22. At the end of the book, he challenged his imagined reader as if the preceding pages constituted one of these Sunday-school discussions: "While I indulge myself, he may feel free to hide behind a door in panic, to grope for another chair and come after me, or if such is his nature, to get himself as rapidly as possible out of the church basement" (347).

2. Ardrey lived in at least two apartments in Rome that fit this description: Piazza dei Mercanti 25 and, later, Via Garibaldi 89: Robert Ardrey Folders, Mss.B.M722, Henry Allen Moe Papers, American Philosophical Society, Philadelphia (hereafter cited as Moe Papers).

3. Ardrey, *African Genesis*, 9.

4. "Aggression and Violence in Man: Dialogue between Dr. L.S.B. Leakey and Mr. Robert Ardrey," *Munger Africana Library Notes* no. 9 (1971): 4.

5. Richard Rhodes, "Goodbye to Darkest Africa," *Playboy*, November 1973, 120; quote on 120, 122.

6. For scholarship on the history of stereotypes about Africa as a "timeless" "primitive" land,

see David Lowenthal, "Past Time, Present Places: Landscape and Memory," *Geographical Review* 65 (1975): 1–36; Melanie G. Wiber, *Erect Men / Undulating Women: The Visual Imagery of Gender, "Race" and Progress in Reconstructive Illustrations of Human Evolution* (Waterloo, ON: Wilfred Laurier University Press, 1977); Curtis A. Keim, *Mistaking Africa: Curiosities and Inventions of the American Mind* (Boulder, CO: Westview Press, 1999); Eric Hobsbawm and Terence Ranger, eds., *The Invention of Tradition* (New York: Cambridge University Press, 1983); Jeremy Rich, *Missing Links: The African and American Worlds of R. L. Garner, Primate Collector* (Athens: University of Georgia Press, 2012).

7. Rhodes, "Goodbye to Darkest Africa," 122.

8. Donna Haraway, "Teddy Bear Patriarchy: Taxidermy in the Garden of Eden, New York City, 1908–36," in *Primate Visions: Gender, Race, and Nature in the World of Modern Science* (New York: Routledge, 1989), 26–58.

9. Robert Ruark, "Far-Out Safari," *Playboy*, March 1965, 172, quote on 173; See also Gregg Mitman, "Cinematic Nature: Hollywood Technology, Popular Culture, and the American Museum of Natural History," *Isis* 84, no. 4 (1993): 637–661.

10. Robert Ardrey, "The Sweetest Boy in All the World: A Short Story about Kenya," *Reporter*, 7 April 1955, 36–39; "A Slight (Archaic) Case of Murder," *Reporter*, 5 May 1955, 34–36; "The Eagles of Swaziland," *Reporter*, 16 June 1955, 31–34; "What's Wrong with Gold? Mr. Marbly Knows," *Reporter*, 14 July 1955, 40–44; and "South Africa: A Personal Report," *Reporter*, 27 November 1958, 22–27.

11. George Hunt, "Provocateur in Anthropology," *Life*, 26 August 1966, 3; "Books: Born in Violence," review of *African Genesis*, by Robert Ardrey, *Time*, 15 December 1961.

12. Ardrey, *African Genesis*, 22; Ardrey described his first meeting with Dart in slightly less laudatory terms in "A Slight (Archaic) Case of Murder," *Reporter*, 5 May 1955, 34–36.

13. Ardrey to Ashley Montagu, 30 July 1961, Series 1, Box 2, Folder: Ardrey, Robert, American Philosophical Archives, Philadelphia, PA, Mss. Coll. 109, Ashley Montagu Papers (hereafter, cited as Montagu Papers).

14. Ardrey to Montagu, 30 July 1961, Montagu Papers.

15. "Vexed Problem of Colour: Woman Who Had Jar of Sherry, Professor Raymond Dart's Evidence," newspaper clipping, University of Witwatersrand Archive, Correspondence Bundle 1, Raymond Dart Papers, as cited by Adrian Young, *Paleoanthropology and Empire: European Colonialism and the Evolution of a Science, 1884–1945* (Ohio State University undergraduate honors thesis, 2008), 161. I'm indebted to Adrian for several illuminating conversations about Dart's politics and history. For a thorough account of Dart's political views, see Christa Kuljian, *Darwin's Hunch: Science, Race, and the Search for Human Origins* (Auckland Park, South Africa: Jacana, 2016), 49–75.

16. Alan G. Morris, "Biological Anthropology at the Southern Tip of Africa," *Current Anthropology* 53, no. S5 (2012): S152–S160. Morris argues that the scientific underpinning of apartheid in South Africa was based on ethnology, not physical anthropology. According to the historian Saul Dubow, however, Dart's research in physical anthropology was later used to shore up the scientific racism on which apartheid had been founded, see, e.g., "Human Origins, Race Typology and the Other Raymond Dart," *African Studies* 55 (1996): 1–30; and *Illicit Union: Scientific Racism in Modern South Africa* (Cambridge: Cambridge University Press, 1995).

17. Raymond Dart, "Rhodesian Engravers, Painters and Pigment Miners of the Fifth Millen-

nium B.C.," *South African Archeological Bulletin* 8, no. 32 (1953): 91–96; Dart, "Rock Engravings in Southern Africa and Some Clues as to Their Significance," *South African Journal of Science* 28 (1931): 475–486; Dart, "The Ancient Iron-Smelting Cavern at Mumbwa," *Transactions of the Royal Society of South Africa* 19 (1931): 379–427; Henrika Kuklick, "Contested Monuments: The Politics of Archaeology in Southern Africa," in *Colonial Situations*, ed. George W. Stocking Jr. (Madison: University of Wisconsin Press, 1991), 135–169.

18. Questions about Dart's interpretation of the site began rather quickly; see J. Desmond Clark, "The Chifubwa Stream Rock Shelter, Solwezi, Northern Rhodesia," *South African Archeological Bulletin* 13, no. 49 (1958): 21–24.

19. On the history of paleoanthropology in South Africa (including Dart and his theories), see Kuljian, *Darwin's Hunch*.

20. Raymond Dart, "The Predatory Transition from Ape to Man," *International Anthropological and Linguistic Review* 1 (1953): 201–218, quote on 207–208.

21. Dart, "Predatory Transition from Ape to Man," 204.

22. Ardrey, *African Genesis*, 315.

23. Ardrey, *African Genesis*, 29.

24. Ardrey, *African Genesis*, 357.

25. Published sporadically by the International Anthropological and Linguistic Circle in Miami, Florida, from 1953 to 1958. De Montigny appears to be remembered today primarily as part of a debate about Atlantis.

26. Ardrey, *African Genesis*, 29; Alan H. Kelso de Montigny, "Editor's Note to the Above," *International Anthropological and Linguistic Review* 1 (1953): 218–219, quote on 218. Italics in original.

27. De Montigny, "Editor's Note," 219.

28. Ardrey, *African Genesis*, 31.

29. Ardrey, *African Genesis*, see esp. 13–15.

30. Ardrey, *African Genesis*, 179.

31. Ardrey, *African Genesis*, 184.

32. Ardrey, *African Genesis*, 19.

33. Ardrey, *Territorial Imperative*, 22. On the complicated relationship of field, farm, and laboratory studies of animal behavior, see Richard W. Burkhardt Jr., *Patterns of Behavior: Konrad Lorenz, Niko Tinbergen, and the Founding of Ethology* (Chicago: University of Chicago Press, 2005).

34. On the complicated relationship between Freudian theories and the "biological revolution" in psychiatry, see Jonathan Metzl, *Prozac on the Couch: Prescribing Gender in the Era of Wonder Drugs* (Durham, NC: Duke University Press, 2003). See also Ardrey's favorable review of Anthony Storr, *Human Aggression* (New York: Penguin Books, 1968) for an account of where he thought psychoanalysts should be directing their attention: "Accomplices to Violence," *New York Times*, 14 July 1968, BR1.

35. He expanded on his animosity toward Freud in *The Territorial Imperative*, 162–163, 292.

36. Ardrey, *African Genesis*, 147. It is additionally worth noting that in this passage Ardrey appealed to *reason* as the basis of humanity. He was remarkably inconsistent in this assertion. In a 1969 letter to Ashley Montagu, Ardrey wrote: "You conceive of the dignity of man as his teachability. I do not. To me man loses all dignity if he is a vulnerable material subject to end-

less manipulation. To me his dignity rests on the three billion years of evolution that brought us forth. To me the iron of man lies in his natural history—an iron malleable, yes—but still iron." Ardrey to Montagu, 29 June 1969, Series 1, Box 2, Folder: Ardrey, Robert, Montagu Papers.

37. . Ardrey, *African Genesis*, 158. Among social scientists, Ardrey would come under fire for his use of biological principles to shore up gender stereotypes, a point to which we will return in Part Three.

38. Ardrey, *African Genesis*, 158.

39. Kenneth MacLeish, "In Deepest Africa: A Startling Link to Mankind," *Life*, 24 November 1961, 85–97, quote on 97.

40. MacLeish, "Deepest Africa," 93.

41. Louis S. B. Leakey, "Exploring 1,750,000 Years into Man's Past," *National Geographic*, October 1961, 564–589.

42. Ardrey to David Pilbeam, 14 January 1969, Box 1, Folder 26, MC 190, Ardrey Papers.

43. Ardrey-Leakey correspondence, Box 2, Folder 8, Ardrey Papers.

44. George Gaylord Simpson to Mr. Simon Michael Bessie, 22 May 1961, Box 1, Folder 7, Ardrey Papers. Simpson was horrified to discover that Bessie had forwarded his letter to Ardrey and called this a "flagrant breach of confidence," as quoted in Ardrey to Simpson, 14 November 1966, Box 1, Folder 16, Ardrey Papers.

45. Clarence Ray Carpenter to Bessie, 29 May 1961, Box 1, Folder 7, Ardrey Papers. When Ardrey responded to Carpenter, this sparked their later friendship.

46. Carleton Coon to Bessie, 30 August 1961, Box 1, Folder 7, Ardrey Papers.

47. Montagu to Ardrey, 19 July 1961, Box 1, Folder 7, Ardrey Papers.

48. Ardrey to Adriaan Kortlandt, 1 December 1962, Box 1, Folder 8, Ardrey Papers.

49. Sherwood Washburn to Bessie, 26 June 1961, Box 1, Folder 7, Ardrey Papers.

50. A review in an academic science journal, *Man*, notes at the beginning that it was hard to find a scientist to review the book and so the current reviewer had no such training and was unfit to comment on the validity of the theses presented: John Layard, [untitled], *Man* 62 (September 1962): 141. *Science News Letter* received a copy of the book for review, which it did not do, merely posting it among the many Books of the Week received and noting in a single sentence: "Controversial theory of man's origin and human behavior in the light of recent evidence": *Science News Letter* 80, no. 24 (December 1961): 388. Other than an occasional oblique reference, the academy remained silent.

51. Sheldon Weeks, "Not Much Higher Than the Apes," *Africa Today* 8, no 10 (1961): 18–19. According to the anthropologist Robin Fox, Ardrey would later recall, "My first book got nothing but lousy reviews. I only took off with the second. Now can you tell me the name of any of the lousy reviewers? . . . Nor can anyone I ask." Robin Fox, *Participant Observer* (New Brunswick, NJ: Transaction Publishers, 2004): 522.

52. Ardrey, *African Genesis*, 16.

53. "Born in Violence," *Time*, 15 December 1961, 88, quote from Ardrey, *African Genesis*, 333.

54. J. E. Havel, [untitled], *American Anthropologist*, n.s., 66, no. 2 (1964): 435–436, quote on 435.

55. Ardrey, *Territorial Imperative*, see especially 32–33.

56. Ardrey to Carpenter, 23 January 1965, Box 1, Folder 13, Ardrey Papers.

57. Ardrey, *Territorial Imperative*, 17.

58. Ardrey, *Territorial Imperative*, 18, 26.

59. Ardrey, *Territorial Imperative*, 18.

60. Ardrey, *Territorial Imperative*, 5.

61. Ardrey, *Territorial Imperative*, 101.

62. Ardrey, *Territorial Imperative*, 69.

63. "Conscience may direct the Christian martyr to die for the brotherhood of man. But that same conscience directs Christian armies to go forth and slaughter the same fellow man." Ardrey, *African Genesis*, 349.

64. Jean-Jacques Petter, *Researche sur l'ecologie et l'etholgie des lémuriens malgaches* (Paris: Editions du Muséum, 1962). See also his essay in Irven DeVore, ed., *Primate Behavior: Field Studies of Monkey and Apes* (New York: Holt Rinehart and Winston, 1965).

65. Ardrey, *Territorial Imperative*, 167. In using Petter's term, Ardrey avoided engaging with traditional sociological categories like *Gemeinschaft* and *Gesellschaft*.

66. Ardrey, *Territorial Imperative*, 174.

67. Ardrey, *Territorial Imperative*, 270.

68. Peter M. Driver, "Toward an Ethology of Human Conflict: A Review," *Journal of Conflict Resolution* 11, no. 3 (1967): 361–374; W. Montgomery Watt, "Traditional Arab Communities in the Modern World," *International Affairs* 44, no. 3 (1968): 494–500, on 496–497.

69. Ardrey, *Territorial Imperative*, 351.

70. Ardrey, *Territorial Imperative*, 225–226.

71. The idea that animal species acted to limit their own population growth for the good of the species would eventually land Wynne-Edwards in intellectual hot water: Mark Borello, *Evolutionary Restraints: The Contentious History of Group Selection* (Chicago: University of Chicago Press, 2010).

72. V. C. Wynne-Edwards, "Population Control in Animals," *Scientific American* (August 1964): 68–74, quote on 71. Ethologists had long believed that male animals engaged in "ritualized" fights rather than killing each other; see, e.g., A. David Blest, "The Concept of 'Ritualisation,'" in *Current Problems in Animal Behaviour*, ed. W. H. Thorpe and O. L. Zangwill (Cambridge: Cambridge University Press, 1961): 102–124; Julian S. Huxley, "A Discussion on Ritualization of Behaviour in Animals and Man," *Philosophical Transactions of the Royal Society of London, Series B* 251 (1966): 249–271.

73. Wynne-Edwards, "Population Control in Animals," 71.

74. Ardrey, *Territorial Imperative*, 48.

75. Ardrey, *Territorial Imperative*, 55.

76. On the fate of Darwin's theory of sexual selection in the twentieth century, see Erika Lorraine Milam, *Looking for a Few Good Males: Female Choice in Evolutionary Biology* (Baltimore: Johns Hopkins University Press, 2010).

77. At the time, occasional hostile accusations circulated that manly characters such as Batman and Robin, the men of *Bonanza*, and even James Bond (due to his utilitarian engagements with women) reflected a growing homosexual tendency in American culture: Carol L. Tilley, "Seducing the Innocent: Frederic Wertham and the Falsifications That Helped Condemn Comics," *Information & Culture* 47 (2012): 383–413; Wendall Hall, "The Fag-Jag on the Boob-Tube," *Fact* 4 (1967): 16–23.

78. Driver, "Toward an Ethology of Human Conflict," 366.

79. Loren Eiseley, "A Script Written in the Bones," *New York Times*, 11 September 1966, 409.

80. Earl W. Count, "Beyond Anthropology: Toward a Man-Science," *American Anthropologist* 74, no. 6 (1972): 1355–1365.

81. On debates over strategies of silent or vocal opposition in the sciences, see Michael Gordin, *Pseudoscience Wars: Immanuel Velikovsky and the Birth of the Modern Fringe* (Chicago: University of Chicago Press, 2012).

82. Ardrey to Harry Frank Guggenheim, 4 November 1967, Folder: Guggenheim, Harry Frank #5, Moe Papers.

83. Robert Ardrey, *The Hunting Hypothesis: A Personal Conclusion Concerning the Evolutionary Nature of Man* (New York: Atheneum, 1976).

Chapter Five. The Human Animal

1. Konrad Lorenz, *Das sogenannte Böse; zur Naturgeschichte der Agression* [The so-called evil, on the natural history of aggression] (Vienna: Schoeler Verlag, 1963), trans. Marjorie Kerr Wilson as *On Aggression* (New York: Harcourt, Brace & World, 1966).

2. Edward Sheehan, "Conversations with Konrad Lorenz," *Harper's Magazine*, May 1968, 69–77, quote on 69.

3. J. Alsop, "Profiles: A Condition of Enormous Probability," *New Yorker*, March 1969, 39, quote on 40.

4. Richard W. Burkhardt Jr., "Ethology, Natural History, the Life Science, and the Problem of Place," *Journal of the History of Biology* 32, no. 3 (1999): 489–508, quotes on 501.

5. Richard W. Burkhardt Jr., *Patterns of Behavior: Konrad Lorenz, Niko Tinbergen, and the Founding of Ethology* (Chicago: University of Chicago Press, 2005), 136–160.

6. Lorenz, *On Aggression*, 3.

7. Lorenz, *On Aggression*, 5.

8. Lorenz, *On Aggression*, 11.

9. Lorenz, *On Aggression*, 11.

10. Lorenz, *On Aggression*, ix.

11. Lorenz, *On Aggression*, 39–41, quote on 41.

12. Tania Munz, " 'My Goose Child Martina': The Multiple Uses of Geese in the Writings of Konrad Lorenz," *Historical Studies in the Natural Sciences* 41, no. 4 (2011): 405–446.

13. Lorenz, *On Aggression*, 220.

14. Konrad Lorenz, "On Killing Members of One's Own Species," *Bulletin of the Atomic Scientists* (October 1970): 2–5, 51–56; Nikolaas Tinbergen, "On War and Peace in Animals and Men," *Science* 160, no. 3835 (1968): 1411–1418; Geoffrey Gorer, "Man Has No 'Killer' Instinct," *New York Times*, 27 November 1966, SM24.

15. Konrad Lorenz, "The Plant Eaters" (excerpt from *Civilized Man's Eight Deadly Sins*), *Vogue*, April 1974, 153, quote from editor on 153.

16. Sheehan, "Conversations with Konrad Lorenz," 73.

17. Ernest Callenbach, *Ecotopia: The Notebooks and Reports of William Weston* (Berkeley, CA: Banyan Tree Books, 1975). Success at the Olympics, of course, did serve as an important symbolic marker during the Cold War. See, for example, Stephen Wagg and David Andrews, eds.,

East Plays West: Sport and the Cold War (New York: Routledge, 2007); and Nicholas Evan Sa-rantakes, *Dropping the Torch: Jimmy Carter, the Olympic Boycott, and the Cold War* (New York: Cambridge University Press, 2011). Ardrey, too, saw the morality of sport as residing in the innate aggression of athletes competing in groups: *Territorial Imperative*, 279.

18. Lorenz, *On Aggression*, 276–278.

19. Eugene Rabinowitch, "Open Letter to Konrad Lorenz," *Bulletin of the Atomic Scientists*, November 1966, 2–3, quote on 3; Harold C. Urey, letter to the editor in response to Rabinow-itch, "Open Letter to Konrad Lorenz," *Bulletin of the Atomic Scientists* (April 1967): 25–26.

20. On Lorenz's complicated engagement with Freud and the resulting public conversation in West Germany about aggression and sexuality, see Dagmar Herzog, *Cold War Freud: Psycho-analysis in an Age of Catastrophes* (Cambridge: Cambridge University Press, 2017), 125–148.

21. Lorenz, *On Aggression*, 38–39.

22. Lorenz, *On Aggression*, 42; see also Lorenz, "On Killing." This article is the first published English translation of a piece Lorenz had published in German in 1955.

23. Sheehan, "Conversations with Konrad Lorenz," 70.

24. Ashley Montagu, ed., *Man and Aggression* (New York: Oxford University Press, 1968), 2nd ed. (New York: Oxford University Press, 1973); Montagu, *The Nature of Human Aggression* (New York: Oxford University Press, 1976).

25. Earl W. Count, "Beyond Anthropology: Toward a Man-Science," *American Anthropologist* 74, no. 6 (1972): 1355–1365, quote on 1360.

26. Marston Bates, "A Naturalist At Large," *Natural History*, April 1967, 14–18; Peter M. Driver, "Toward an Ethology of Human Conflict: A Review," *Journal of Conflict Resolution* 11, no. 3 (1967): 361–374.

27. B. F. Skinner, "The Phylogeny and Ontogeny of Behavior," review of *On Aggression*, by Konrad Lorenz, *Science* 153, no. 3741 (1966): 1205–1213, quote on 1212.

28. Skinner, "Phylogeny and Ontogeny of Behavior," 1212.

29. Theodore Schneirla, "Instinct and Aggression," *Natural History*, December 1966, 16–20; Ethel Tobach, "Evolution of Behavior and the Comparative Method," *International Journal of Psychology* 11 (1976): 185–201.

30. Bates, "Naturalist at Large," 14.

31. On Lorenz's intellectual and political commitments, as well as a nuanced discussion of complicity and responsibility during the Second World War, see Burkhardt, *Patterns of Behavior*, 231–280.

32. On Lorenz's excitement over Germany's annexation of Austria in 1938 and his incorporation of Nazi rhetoric into his publications during the war, see Munz, " 'My Goose Child Martina.' "

33. Burkhardt, *Patterns of Behavior*, 278.

34. Daniel S. Lehrman, "A Critique of Konrad Lorenz's Theory of Instinctive Behavior," *Quarterly Review of Biology* 28, no. 4 (1953): 337–363. See also Paul E. Griffiths on the skepticism with which Lorenz's conception of "instinct" was greeted in Britain, "Instinct in the '50s: The British Reception of Konrad Lorenz's Theory of Instinctive Behavior," *Biology and Philosophy* 19, no. 4 (2004): 609–631.

35. Jay S. Rosenblatt, "Daniel S. Lehrman, 1919–1972," *Biographical Memoirs of the National Academy of Sciences* 66 (1995): 226–245, esp. 235.

36. Burkhardt, *Patterns of Behavior*, 384–407.

37. Nikolaas Tinbergen, "On the Aims and Methods of Ethology," *Zeitschrift für Tierspsychologie* 20 (1963): 410–433. The paper was dedicated to Konrad Lorenz on the occasion of his sixtieth birthday, but Lorenz never fully agreed with the new direction for the field Tinbergen developed in its pages; see Burkhardt, *Patterns of Behavior*, 409–446.

38. Robert Hinde, "The Nature of Aggression," *New Society*, 2 March 1967, 302–304, quote on 303. Emphasis in original.

39. Hinde, "Nature of Aggression," quotes on 303, 304.

40. S. Anthony Barnett, review of *On Aggression* by Konrad Lorenz, *Scientific American*, February 1967, 135–138, quote on 135.

41. Ardrey to Piel, 23 February 1967, Ardrey to Morris, 27 February 1967, Box 1, Folder 18, Robert Ardrey Papers, 1955–1980, Special Collections and University Archives, Rutgers University Libraries, New Brunswick, NJ. Ardrey and Lorenz would not meet until May 1973.

42. Marshall Sahlins, "Books," *Scientific American*, July 1962, 169–174. Sahlins also included a catchy limerick mocking Ardrey's reconstruction of prehuman history: "There was a young man named *Zinjanthropus*, / Who never became *Pithecanthropus*, / By his small cousin Cain / He was dastardly slain, / And that is what saved my hypothosus" (172).

43. Solly Zuckerman, "The Human Beast," *Nature* 212, no. 5062 (1966): 563–564, quote on 563. On Zuckerman's prewar primatological research (especially *The Social Life of Apes and Monkeys*, published in 1932) and his postwar fall from grace among ethologists, see Jonathan Burt, "Solly Zuckerman: The Making of a Primatological Career in Britain, 1925–1945," *Studies in History and Philosophy of Biology and Biomedicine* 37 (2006): 295–310. Although a British publication, *Nature* was positioning itself as an international science journal by publishing the work of an international array of scientists, so much so that in 1970 the journal opened an office in Washington, DC: Melinda Baldwin, *Making 'Nature': The History of a Scientific Journal* (Chicago: University of Chicago Press, 2015).

44. Zuckerman, "Human Beast," 564.

45. Ashley Montagu, "Aggressive Behavior of Man," *Chicago Tribune*, 17 July 1966, O10. The real touchstone for debates over nature (genetics) and nurture (development) in the late 1960s and early 1970s, however, was the question of an IQ gap between blacks and whites. Montagu similarly argues that any measured differences in IQ between races are due to environmental factors, not hereditary ones; e.g., Ashley Montagu, ed., *Race and IQ* (New York: Oxford University Press, 1975).

46. Ashley Montagu, "What Comes Naturally?" *New York Times*, 25 February 1968, BR14. For a similar definition of human nature as culture, see Margaret Mead, "What Is Human Nature," *Look*, 19 April 1955, 56–62.

47. For example, "The Improvement of Human Relations through Education," *School and Society* 65 (28 June 1947): 465–469; "Social Instincts," *Scientific American*, April 1950, 54–56; "Man Is a Creature Who Can Make or Unmake Himself," *New York Times*, 9 February 1964, BR6; "The New Litany of 'Innate Depravity,' or Original Sin Revisited," in *Man and Aggression*, ed. Ashley Montagu (Oxford University Press, 1968): 3–18; "Is Man Born as Mean as He Is?," *Washington Post*, 19 May 1968, B1 (also printed on the same day in the *Times Herald* and on 26 May in the *Los Angeles Times* as ". . . Or a Child of an Imperfect Society?"). Some of his books include, *Man in Process* (Cleveland: World Publication, 1961), *The Human Revolution* (Cleve-

land: World Publication, 1965), as editor, *Culture: Man's Adaptive Dimension* (New York: Oxford University Press, 1968), *The Concept of the Primitive* (New York: Free Press, 1968), and with C. Loring Brace, *Man's Evolution: An Introduction to Physical Anthropology* (New York: Macmillan, 1965).

48. Montagu, "Is Man as Mean as He Is?," B1.

49. Sandra McPherson, letter to the editor in response to Rabinowitch, "Open Letter to Konrad Lorenz," *Bulletin of the Atomic Scientists* (April 1967): 36.

50. Peter M. Driver, letter to the editor, *Scientific American*, May 1967, 8.

51. Geoffrey Gorer, "Man Has No 'Killer' Instinct," *New York Times*, November 27, 1966, quote on SM24.

52. Driver, "Toward an Ethology of Human Conflict," 363; Bates, "Naturalist at Large," 18. Other scientists who wrote favorably of Lorenz's perspective before the publication of *On Aggression* were also cited. See, for example, Anatol Rapoport, "Is Warmaking a Characteristic of Human Beings or of Culture?," review of *The Natural History of Aggression*, ed. J. D. Carthy and F. J. Ebling, *Scientific American*, October 1965, 115–118.

53. Sheehan, "Conversations with Konrad Lorenz," 72. In future publications, Lorenz repeated this point; see, for example, coverage of his "Eight Deadly Sins," *Newsweek*, 6 August 1973, 58.

54. Lorenz, "On Killing," 56.

55. B. Lawren, "Interview with Konrad Lorenz," *Omni*, April 1987, 86.

56. Frédéric de Towarnicki and Konrad Lorenz, "A Talk with Konrad Lorenz," trans. Stanley Hochman, *New York Times*, 5 July 1970, 121.

Chapter Six. Man and Beast

1. The book was released earlier in the UK, on 12 October 1967.

2. Henry Raymont, "Review of 'Naked Ape' Causes 2 Papers to Call Back Sections," *New York Times*, 21 January 1968, 76. In *The Naked Ape*, Morris described the clitoris as "the female counterpart of the male penis" (58), a word that would have drawn far more consternation.

3. "Programming: Reasonable v. Raunchy," *Time*, 9 February 1968, 67–68. See also Desmond Morris, *Watching: Encounters with Humans and Other Animals* (London: Max Press, 2006), 313.

4. Morris, *Watching*, 314–315.

5. Despite the 1967 copyright date on the book, *The Naked Ape* became available for sale in the United States in the third week of January 1968. For details on the movie contract, see A. H. Weiler, "Everybody's Going Ape," *New York Times*, 14 April 1968, D15; and Bill Edwards, "Bufman, Driver Strike It Rich in Univ. Deal," *Daily Variety*, 26 February 1969, 1.

6. *Daily Variety*, 27 February 1974, 20.

7. Morris, *Watching*, 41.

8. Morris, *Watching*, 57.

9. Morris *Watching*, 60.

10. Morris, *Watching*, 65–66, 112, 123–124. A geneticist with interests in animal behavior, Spurway was married to the geneticist J.B.S. Haldane at the time of the affair. By the time Morris published, both Lorenz and Spurway had passed away.

11. Morris, *Watching*, 63.

12. Alfred C. Kinsey, Wardell B. Pomeroy, and Clyde E. Martin, *Sexual Behavior in the Human Male* (Philadelphia: W. B. Saunders, 1948); Alfred C. Kinsey and the Institute for Sex Research, *Sexual Behavior in the Human Female* (Philadelphia: W. B. Saunders, 1953); Masters and Johnson, *Human Sexual Response*. Morris's doctoral research focused on the sexual behavior of birds and fishes, with particular attention to "pseudo-male" and "pseudo-female" courtship behavior.

13. Morris, *Naked Ape*, 63.

14. John Hurrell Crook, "Sexual Selection, Dimorphism, and Social Organization in the Primates," in *Sexual Selection and the Descent of Man*, ed. Bernard Campbell (Chicago: Aldine, 1972): 231–281, esp. 249–250.

15. Morris, "Sex," in *Naked Ape*, 50–102.

16. Less risqué portions of the book were excerpted in *Life* magazine (as with Ardrey's *Territorial Imperative*); Desmond Morris, "The Naked Ape," *Life*, 22 December 1967, 94–108.

17. Ardrey to HFG, 13 January 1969, Folder: Guggenheim, Harry Frank, Foundation #2, Mss.B.M722, Henry Allen Moe Papers, American Philosophical Society, Philadelphia.

18. Desmond Morris, *The Human Zoo* (New York: McGraw-Hill, 1969); *Intimate Behaviour* (New York: Random House, 1971); *Patterns of Reproductive Behaviour* (London: Cape, 1970).

19. Desmond Morris, "Status and Superstatus in the Human Zoo," *Playboy*, September 1969, 122.

20. Morris, "Status and Superstatus," 124, 202–204. Each point was followed by multiple paragraphs of justification.

21. Morris, "Status and Superstatus," 204.

22. Morton Hunt, "Man and Beast," in *Man and Aggression*, 2nd ed., ed. Ashley Montagu (Oxford: Oxford University Press, 1973), 19–38, 80–81.

23. Hunt, "Man and Beast," 81.

24. Morton Hunt, *Her Infinite Variety: The American Woman as Lover, Mate, and Rival* (New York: Harper & Row, 1962), 36. See also Hunt, *The Natural History of Love* (1959; New York: Minerva Press, 1970), *The Affair: A Portrait of Extra-Marital Love in Contemporary America* (New York: World Publishing, 1969), and *Sexual Behavior in the 1970s* (Chicago: Playboy Press, 1974). The last of these, *Sexual Behavior in the 1970s*, served as a popular contextualization of Kinsey's results, updated for a new era, and he cited none of the contemporary colloquial work on human nature.

25. Hunt, "Man and Beast," 81. See also, Evelyn Reed, "Is Man[NB] an 'Aggressive Ape'?" *International Socialist Review* 31, no. 8 (1970): 27–31, 40–42; George Gaylord Simpson, *The Meaning of Evolution* (New Haven, CT: Yale University Press, 1969).

26. Hunt, "Man and Beast," 82.

27. Hunt, "Man and Beast," 181–182.

28. Hunt, "Man and Beast," 179.

29. Hunt, "Man and Beast," 183.

30. Dear Playboy, letters to the editor, *Playboy*, October 1970, 14.

31. Hunt, "Man and Beast," 19–38.

32. Loren Eiseley, "The Intellectual Antecedents of *The Descent of Man*," in *Sexual Selection and the Descent of Man*, ed. Bernard Campbell (London: Heinemann, 1972), 1–16, quotes on 14, 15.

33. George Gaylord Simpson, "The Evolutionary Concept of Man," in *Sexual Selection and the Descent of Man*, ed. Campbell, 17–39, quote on 31.

34. Simpson, "Evolutionary Concept of Man," 20, 31, 36.

35. J. H. Crook, "Sexual Selection, Dimorphism, and Social Organization in the Primates," in *Sexual Selection and the Descent of Man*, ed. Campbell, 231–281.

36. George Schaller, "The Social Kingdom," in *The Marvels of Animal Behavior*, ed. Thomas B. Allen (Washington, DC: National Geographic Society, 1972), 66–87, quote on 86.

37. S. Dillon Ripley, preface to *Man and Beast: Comparative Social Behavior*, ed. J. F. Eisenberg and W. S. Dillon (Washington, DC: Smithsonian Institution Press, 1971), 5.

38. Gorer to Mead, 30 November 1966; and Mead to Gorer, 2 December 1966; Series B. Special Correspondence, Box B7, Folder 1, Geoffrey Gorer, 1965–1969, Margaret Mead Papers and South Pacific Ethnographic Archives, 1838–1996, MSS32441, Library of Congress, Washington, DC (hereafter cited as Mead Papers).

39. Margaret Mead, "Innate Behavior and Building New Cultures: A Commentary," in *Man and Beast*, ed. Eisenberg and Dillon, 369–381, quote on 371.

40. Mead, "Innate Behavior," 371, 373.

41. Mead to Gorer, 26 March 1967, Series B. Special Correspondence, Box B7, Folder 1, Geoffrey Gorer, 1965–1969, Mead Papers.

42. Richard Evans, "A Conversation with Konrad Lorenz about Aggression, Homosexuality, Pornography, and the Need for a New Ethic," *Psychology Today* 8 (November 1974): 82–87," 86.

43. Evans, "Conversation with Konrad Lorenz," 86. The later work he referred was likely *The Social Contract* (New York: Atheneum, 1970).

44. As quoted in "Behavior: The Watchers," *Time*, 22 October 1973, 68.

45. Konrad Lorenz, *Civilized Man's Eight Deadly Sins*, trans. Marjorie Kerr Wilson (New York: Harcourt Brace Jovanovich, 1974).

46. Bill Lawren, "Interview: Konrad Lorenz," *Omni*, April 1987, 84–90, quote on 84.

47. Peter Marler and Donald R. Griffin, "The 1973 Nobel Prize for Physiology or Medicine," *Science* 182, no. 4111 (1973): 464–466; see also "From Bees, Geese and Wasps to Man," *Science News* 104, no. 16 (October 1973): 244; "Learning from the Animals," *Newsweek*, 22 October 1973, 102.

48. Morris's conception of an "imperfect" pair-bond differed from Hefner's earlier views of sex relations in early man. In a 1962 interview, Hefner suggested that sex roles used to be quite easy and only became complicated with the influence of modern civilization: "You know it goes back to the beginning of time. The man goes out and kills a saber-toothed tiger while the woman stays at home and washes out the pots." See Simon Nathan, "About the Nudes in *Playboy*," *U.S. Camera*, April 1962, 69–70, as quoted in Elizabeth Fraterrigo, Playboy *and the Making of the Good Life in Modern America* (New York: Oxford University Press, 2009), 34.

49. Murf. [Arthur Murphy], "The Naked Ape," *Variety*, 15 August 1973, 12.

50. Vernon Scott, "The Girl Who Has Everything: Victoria Principal," *Good Housekeeping*, June 1989, 112, quote on 114.

51. D. Morris, *Watching*, 315.

52. *Daily Variety*, 27 February 1974, 20; Frank Brady, *Hefner: An Unauthorized Biography* (New York: Macmillan, 1974), 212.

Part Three. Unmaking Man

Margaret Mead to Konrad Lorenz, reacting to *On Aggression*, 22 December 1966, Folder 1966 L, Box C77, Margaret Mead Papers and South Pacific Ethnographic Archives, 1838–1996, MSS32441, Library of Congress, Washington, DC.

1. Lois Banner and Mary Hartman, program co-chairwomen, 1973, memo to Members of the Berkshire Conference, Re: Conference, "Historical Perspectives on Women," Douglass College, March 2–3, 1973," Box 13, Folder 16: "First Conference, Douglass College (March 2–3, 1973), 1972–1973," MC606 Berkshire Conferences on the History of Women, Schlesinger Library, Radcliffe Institute for Advanced Study, Harvard University (hereafter cited as BCHW Papers). Thanks to Sally Gregory Kohlstedt for bringing Tiger's talk at the Berks to my attention.

2. Lionel Tiger and Robin Fox, "The Zoological Perspective in Social Science," *Man*, n.s., 1, no. 1 (1966): 75–81.

3. Lionel Tiger, *Men in Groups* (New York: Random House, 1969).

4. Lionel Tiger, "But I Never Said Women are Inferior," *Maclean's*, 1 January 1970, 61–63.

5. Tiger, "But I Never Said," 63.

6. Constance Ashton Myers, "Historical Perspectives on Women," Box 13, Folder 16: "First Conference, Douglass College (March 2–3, 1973), 1972–1973," BCHW Papers.

7. On the power of anti-Lysenkoism as anti-communist rhetoric in the United States, see the special issue "The Lysenko Controversy and the Cold War," ed. William DeJong-Lambert and Nikolai Krementsov, *Journal of History of Biology* 45, no. 3 (2012): 373–556.

8. As quoted and described in Myers, "Historical Perspectives on Women."

9. Banner and Hartman, "Conference, 'Historical Perspectives on Women.' "

10. Carolyn Merchant, *The Death of Nature: Women, Ecology, and the Scientific Revolution* (San Francisco: Harper & Row, 1980); Sherry Ortner, "Is Female to Male as Nature Is to Culture?," in *Woman, Culture, and Society*, ed. Michelle Rosaldo and Louise Lamphere (Palo Alto, CA: Stanford University Press, 1974), 67–88.

11. Michelle Murphy, *Seizing the Means of Reproduction: Entanglements of Feminism, Health, and Technoscience* (Durham, NC: Duke University Press, 2012); Daniel T. Rodgers, *Age of Fracture* (Cambridge, MA: Belknap University Press of Harvard University Press, 2011).

12. Noam Chomsky et al., *The Cold War and the University—Toward an Intellectual History of the Postwar Years* (New York: New Press, 1997), xiv.

13. Dorothy Zinberg, "Past Decade for Women Scientists—Win, Lose, or Draw?," *Trends in Biochemical Science* 2, no. 6 (1977): N123–N126; Harriet Zuckerman and Jonathan R. Cole, "Women in American Science," *Minerva* 13, no. 1 (1975): 82–102; Ann Fischer and Peggy Golde, "The Position of Women in Anthropology," *American Anthropologist* 70, no. 2 (1968): 337–344.

14. *Climbing the Academic Ladder: Doctoral Women Scientists in Academe* (Washington, DC: National Academy of Sciences, 1979): xiv.

15. On mid-century politics over homosexuality, see Jennifer Terry, *An American Obsession: Science, Medicine, and Homosexuality in Modern Society* (Chicago: University of Chicago Press, 1999); Margot Canaday, *The Straight State: Sexuality and Citizenship in Twentieth-Century America* (Princeton, NJ: Princeton University Press, 2009). On the connections

between contemporary discussions of anthropology, race, and sexuality, see Joanne Meyerowitz, " 'How Common Culture Shapes the Separate Lives': Sexuality, Race, and Mid-Twentieth Century Social Constructivist Thought," *Journal of American History* 96, no. 4 (2010): 1057–1084.

16. Margaret W. Rossiter, "The Path to Liberation: Consciousness Raised, Legislation Enacted," in *Women Scientists in America: Before Affirmative Action, 1940–1972* (Baltimore: Johns Hopkins University Press, 1995), 361–382; and on the more recent decades, Margaret W. Rossiter, *Women Scientists in America: Forging a New World since 1972* (Baltimore: Johns Hopkins University Press, 2012).

17. Laura Nader, "Up the Anthropologist—Perspectives Gained from Studying Up," in *Reinventing Anthropology*, ed. Dell H. Hymes (New York: Pantheon Press, 1972), 284–311, quote on 289. (Yes, Ralph Nader is her brother.)

18. Robert A. Nye, "Kinship, Male Bonds, and Masculinity in Comparative Perspective," *American Historical Review* 105, no. 5 (2000): 1656–1666; and Nye, "Medicine and Science as Masculine 'Fields of Honor,' " *Osiris* 12 (1997): 60–79.

19. Daniel Wickberg, *The Senses of Humor: Self and Laughter in Modern America* (Ithaca, NY: Cornell University Press, 1998), 96, 170.

20. James Secord examined one of the well-documented cases of this phenomenon in history of science in *Victorian Sensation: The Extraordinary Publication, Reception, and Secret Authorship of Vestiges of the Natural History of Creation* (Chicago: University of Chicago Press, 2000).

21. Kate Millet, *Sexual Politics* (Boston: New England Free Press, 1968): 11.

22. Millet, *Sexual Politics*, 12.

23. Robert Ardrey, *The Territorial Imperative: A Personal Inquiry into the Animal Origins of Property and Nations*, (New York: Atheneum, 1966).

24. *Annual Behavior Society Newsletter* 13, no. 4 (November 1968): 1. Ardrey wrote to Harry Frank Guggenheim on 13 January 1969: "Charles Southwick, the president of the ABS, said, 'Why should you be astounded? You did it.' I said they weren't here to hear me. 'Maybe not,' he said, 'but they wouldn't know about us except for you.' It was a good feeling Harry. I'd like to think it was true." Box 1, Folder 25, Robert Ardrey Papers, 1955–1980, Special Collections and University Archives, Rutgers University Libraries, New Brunswick, NJ (hereafter cited as Ardrey Papers).

25. Elaine Morgan, *The Descent of Woman* (New York: Stein and Day, 1972).

26. Antony Jay, *Corporation Man* (New York: Random House, 1971), paperback by Pocket (1973). Jay's book was equally an homage to William H. Whyte's *The Organization Man* (New York: Simon & Schuster, 1956).

27. This fluid intellectual community with multiple centers of concentration allowed for fairly amicable relations—you do your thing, I'll do mine—without the need for hard demarcation within a single professional association or department. This flexibility can be seen as a sign of intellectual health and vigor, contra Thomas Kuhn's vision of mature science as operating within iterative, single paradigms. Kuhn's articulation of science as paradigmatic became popular among many scientists at the exact moment that anthropology (and other social and natural sciences) settled on a multiple-paradigm approach within their discipline. For an alternative pluralistic vision of "normal" science, see Helen Longino, *Studying Human Behavior: How*

Scientists Investigate Aggression and Sexuality (Chicago: University of Chicago Press, 2013) and Hasok Chang, *Is Water H₂O? Evidence, Pluralism and Realism* (Dordrecht: Springer, 2012).

28. Chomsky et al., *Cold War and the University*, xiii.

29. Richard I. Evans, Konrad Lorenz interview in *Psychology Today*, November 1974, 83.

30. Ethel Tobach, ". . . Personal Is Political Is Personal Is Political . . . ," *Journal of Social Issues* 50, no. 1 (1994): 221–244.

31. Harriet Zuckerman and Jonathan R. Cole, "Women in American Science," *Minerva* 13, no. 1 (1975): 82–102, on 84.

32. Zuckerman and Cole, "Women in American Science"; Robert K. Merton, "The Matthew Effect in Science," *Science* 159, no. 3810 (1968): 56–63; Margaret Rossiter, "The ~~Matthew~~ Mathilda Effect in Science," *Social Studies of Science* 23, no. 2 (1993): 325–341.

33. Lionel Tiger, interview by author, 10 November 2011.

34. Tiger to Ardrey, 12 December 1969, Box 1, Folder 26, Ardrey Papers. Tiger still recalled the indignity of this event when I interviewed him forty years later. See also his account of a television appearance with Betty Friedan "and 5 of the nation's most militant feminists," Tiger to Ardrey 24 June 1969, Box 1, Folder 26, Ardrey Papers. Tiger later coordinated a conference and a co-edited collection, with Heather Fowler, entitled *Female Hierarchies* (Chicago: Beresford Book Service, 1978).

35. Tiger, interview by author.

Chapter Seven. Woman the Gatherer

1. For a fuller account of the importance of women in the development of primatology and theories of human nature, see Shirley C. Strum and Linda Marie Fedigan, eds., *Primate Encounters: Models of Science, Gender, and Society* (Chicago: University of Chicago Press, 2000).

2. Sally Slocum, "Woman the Gatherer: Male Bias in Anthropology," in *Towards an Anthropology of Women*, ed. Rayna Reiter (New York: Monthly Review Press, 1975), 36–50; M. Kay Martin and Barbara Voorhies, *Female of the Species* (New York: Columbia University Press, 1975); Carol B. Stack, Mina Davis Caulfield, Valerie Estes, Susan Landes, Karen Larson, Pamela Johnson, Juliet Rake, and Judith Shirek, "Anthropology," *Signs* 1, no. 1 (1975): 147–159; Nancy Tanner and Adrienne Zihlman, "Women in Evolution. Part I: Innovation and Selection in Human Origins," *Signs* 1, no. 3 (1976): 585–608; Adrienne Zihlman, "Women in Evolution. Part II: Subsistence and Social Organization among Early Hominids," *Signs* 4, no. 1 (1978): 4–20; Frances Dahlberg, ed., *Woman the Gatherer* (New Haven, CT: Yale University Press, 1981).

3. Mary Douglas, "Animals in Lele Religious Symbolism," *Africa* 27, no. 1 (1 January 1957): 46–58, on 56; reprinted in in *Implicit Meanings: Selected Essays in Anthropology*, 2nd ed. (New York: Routledge, 1999).

4. Konrad Lorenz, *King Solomon's Ring: New Light on Animal Ways*, with a foreword by Julian Huxley, trans. Marjorie Kerr Wilson (New York: Crowell, 1952).

5. Rayna Reiter, ed. *Toward an Anthropology of Women* (New York: Monthly Review, 1975); Michelle Rosaldo and Louise Lamphere, eds., *Women, Culture, and Society* (Stanford, CA: Stanford University Press, 1974). Rosaldo and Lamphere earned their PhDs in anthropology from Harvard University (in 1972 and 1968, respectively). Rosaldo was working at Stanford when the edited collection was published, Lamphere at Brown. When Lamphere was denied tenure in

1974, she sued and was awarded tenure retroactively in 1978. For an online exhibit at Brown regarding the case and its legacy, see "The Lamphere Case" at https://pembrokeexhibits .squarespace.com (accessed 26 January 2018). The author biographies at the back of the volume reveal only one contributor with tenure at the time of publication (Peggy Sanday).

6. Naomi Quinn, "The Divergent Case of Cultural Anthropology," in *Primate Encounters*, ed. Strum and Fedigan, 223–242.

7. Sally Slocum, "Woman the Gatherer: Male Bias in Anthropology," originally published under Sally Linton, in *Women in Cross-Cultural Perspectives*, ed. Sue Ellen Jacobs (Urbana: University of Illinois Press, 1971). An anthropology graduate student at University of California, Berkeley, Slocum ultimately earned her PhD at the University of Colorado, Boulder, in 1975. After teaching for a number of years, she joined the State Department in 1985.

8. Sherwood L. Washburn and Curt S. Lancaster, "The Evolution of Hunting," in *Man the Hunter*, ed. Richard B. Lee and Irven DeVore (Chicago: Aldine, 1968), 293–303, quote on 303. The volume emerged from a 1966 symposium by the same name sponsored by the Wenner-Gren Foundation for Anthropological Research.

9. Rayna Reiter, "Introduction," in *Toward an Anthropology of Women*, ed. Reiter, 16. Now Rayna Rapp, she completed her PhD in 1973 at the University of Michigan. At the time of publication, she was an assistant professor in the Department of Anthropology at the New School for Social Research in New York.

10. See, e.g., Peggy Golde, ed., *Women in the Field: Anthropological Experiences* (Chicago: Aldine, 1970), a second edition issued in 1986 contained two new essays. This volume served not only to explore the unique circumstances confronting each female anthropologist in the field (its stated goal) but also to establish the street credibility of women as anthropologists. They, too, were field-tested professionals, who had struggled to understand another culture in potentially dire and dangerous circumstances. "*Women in the Field* is touted as the first book to acknowledge how gender influences data collection and analysis, and to recognize the personal, subjective impact of fieldwork immersion." Pierrette Hondagneu-Sotelo, "Gender and Fieldwork," *Women's Studies International Forum* 11, no. 6 (1988): 611–618, quote on 612.

11. For example, Richard B. Lee, "What Hunters Do for a Living, or, How to Make Out on Scarce Resources," in *Man the Hunter*, ed. Lee and DeVore, 30–48.

12. Quinn, "Divergent Case of Cultural Anthropology," quote on 227. Not all important essays calling attention to the relative silence of women in anthropology were authored by women. Edwin Ardener raised the question of male bias in ignoring the status of women in his influential "Belief and the Problem of Women" (1972) and "The 'Problem' Revisited," in *Perceiving Women*, ed. Shirley Ardener (London: Malaby Press, 1975), 1–18 and 19–28.

13. On the contested history of mother love during this era, see Marga Vicedo, *The Nature and Nurture of Love: From Imprinting to Attachment in Cold War America* (Chicago: University of Chicago Press, 2013).

14. Slocum, "Woman the Gatherer," 46.

15. Catherine R. Stimpson, Joan N. Burstyn, Domna C. Stanton, and Sandra M. Whisler, "Editorial," *Signs* 1, no. 1 (1975): v–viii.

16. Carol B. Stack et al. "Anthropology," *Signs* 1, no. 1 (1975): 147–159. All except Juliet Rake earned a PhD in anthropology. Rake left graduate school, became an antique dealer in Virginia and then a photographer.

17. Susan Sperling, "Social Determinants of Attitudes toward Primates and Other Animals" (PhD diss., University of California, Berkeley, 1985), 16–17.

18. Phyllis Jay, "The Social Behaviour of the Langur Monkey" (PhD diss., University of Chicago, 1963). Jay completed her PhD in 1963 under the direction of Sherwood Washburn at the University of Chicago and moved to the University of California, Berkeley, in 1966, where she worked with a great number of female graduate students. Alongside Washburn, she edited the first two in the five-volume series *Perspectives on Human Evolution* for Holt, Rinehart and Winston (1968, 1972). During these years, she also edited *Primate Patterns* (1972) and *Primates: Studies in Adaptation and Variability* (1968). In all her work she strove to emphasize the non-aggressive nature of primates societies studied under natural conditions. On Phyllis Jay Dolhinow's later work, see Amanda Rees, *The Infanticide Controversy: Primatology and the Art of Field Science* (Chicago: University of Chicago Press, 2009).

19. Alison Jolly, "Lemur Social Behavior and Primate Intelligence," *Science* 153, no. 3735 (1966): 501–506; and Jolly, *The Evolution of Primate Behavior* (New York: Macmillan, 1972).

20. Thelma Rowell, *The Social Behaviour of Monkeys* (Harmondsworth: Penguin, 1972); Rowell, "The Concept of Social Dominance," *Behavioral Biology* 11 (1974): 131–154; Vinciane Despret, "Culture and Gender Do Not Dissolve into How Scientists 'Read' Nature: Thelma Rowell's Heterodoxy," in *Rebels, Mavericks, and Heretics in Biology*, ed. Oren Harman and Michael Deitrich (New Haven, CT: Yale University Press, 2009), 338–355.

21. Robert Trivers, review of *The Social Behaviour of Monkeys* by Thelma Rowell, *American Journal of Physical Anthropology* 41, no. 1 (1974): 163–164.

22. Susan Sperling, "Baboons with Briefcases vs. Langurs with Lipstick: Feminism and Functionalism in Primate Societies," in *The Gender/Sexuality Reader: Culture, History, Political Economy*, ed. Roger N. Lancaster and Micaela di Leonardo (New York: Routledge, 1997), 249–264. See also the publications by Ronald Nadler on the sexual behavior of great apes in captivity at the Yerkes Primatological Center in the mid-1970s, e.g., as described by Sarah Blaffer Hrdy in "The Primate Origins of Human Sexuality," in *The Evolution of Sex*, ed. Robert Bellig and George Stevens (New York: Harper & Row, 1988), 101–138. See also the first part of Frederick Wiseman's controversial film *Primate* (Zipporah Films, 1974), 113 min.

23. Slocum, "Woman the Gatherer," 50.

24. Kay Milton, "Male Bias in Anthropological Research," *Man*, n.s., 14, no. 1 (1979): 40–54, quote on 43. Among the articles she singled out for critique were Ardener, "Belief and the Problem of Women," Slocum, "Male Bias in Anthropology," and Sherry Ortner, "Is Female to Male and Nature is to Culture?," in Rosaldo and Lamphere, *Women, Culture, and Society*.

25. Beth Elverdam, Kay Milton, "Correspondence: Male Bias?" *Man*, n.s., 14, no. 4 (1979): 750–752. See also Maria-Barbara Watson-Franke and Kay Milton, "Bias, Male and Female," *Man*, n.s., 15, no. 2 (1980): 377–380.

26. For example, Tracy Teslow, *Constructing Race: The Science of Bodies and Cultures in American Anthropology* (New York: Cambridge University Press, 2014).

27. Ernestine Friedl, *Women and Men: An Anthropologist's View* (New York: Holt, Rinehart and Winston, 1975); See also Ernestine Friedl, "The Position of Women: Appearance and Reality," *Anthropological Quarterly* 40 (1967): 97–108.

28. Evelyn S. Kessler, *Women: An Anthropological View* (New York: Holt, Rinehart and Win-

ston, 1976), 239. See also Kessler, *Anthropology: The Humanizing Process* (Boston: Allyn and Bacon, 1974). The University of South Florida hired Kessler in 1967, and she worked there until her death, in 1977 of pancreatic cancer. Her daughter, Linda S. Cordell, earned her PhD in 1972 from the anthropology department at University of California, Santa Barbara.

29. Kessler, *Women*, 3, 248.

30. Ortner, "Is Female to Male?," 68–87; Carol MacCormack and Marilyn Strathern, eds., *Nature, Culture and Gender* (Cambridge University Press, 1980), esp. MacCormack's introduction to the volume, "Nature, Culture, and Gender: A Critique," 1–24.

31. See especially Misia Landau, "Human Evolution as Narrative," *American Scientist* 72, no. 3 (1984): 262–268; and Landau, *Narratives of Human Evolution* (New Haven, CT: Yale University Press, 1991); Donna Haraway, *Primate Visions: Gender, Race, and Nature in the World of Modern Science* (New York: Routledge, 1989).

32. Lila Abu-Lughod, "Can There Be a Feminist Ethnography?" *Critique of Anthropology* 13 (1993): 17, as quoted in Quinn, "Divergent Case of Cultural Anthropology," 240. See also Mary Hartsock, "Rethinking Modernism: Minority vs. Majority Theories," *Cultural Critique* 7 (1987): 187–206.

33. As quoted in Virginia Morell, "Anthropology: Nature-Cultural Battleground," *Science* 261, no. 5129 (1993): 1798, 1801–1802.

34. Quinn, "Divergent Case of Cultural Anthropology," 241.

35. Lori D. Hager, "Sex and Gender in Paleoanthropology," in *Women in Human Evolution*, ed. Lori D. Hager (New York: Routledge, 1997), 1–28, quote on 2.

36. M. D. Leakey and R. L. Hay, "Pliocene Footprints in the Laetolil Beds at Laetoli, Northern Tanzania," *Nature* 278, no. 5702 (1978): 317–323. Interpreting the footprints remained controversial: see Russell H. Tuttle, "Footprint Clues in Hominid Evolution and Forensics: Lessons and Limitations," *Ichnos* 15 (2008): 3–4; and Donald Johanson, Tim White, and Yves Coppens, "A New Species of the Genus *Australopithecus* (Primates: Hominidae) from the Pliocene of Eastern Africa," *Kirtlandia* 28 (1978): 1–11. Mitochondria are the powerhouses of cells. As eggs (but not sperm) contain mitochondria, they are passed from a mother to all of her offspring (male and female). Mitochondrial Eve, then, represents the most recent female common ancestor of all humans alive today. The research calculating the age of Mitochondrial Eve started in the late 1970s: Rebecca Cann, Mark Stoneking, and Allan Wilson, "Mitochondrial DNA and Human Evolution," *Nature* 325, no. 6099 (1987): 31–36.

37. Hager, "Sex and Gender in Paleoanthropology," 3–4.

38. Carole Vance, "Sexism in Anthropology? The Status of Women in Departments of Anthropology," *AAS Newsletter* 11, no. 9 (1970): 5–6; Roger Sanjek, "The Position of Women in the Major Departments of Anthropology, 1967–1976," *American Anthropologist* 80, no. 4 (1978): 894–904; Sanjek, "The American Anthropological Association Resolution on the Employment of Women: Genesis, Implementation, Disavowal, and Resurrection," *Signs* 7, no. 4 (1982): 845–868.

39. On the legacy of UC-Berkeley and Washburn for field primatology today, see the careful intellectual genealogy created by Elizabeth A. Kelley and Robert W. Sussman, "An Academic Genealogy on the History of American Field Primatologists," *American Journal of Physical Anthropology* 132 (2007): 406–425; Donna Haraway, "Remodeling the Human Way of Life: Sherwood Washburn and the New Physical Anthropology, 1950–1980," in *Primate Visions*, 186–230.

Women who earned their PhDs in this era included Virginia Avis (1958), Phyllis Jay (Dolhinow) (1963), Suzanne Ripley (1965), Adrienne Zihlman (1967), Judy Shirek Ellefson (1967), Jane Lancaster (1967), Suzanne Chevalier-Skolnikoff (1971), Mary Ellen Morebeck (1972), Naomi Bishop (1975), Shirley Strum (1976), Sheila Curtain (1976), Jane Bogess (1976), and Elizabeth McCown (1977).

40. Stephen O. Murray, "A 1978 Interview with Mary R. Haas," *Anthropological Linguistics* 39, no. 4 (1997): 695–722.

41. Laura Nader, "Up the Anthropologist—Perspectives Gained from Studying Up," in *Nature, Culture and Gender*, ed. Carol MacCormack and Marilyn Strathern (Cambridge University Press, 1980), 284–311.

42. Gerald Berreman, "Is Anthropology Alive? Social Responsibility in Social Anthropology," *Current Anthropology* 9, no. 5 (1968): 391–396; Elizabeth Colson, *The Social Consequences of Resettlement: The Impact of the Kariba Resettlement upon the Gwembe Tonga*, vol. 4 of Kariba Studies (Manchester: University of Manchester Press, 1971).

43. David Burner, *Making Peace with the 60s* (Princeton, NJ: Princeton University Press, 1996), 136–147.

44. Adrienne Zihlman, interview by author, 24 October 2011.

45. Adrienne Zihlman, "Human Locomotion: A Reappraisal of the Functional and Anatomical Evidence" (PhD diss., University of California, Berkeley, 1967); Nancy Tanner, "Minangkabu Disputes" (PhD diss., University of California, Berkeley, 1972).

46. Zihlman, interview by author; see also Haraway, "The Paleoanthropology of Sex and Gender," in *Primate Visions*, 331–348.

47. P. B. Medawar, "The Ape Redressed," *New York Review of Books* (8 March 1973), 21; Alexander Alland Jr., *The Human Imperative* (New York: Columbia University Press, 1972).

48. W. G. Berl, "A Brief Prospectus of the 1971 AAAS Annual Meeting," *Science* 174, no. 4011 (1971): 847–856, on 853.

49. Jacob Arlow, "The Role of Aggression in Human Adaptation," *Science* 174, no. 4008 (1971): 526; Arie Y. Lewin, "Women in Academia," *Science* 174, no. 4011 (1971): 859.

50. Zihlman, "Human Locomotion," 80.

51. Tanner, "Minangkabu Disputes"; Nancy Tanner, "Speech and Society among the Indonesian Elite, a Case Study of a Multilingual Community," *Anthropological Linguistics* 9, no. 3 (1967): 15–40; Nancy Tanner, "Disputing and the Genesis of Legal Principles: Examples from the Minangkabu," *Southwestern Journal of Anthropology* 26, no. 4 (1970): 375–401.

52. Tanner and Zihlman, "Women in Evolution. Part I."

53. Zihlman, interview by author.

54. Sarah Blaffer Hrdy, interview by author, 27 October 2011; Sarah Blaffer Hrdy, "Empathy, Polyandry and the Myth of the Coy Female," in *Feminist Approaches to Science*, ed. Ruth Bleier (New York: Pergamon, 1986): 119–146; reprinted in *Conceptual Issues in Evolutionary Biology*, ed. Elliott Sober (Cambridge, MA: MIT Press, 2006), 131–160.

55. Tanner and Zihlman, "Women in Evolution. Part I," 586.

56. Zihlman, interview by author.

57. Zihlman, "Women in Evolution, Part II." See also Adrienne Zihlman, "Women as Shapers of Human Adaptation," in *Woman the Gatherer*, ed. Frances Dahlberg (New Haven, CT: Yale University Press, 1981), 75–120.

58. Nancy Makepeace Tanner, *On Becoming Human* (New York: Cambridge University Press, 1981).

59. Journalists cared most about Sally Slocum's anthropological research when during her 1975 paper at the American Anthropological Association in San Francisco—"Strippers and Their Customers: Interaction at the Bar"— she revealed she had performed as Autumn Lee in order to pay for her graduate studies. The AP wire story from San Francisco appeared in the *Bloomington Pantagraph*, 6 December 1975, 7; the *Pottsdown Mercury*, 6 December 1975, 15; the *Baltimore Sun*, 8 December 1975, B1; and probably other local newspapers. See also Frank W. Martin, "Does Teaching College Have Its Bumps? Is It a Grind? Professor Sally (Autumn Lee) Slocum Knows," *People*, 27 June 1977, 78.

Chapter Eight. The Academic Jungle

1. Desmond Morris, *The Naked Ape: A Zoologist's Study of the Human Animal* (New York: McGraw Hill, 1967).

2. Robin Fox, *Participant Observer: Memoir of a Transatlantic Life* (New Brunswick, NJ: Transaction Publishers, 2004): 329–330. Fox's autobiography is narrated in the third person, as if by an anthropological observer. It reads like a cross between a bildungsroman and a campus novel. Burt Benedict, a London School of Economics colleague and good friend of Fox's, was also present. See also Alex Walter, "An Interview with Robin Fox," *Current Anthropology* 34, no. 4 (1993): 441–452.

3. Lionel Tiger, interview by author, 10 November 2011.

4. Lionel Tiger, "My Life in the Human Nature Wars," *Wilson Quarterly* (Winter 1966): 14–25. It crept back in his later writing, however; see Max Weber, *The Theory of Social and Economic Organization* (Glencoe, IL: Free Press, 1947), 106, as quoted by Tiger, *Men in Groups* (New York: Random House, 1969), 134.

5. Tiger, interview by author.

6. Robin Fox, interview by author, 9 November 2011. On Bowlby's influence in the United States, see Marga Vicedo, "The Social Nature of the Mother's Tie to Her Child: John Bowlby's Theory of Attachment in Post-War America," *British Journal for the History of Science* 44, no. 3 (2001): 401–426.

7. Fox, interview by author.

8. Noam Chomsky, *Aspects of the Theory of Syntax* (Cambridge, MA: MIT Press, 1965); Sigmund Freud, *Totem and Taboo: Some Points of Agreement between the Mental Lives of Savages and Neurotics* (New York: W. W. Norton, 1950).

9. Robert G. W. Kirk, "Between the Clinic and the Laboratory: Ethology and Pharmacology in the Work of Michael Robin Alexander Chance, c. 1946–1964," *Medical History* 53, no. 4 (2009): 513–536.

10. M.R.A. Chance, "Social Structure of a Colony of *Macaca mulatta*," *British Journal of Animal Behavior* 4 (1956): 1–13; Chance, "What Makes Monkeys Sociable?" *New Scientist* (5 March 1959): 520–523; Chance, "The Nature and Social Features of the Instinctive Social Bond of Primates," in *The Social Life of Early Man*, ed. Sherwood Washburn (New York: Viking Fund, 1961), 17–33; M.R.A. Chance and C. J. Jolly, *Social Groups of Monkeys, Apes and Men* (London: Cape, 1970). In these papers Chance first elaborated the idea of a male bond.

11. M.R.A. Chance, "Attention Structure as the Basis of Primate Rank Orders," *Man* 2, no. 4 (1967): 503–518.

12. Fox was also impressed by M.R.A. Chance's, "The Nature and Special Features of the Instinctive Social Bond of Primates," in *The Social Life of Early Man*, ed. Sherwood L. Washburn (London: Methuen, 1962): 17–33.

13. Lionel Tiger and Robin Fox, "The Zoological Perspective in Social Science," *Man*, n.s., 1, no. 1 (1966): 75–81.

14. Tiger, "My Life in the Human Nature Wars," 17.

15. Tiger and Fox, "Zoological Perspective," 76–77, 80. Tiger and Fox's arguments were often read as being biologically determinist. Yet they firmly believed that culture and experience modify behavior, too, and would not have identified as biological "determinists." At issue were differing beliefs regarding the *degree* of constraint.

16. Tiger, *Men in Groups*.

17. Robin Fox, *Kinship and Marriage: An Anthropological Perspective* (Baltimore: Penguin, 1967).

18. Lionel Tiger, "Male Dominance? Yes, Alas. A Sexist Plot? No." *New York Times*, 25 October 1970, SM18.

19. Simone de Beauvoir, *The Second Sex*, trans. H. M. Parshley (New York: Knopf, 1952); Betty Friedan, *The Feminine Mystique* (New York: Norton, 1963); Kate Millett, *Sexual Politics* (Garden City, NY: Doubleday, 1970).

20. Robin Fox, "The Evolution of Sexual Behavior," *New York Times*, 24 March 1968, SM32.

21. Contemporaneous academically oriented collections advanced a similar fascination with the idea. See, e.g., Richard B. Lee and Irven DeVore, eds., *Man the Hunter* (Chicago: Aldine, 1968).

22. Tiger, "Male Dominance?"

23. Tiger, Fox, and Ardrey attacked a simplistic vision of psychological theory more characteristic, perhaps, of the 1950s than of the changing gender dynamics of the profession in the 1960s; Jonathan Metzl, *Prozac on the Couch: Prescribing Gender in the Era of Wonder Drugs* (Durham, NC: Duke University Press, 2003).

24. Robert Ardrey, "A Tiger About to Stir up a Mare's Nest," *Life* , 20 June 1969, 11.

25. Tiger, interview by author.

26. Fox, *Participant Observer*, 321.

27. Fox, *Participant Observer*, 324.

28. Case in point: according to Fox, Ardrey was a small, round man who had told him, "I haven't seen my thing, except in the mirror, for a decade." Fox, *Participant Observer*, 349.

29. Fox, *Participant Observer*, 349.

30. See professional reviews: Rodney Needham, review of *Kinship and Marriage*, by Robin Fox, *Man* 3, no. 2 (1968): 324–325; Eugene A. Hammel, review of *Kinship and Marriage*, by Robin Fox, *American Anthropologist* 70, no. 5 (1968): 972–973.

31. Robin Fox, "Preface to the Transaction Edition," *Encounter with Anthropology*, 2nd ed. (New Brunswick, NJ: Transaction Publishers, 1991), 5.

32. Robin Fox, "The Cultural Animal," in *Encounter with Anthropology*, 311–339, originally published in *Social Science Information* 9, no. 1 (1970): 7–25.

33. Fox, "Cultural Animal," *Social Science Information*, 11.

34. Fox's statement echoes common sentiment, as most anthropologists at the time invoked the inseparability of nature and nurture.

35. Fox, "Cultural Animal," *Social Science Information*, 21.

36. Noam Chomsky, *Aspects of the Theory of Syntax* (Cambridge, MA: MIT Press, 1965); Chomsky, *Syntactic Structures* (1957; The Hague: Mouton, 1965). Fox, "Cultural Animal," *Social Science Information*, 13–15, quote on 15.

37. Robin Fox, "Primate Kinship and Human Kinship," in *Biosocial Anthropology* (New York: John Wiley & Sons, 1975): 9–35; Fox, "Alliance and Constraint: Sexual Selection in the Evolution of Human Kinship Systems," in *Sexual Selection and Descent of Man*, ed. Bernard Campbell (Chicago: Aldine, 1972).

38. He claimed inspiration from both Konrad Lorenz and Charles Darwin, especially Darwin's *The Expression of Emotions in Man and Animals* (London: John Murray, 1872). *Expression* has attracted far less historical attention than either *On the Origin of Species* or *Descent of Man and Selection in Relation to Sex*, but it was a particular favorite of scientists interested in the evolution of behavior. Lorenz, for example, wrote an introduction for a 1965 reprint, in which he hailed Darwin as the "patron saint" of ethology. Charles Darwin, *The Expression of Emotions in Man and Animals* (Chicago: University of Chicago Press, 1965), xi.

39. Robin Fox, "The Evolution of Human Sexual Behavior," *New York Times Magazine*, 24 March 1968, 32.

40. Fox, "Evolution of Human Sexual Behavior," quotes on 87, 97, 79, 100, 92.

41. Fox, "Cultural Animal," *Social Science Information*, 12.

42. After publication he occasionally claimed he had no initial hope that the book would reach such a wide audience, a defense against nonprofessional interpretations of his arguments; Lionel Tiger, "But I Never Said Women are Inferior," *Maclean's*, 1 January 1970, 61–63.

43. Quotes respectively from Joan Cook, "Explaining Why Men Love to Be One of the Boys," *New York Times*, 21 June 1969, 14; and Marylin Bender, "No Time for Dandies?" *New York Times*, 14 September 1969, SM2A2. Tiger also noted that "when I wrote my book *Men in Groups*, everyone assumed I was a homosexual. One of the bases was that homosexuals wore colorful clothes. I realized how irrelevant that was to the issue, and I would never answer the question or do anything that would make my own person relative to it." Cynthia Fuchs Epstein and Lionel Tiger, "[Debate] Will Women's Lib Really Help Women," *Sexual Behavior* 1, no. 6 (1971): 56–67, quote from Tiger on 67.

44. Tiger, *Men in Groups*, xiii. These questions were also printed on the cover of his book.

45. Tiger, *Men in Groups*, 196. Within a few years, "unisexual" selection would become known by its Darwinian name, intrasexual selection (selection within a single sex), as typically manifested by male-male competition.

46. Tiger, *Men in Groups*, 202–203.

47. Tiger, *Men in Groups*, 214.

48. For a historical perspective, see Erika Lorraine Milam and Robert Nye, eds., *Scientific Masculinities*, *Osiris* 30 (2015).

49. Elaine Morgan, *The Descent of Woman* (New York: Stein and Day, 1972), 190–193.

50. Christopher Lehmann-Haupt, "The Disturbing Rediscovery of the Obvious," *New York Times*, 18 June 1969, 45.

51. Millett, *Sexual Politics*, 32.

52. Sherwood Washburn, "Does Biology Account for the Men's Club?" *New York Times*, 27 July 1969, BR10.

53. Ardrey, "Tiger About to Stir up a Mare's Nest," 12.

54. Ardrey, "Tiger About to Stir up a Mare's Nest," 12.

55. Lionel Tiger and Robin Fox, *The Imperial Animal* (1971; New York: Henry Holt, 1989). The reprinted version includes a foreword by Konrad Lorenz and a new introduction by the authors.

56. Tiger and Fox, *Imperial Animal*, 55.

57. Tiger and Fox, *Imperial Animal*, 236.

58. Fox, *Participant Observer*, 437.

59. Elizabeth Fisher, "Nature and the Animal Determinists," *Nation*, 17 January 1972, 89–90.

60. Adrienne Zihlman, "The Imperial Animal," *American Journal of Physical Anthropology* 39, no. 1 (1973): 145–146.

61. Fox, *Participant Observer*, 395.

62. Robin Fox, *Encounter with Anthropology* (New York: Harcourt, Brace, Jovanovich, 1973), 39–40.

63. Lionel Tiger and Joseph Shepher, *Women in the Kibbutz* (New York: Harcourt Brace Jovanovich, 1975).

64. See Ullica Segerstråle's description of the conference in *Defenders of the Truth: The Battle for Science in the Sociobiology Debate and Beyond* (New York: Oxford University Press, 2000): 90–94.

65. Quotes from Tiger, author interview.

66. Fox, *Participant Observer*, 395.

67. Fox, *Participant Observer*, 559.

Chapter Nine. The Edge of Respectability

1. Elaine Morgan, *The Descent of Woman* (New York: Stein and Day, 1972), 57.

2. Morgan, *Descent of Woman*, 3.

3. Antony Jay, *Corporation Man: Who He Is, What He Does, Why His Ancient Tribal Impulses Dominate the Life of the Modern Corporation* (New York: Random House, 1971). Forty years later, he continued to use anthropological comparisons to describe the importance of institutions to the stability of civilization: Antony Jay, "Confessions of a Reformed BBC Producer" (London: Center for Policy Studies, 2007). He is most well known for his later work on *Yes, Minister*.

4. Antony Jay, *Management and Machiavelli* (New York: Holt, Rinehart and Winston, 1967).

5. Antony Jay to Robert Ardrey, 19 September 1972, Box 2, Folder 7, Robert Ardrey Papers, 1955–1980, Special Collections and University Archives, Rutgers University Libraries, New Brunswick, NJ (hereafter cited as Ardrey Papers).

6. Sloan Wilson, *Man in the Gray Flannel Suit* (New York: Simon & Schuster, 1955)—it was almost immediately made into a movie starring Gregory Peck and Jennifer Jones (Twentieth

Century Fox, 1956), 153 min. David Riesman, in collaboration with Reuel Denney and Nathan Glazer, *The Lonely Crowd: A Study of the Changing American Character* (New Haven, CT: Yale University Press, 1950).

7. Elizabeth Fraterrigo, Playboy *and the Making of the Good Life in Modern America* (Oxford: Oxford University Press, 2009); Erika Lorraine Milam, "Science of the Sexy Beast: Biological Masculinities and the *Playboy* Lifestyle," in *Groovy Science: The Counterculture's Embrace of Science in the Long 1970s*, ed. David Kaiser and W. Patrick McCray (Chicago: University of Chicago Press, 2016), 270–302.

8. Helen Gurley Brown, *Sex and the Single Girl* (New York: B. Geis Associates, 1962).

9. Jay, *Corporation Man*, 8–11.

10. Antony Jay, "The White-Collar Ape," *Time*, 27 September 1971, 100.

11. See also Timothy Colton, "The 'New Biology' and the Causes of War," *Canadian Journal of Political Science / Revue Canadienne de Science Politique* 2, no. 4 (1969): 434–447. The term "new biology" was also used at the time to refer to molecular biology; see, for example, Garland E. Allen, "The New Biology: The Meaning behind the Recent Revolution," *Bios* 40, no. 2 (1969): 43–57.

12. Jay, *Corporation Man*, 19.

13. Morgan's semantic point was later adopted by anthropologists and linguists; Virginia L. Warren, "Guidelines for the Nonsexist Use of Language," *Proceedings and Addresses of the American Philosophical Association* 59, no. 3 (1986): 471–484.

14. Morgan, *Descent of Woman*, 56.

15. Morgan, *Descent of Woman*, 156.

16. Morgan, *Descent of Woman*, 165.

17. Ardrey as excerpted and discussed in Morgan, *Descent of Woman*, 180–181. For the (slightly different) original wording, see Robert Ardrey, *Territorial Imperative: A Personal Inquiry into the Animal Origins of Property and Nations* (New York: Atheneum, 1966), 227.

18. Morgan, *Descent of Woman*, 159.

19. Morgan, *Descent of Woman*, 190–193.

20. Morgan, *Descent of Woman*, 160–163.

21. Morgan, *Descent of Woman*, 179.

22. Margaret Mead, "On the Family and Growing Up," recorded 26 September 1972, New York, sound cassette (New York: Encyclopedia Americana/CBS News Audio Resource Library), 21 min.

23. Desmond Morris, *The Naked Ape: A Zoologist's Study of the Human Animal* (New York: McGraw Hill, 1967), 43–45; Morgan, *Descent of Woman*, 24–25.

24. Nadine Weidman, "Popularizing the Ancestry of Man: Robert Ardrey and the Killer Instinct," *Isis* 102, no. 2 (2011): 269–299, 288–289.

25. N. B. Marshall, "Alister Clavering Hardy, 10 February 1896–22 May 1985," *Biographical Memoirs of Fellows of the Royal Society* 32 (1986): 222–273.

26. Sir Alister Hardy, "Was Man More Aquatic in the Past?," *New Scientist*, 17 March 1960, 642–645.

27. A. C. Hardy, "Was There a *Homo aquaticus*?," *Zenith* (Oxford University Scientific Society) 15, no. 1 (1977), 4–6.

28. Desmond Morris, "Biologist with a Broader Outlook: An Appreciation of Sir Alister Hardy," *Oxford Times*, 31 May 1985, reprinted in Michael G. Hardy, "Sir Alister Hardy and the Aquatic Ape Theory: A Brief Account of His Life," *Nutrition and Health* 9 (1993): 161–164, quote on 163.

29. Elaine Morgan to Ernest Hecht, Souvenir Press, 21 November 1971, Series 2, Box 18, Folder 1, Stein and Day Publisher Records, Rare Book and Manuscript Library, Columbia University, New York (hereafter cited as Stein and Day Records).

30. Sol Stein to Ernest Hecht, 29 November 1971, Box 18, Folder 1, Series 2, Stein and Day Records.

31. Ardrey to David Wolper, 19 June 1972, Box 2, Folder 14, Ardrey Papers.

32. Memo Vicki to Jethro, cc: Sheri Sol, "DESCENT OF WOMAN—Morgan," 9 August 1972, Box 18, Folder 1, Series 2, Stein and Day Records.

33. "Books," *Playboy*, June 1972, 20.

34. Hugh Kenner, "Reviewer's Choice," *Life*, 14 July 1972, 20; see also Mary Morain, "The Endocentric [sic] Bias," *Humanist* 33, no. 2 (1973): 44.

35. Jurate Kazickas, "Is the Missing Link a Mermaid?," *Stars and Stripes*, Thursday, 6 July 1972, 14–15; Morain, "Endocentric Bias," 44.

36. John Pfeiffer, "The Descent of Woman," *New York Times*, 25 June 1972, BR6.

37. For example, Richard B. Lee, "Politics, Sexual and Non-Sexual in an Egalitarian Society," *Social Science Information* 17, no. 6 (1978): 871–895, 871; Marie Withers Osmond, "Cross-Societal Family Research: A Macrosociological Overview of the Seventies," *Journal of Marriage and Family* 42, no. 4 (1980): 995–1016, 1004; Martin J. Waterhouse, review of *On Becoming Human*, by Nancy Tanner, *Man* 17, no. 2 (1982): 352; Margaret W. Conkey and Janet D. Spector, "Archeology and the Study of Gender," *Advances in Archeological Method and Theory* 7 (1984): 1–38, 7 and 14; Hilary Callan, "The Imagery of Choice in Sociobiology," *Man* 19, no. 3 (1984): 404–420, 418; Donna J. Haraway, "Primatology Is Politics by Other Means," *PSA: Proceedings of the Biennial Meeting of the Philosophy of Science Association* 2 (1984): 489–524, 503; Nancy Tuana, "Re-Presenting the World: Feminism and the Natural Sciences," *Frontiers: A Journal of Women Studies* 8, no. 3 (1986): 73–75; Matt Cartmill, "Paleoanthropology: Science or Mythological Charter?" *Journal of Anthropological Research* 58, no. 2 (2002): 183–201.

38. Adrienne Zihlman, "Gathering Stories for Hunting Human Nature," *Feminist Studies* 11, no. 2 (Summer 1985): 365–377, on 367; see also Robert Attenborough, "Between Men and Women," *Nature* 319 (1986): 271–272; and Helen E. Longino, "Science, Objectivity, and Feminist Values," *Feminist Studies* 14, no. 3 (1988): 561–574.

39. In "Gathering Stories," Zihlman mentions C. Owen Lovejoy, "The Origin of Man," *Science* 211, no. 4480 (1981): 341–350.

40. Jerold M. Lowenstein and Adrienne L. Zihlman, "The Wading Ape: A Watered-Down Version of Human Evolution," *Oceans* 13 (May–June 1980): 3–6, 6.

41. Morgan, *Descent of Woman*, 17, 14, 15, 17.

42. Morgan, *Descent of Woman*, 21–22.

43. Morgan, *Descent of Woman*, 113–130.

44. Morgan, *Descent of Woman*, 131.

45. Robert Ardrey and Antony Jay used similar tactics: Ardrey, *African Genesis: A Personal*

Investigation into the Animal Origins and Nature of Man (1961; New York: Atheneum, 1968), 252–257.

46. Morgan, *Descent of Woman*, 28, 30.

47. Jordan Bonafonte, "The Naked Ape Is All Wet, Says a Liberated Lady," *Life*, 21 July 1973, 77.

48. Pfeiffer, "Descent of Woman," BR6.

49. Robert Ardrey to David Wolper, 19 June 1972, Folder 14, Box 2, MC 190, Ardrey Papers. Ardrey refers here to Germaine Greer, *The Female Eunuch* (London: MacGibbon & Kee, 1970) and Kate Millett, *Sexual Politics* (Garden City, NY: Doubleday, 1970).

50. Lowenstein and Zihlman, "Wading Ape," 3.

51. Lowenstein and Zihlman, "Wading Ape," 6; Erich von Däniken, *Chariots of the Gods? Unsolved Mysteries of the Past*, trans. Michael Heron (New York: Putnam, 1968). A couple of years later, Zihlman also compared the aquatic ape to Immanuel Velikovsky's *Worlds in Collision* (New York: Macmillan, 1950); Adrienne Zihlman, review of *The Monkey Puzzle: Reshaping the Evolutionary Tree*, by John Gribbin and Jeremy Cherfas (New York: Pantheon Books, 1982), *American Anthropologist* 85, no. 2 (June 1983): 458–459. See also Ian Tattersall and Niles Eldredge, "Fact, Theory, and Fantasy in Human Paleontology," *American Scientist* 65, no. 2 (1977): 204–211, 207.

52. Elaine Morgan, *The Aquatic Ape* (New York: Stein and Day, 1982); Barbara Miner, "Author Waters Down Anthropological Views," *Hutchinson News*, 2 April 1983, 10; Kate Douglas, "Natural Optimist," interview with Elaine Morgan, *New Scientist*, 23 April 2005, 50–53.

53. Jennifer Rees, "Trigger of Change," *New Scientist*, 27 May 1982, 592.

54. Jerold M. Lowenstein, "Swimmers or Swingers," *Oceans* 17 (July–August 1984): 72.

55. Elaine Morgan, *Scars of Evolution* (London: Souvenir Press, 1990); Morgan, *Descent of the Child: Human Evolution from a New Perspective* (Oxford University Press, 1995).

56. Academic reviews of the book are sparse: Dwight E. Robinson called it a "social-biologic fling" in his review in *Academy of Management Journal* 15, no. 3 (1972): 345–353, on 351. Glendon Schubert mentions the book only in a footnote as working out Tiger and Fox's suggestion in *The Imperial Animal* that evolution could be used to understand entrepreneurial culture, *Polity* 6, no. 2 (1973): 240275, on 274.

57. Robert Ardrey, "The Hunter Home from the Office," *Life*, 22 October 1971, 10.

58. Judith Shapiro, letter to the editor, *New York Magazine*, 11 October 1971, 5. Shapiro had a long and impressive career, serving as the tenth president of Barnard College from 1994 to 2008.

59. Susan Kreisler, letter to the editor, *New York Magazine*, 11 October 1971, 5.

60. Jay to Ardrey, 12 May 1972, Box 2, Folder 7, Ardrey Papers.

61. On demarcation, see Thomas Geiryn, *Cultural Boundaries of Science: Credibility on the Line* (Chicago: University of Chicago Press, 1999); James Rodger Fleming, *Fixing the Sky: The Checkered History of Weather and Climate Control* (New York: Columbia University Press, 2010); Naomi Oreskes and Erik Conway, *Merchants of Doubt: How a Handful of Scientists Obscured the Truth on Issues from Tobacco Smoke to Global Warming* (New York: Bloomsbury, 2010); Michael Gordin, *The Pseudoscience Wars: Immanuel Velikovsky and the Birth of the Modern Fringe* (Chicago: University of Chicago Press, 2012). In terms of evolutionary theory specifically, see Philip Kitcher's *Abusing Science: The Case against Creationism* (Cambridge, MA: MIT Press, 1982).

62. Gordin, "Experiments in Rehabilitation," in *Pseudoscience Wars*, 106–134.

63. Zihlman, interview by author.

Part Four. Political Animals

Martin Luther King Jr. acceptance speech for the Nobel Peace Prize, Oslo, Sweden, 10 December 1964.

1. "Arizona Shootout," *Time*, 20 September 1976, 24.

2. Cong. Rec. H9497 (9 April 1975) (statement of Rep. John B. Conlan). Conlan quoted from an article included in the MACOS teaching materials: Geoffrey Gorer, "Man Has No 'Killer' Instinct," *New York Times*, 27 November 1966, reprinted in MACOS, *Seminars for Teachers* (Washington, DC: Curriculum Development Associate, 1970), 52–55, quote on 54. Teachers were asked to read Gorer's article alongside materials authored by Irenaus Eibl-Eibesfeldt, Leonard Berkowitz, Anthony Storr, and Edmund Leach, all trying to answer the question "Is man innately aggressive?" Teachers were then instructed to ask themselves "What do various fields of study contribute to the question?" and "What are the consequences of the different approaches?" (40).

3. On the fate of scientific concepts during these decades, see David Kirby, "Censoring Science in 1930s and 1940s Hollywood Cinema," in *Hollywood Chemistry: When Science Met Entertainment*, ed. Donna Nelson, Kevin Grazier, Jamie Paglia, and Sidney Perkowitz (New York: American Chemical Society, 2013), 229–240. Comic books fell under similar scrutiny in the 1950s thanks to Frederic Wertham's sensationalist *Seduction of the Innocent* (New York: Rinehart, 1954).

4. David Kirby, "Darwin on the Cutting Room Floor: Evolution, Religion, and Film Censorship," (unpublished manuscript, 2017); see also David Kirby, "The Devil in Our DNA: A Brief History of Eugenics in Science Fiction Films," *Literature and Medicine* 26, no. 1 (2007): 83–108.

5. Stephen Weinberger, "Joe Breen's Oscar," *Film History* 17, no. 4 (2005): 380–391.

6. Jerold Simmons, "The Production Code under New Management: Geoffrey Shurlock, 'The Bad Seed,' and 'Tea and Sympathy,'" *Journal of Popular Film and Television* 22, no. 1 (1994): 2–10.

7. Gene Siskel, "The Movies: See No Evil," *Chicago Tribune*, 22 April 1970, C5; Gary Arnold, "Liberation of a Censor," *Washington Post*, 14 June 1970, E2.

8. Andrew Sarris, "A Hollywood Censor Reasons and Remembers," *New York Times*, 19 July 1970, 193; John J. O'Connor, "What's Off at the Pictures," *Wall Street Journal*, 30 November 1970, 16. Christian conservatives worried about both the content and political messages of films: see Frank R. Walsh, *Sin and Censorship: The Catholic Church and the Motion Picture Industry* (New Haven, CT: Yale University Press, 1996); and William Romanowski, *Reforming Hollywood: How American Protestants Fought for Freedom at the Movies* (New York: Oxford University Press, 2012). Concerns over a left liberal bias in Hollywood have not disappeared: e.g., Ronald Radosh and Allis Radosh, *Red Star over Hollywood: The Film Colony's Long Romance with the Left* (New York: Encounter, 2006); Ben Shapiro, *Primetime Propaganda: The True Hollywood Story of How the Left Took Over Your TV* (New York: Broadside, 2011).

9. On violence in film, see J. David Slocum, ed., *Violence and American Cinema* (New York:

Routledge, 2001); Stephen Prince, *Savage Cinema: Sam Peckinpah and the Rise of Ultraviolent Movies* (Austin: University of Texas Press, 1998).

10. Michael Anderegg, ed., *Inventing Vietnam: The War in Film and Television* (Philadelphia: Temple University Press, 1991); Gary Gerstle, *American Crucible: Race and Nation in the Twentieth Century* (Princeton, NJ: Princeton University Press, 2001); Daniel C. Hallin, *The "Uncensored War": The Media and Vietnam* (Berkeley: University of California Press, 1986); Susan Sontag, *Regarding the Pain of Others* (New York: Penguin Books, 2004).

11. Thomas Sugrue, *The Origins of the Urban Crisis: Race and Inequality in Postwar Detroit* (Princeton, NJ: Princeton University Press, 1996), xvii. On conservative reactions to such "race riots," see Dan Carter's *The Politics of Rage: George Wallace, the Origins of the New Conservatism, and the Transformation of American Politics*, 2nd ed. (Baton Rouge: Louisiana State University Press, 2000).

12. Other assassinations in the United States during the 1960s included the civil rights activist Medgar Evers (June 1963), Malcolm X (February 1965), and George Rockwell, leader of the American Nazi Party (August 1967).

13. Clay Risen, *A Nation on Fire: America in the Wake of the King Assassination* (Hoboken, NJ: John Wiley & Sons, 2009).

14. Michael Couzens (of the Brookings Institution), "Reflections on the Study of Violence," *Law & Society Review* 5, no. 4 (1971): 583–604; Julian E. Zelizer, introduction to *The Kerner Report: National Advisory Commission on Civil Disorders* (1968; Princeton, NJ: Princeton University Press, 2016), xiii–xxxvi.

15. *Kerner Report*, 1.

16. Zelizer, introduction to *Kerner Report*.

17. National Commission on the Causes and Prevention of Violence, Task Force Reports (Washington, DC: US Government Printing Office, 1969): H. D. Graham and T. R. Gurr, eds., *Violence in America: Historical and Comparative Perspectives*; J. H. Skolnick, *The Politics of Protest*; J. R. Sahid, director, *Rights in Concord: Response to the Counter-Inaugural Protest Activities in Washington, D.C., January 18–20, 1969*; L. H. Masotti and J. R. Corsi, *Shoot-Out in Cleveland: Black Militants and the Police*; W. H. Orrick Jr., dir., *Shut It Down! A College in Crisis: San Francisco State College, October 1968–April 1969*; G. D. Newton Jr. and F. E. Zimring, *Firearms and Violence in American Life*; J. F. Kirkham, S. G. Levy, and W. J. Crotty, *Assassination and Political Violence*; R. K. Baker, and S. J. Ball, *Mass Media and Violence*; J. S. Campbell, J. R. Sahid, and D. P. Stand, *Law and Order Reconsidered*; D. Walker, Director, *Rights in Conflict* (1968; published additionally by Bantam Books); L. J. Hector and P. L. E. Helliwell, directors, *Miami Report*; D. J. Mulvihill, M. M. Tumin, and L. A. Curtis, *Individual Acts of Violence* (1970). On the commission's contention that Americans suffered from a "historical amnesia" regarding violence and race in the country, see Richard Hofstadter and Michael Wallace, eds., *American Violence: A Documentary History* (New York: Knopf, 1970).

18. On the history of debates over the content of nature films and television shows, see Gregg Mitman, *Reel Nature: America's Romance with Wildlife on Film* (Cambridge, MA: Harvard University Press, 1999). In his analysis, after the Second World War television producers domesticated nature for younger audiences and carefully crafted their shows to reflect Christian morals. On the one hand, depictions of violence in nature continued to convey a sense of authenticity and garnered a wide viewership but were typically deemed inappropriate for

educational shows; on the other, too-frequent violent interactions between animals risked overly dramatizing nature to the point of fakery.

19. Anthropologists who fought against the killer ape conception of humanity repeatedly expressed this concern in their publications. See, e.g., Robert Hinde, "The Nature of Aggression," *New Society* 9 (1 March 1967), 302–304; Ashley Montagu, ed., *Man and Aggression* (New York: Oxford University Press, 1968); and David Pilbeam, "The Fashionable View of Man as a Naked Ape is 1. An Insult to Apes, 2. Simplistic, 3. Male-Oriented, 4. Rubbish," *New York Times,* 3 September 1972, SM10.

20. Garth S. Jowett, Ian C. Jarvie, and Kathryn H. Fuller, *Children and the Movies: Media Influences and the Payne Fund Controversy* (New York: Cambridge University Press, 1996). According to the historian Mark Lynn Anderson, the published reports never garnered a wide public readership but through discussion in articles in newspapers and magazines shaped how intellectuals thought about media; see "Taking Liberties: The Payne Fund Studies and the Creation of the Media Expert," in *Inventing Film Studies*, ed. Lee Grieveson and Haidee Wasson (Durham, NC: Duke University Press, 2008), 38–65.

21. Shearon Lowery and Melvin L. DeFleur, *Milestones in Mass Communication Research: Media Effects* (New York: Longman, 1983).

22. Meg Jacobs, *Panic at the Pump: The Energy Crisis and the Transformation of American Politics in the 1970s* (New York: Farrar, Straus and Giroux, 2016).

23. Mark Solovey, *Shaky Foundations: The Politics–Patronage–Social Science Nexus in Cold War America* (New Brunswick, NJ: Rutgers University Press, 2013).

24. John Denton Carter, *The Warren Court and the Constitution: A Critical View of Judicial Activism* (Nashville, TN: Pelican Publishing, 1973); John P. Jackson Jr., *Social Scientists for Social Justice: Making the Case Against Segregation* (New York: New York University Press, 2001).

25. C. R., "Scientists Defend Studies of Foreign Frogs and Lizards," *BioScience* 24, no. 10 (1974): 607.

26. Daniel Rodgers, *Age of Fracture* (Cambridge, MA: Harvard University Press, 2011).

27. For example, John C. Whitcomb and Henry M. Morris, *The Genesis Flood: The Biblical Record and Its Scientific Implications* (Philadelphia: Presbyterian and Reformed Publishing, 1961); Immanuel Velikovsky, *Worlds in Collision* (New York: Doubleday & Company, 1950), *Ages of Chaos: From the Exodus to King Akhnaton* (Garden City, NY: Doubleday, 1952), and *Earth in Upheaval* (Garden City, NY: Doubleday, 1955); Hal Lindsey and C. C. Carlson, *The Late Great Planet Earth* (Grand Rapids, MI: Zondervan, 1970), reissued in 1973 by Bantam Books. See Matthew Sutton, *American Apocalypse: A History of Modern Evangelicalism* (Cambridge, MA: Harvard University Press, 2014); and Michael Gordin, *The Pseudoscience Wars: Immanuel Velikovsky and the Birth of the Modern Fringe* (Chicago: University of Chicago Press, 2012).

Chapter Ten. The White Problem in America

1. Joseph A. Towles and Colin M. Turnbull, "The White Problem in America," *Natural History,* June–July 1968, 6–18. Although they thought of their partnership in both emotional and intellectual terms, this was their sole joint publication and Towles's only published foray into anthropology. On their stormy relationship, see Roy Richard Grinker, *In the Arms of Africa: The*

Life of Colin M. Turnbull (New York: St. Martin's Press, 2000). On the difficulties created by unequal professional status in married couples, see Helena M. Pycior, Nancy G. Slack, and Pnina Abir-Am, eds., *Creative Couples in the Sciences* (New Brunswick, NJ: Rutgers University Press, 1996).

2. Grinker suggests that Towles instigated their joint authored paper on urban violence and that the religious imagery in the article belonged to him, not Turnbull; Grinker, *In the Arms of Africa*, 120, 191–192.

3. These studies existed alongside sociological inquiries into the relationship between collective violence, poverty, and crime, such as the twelve-volume report of the National Commission on the Causes and Prevention of Violence; see note 17 in "Part 4: Political Animals."

4. *The Kerner Report: National Advisory Commission on Civil Disorders* (1968; Princeton, NJ: Princeton University Press, 2016), 2. See also Kevin Kruse, *White Flight: Atlanta and the Making of Modern Conservatism* (Princeton, NJ: Princeton University Press, 2005).

5. James Baldwin, "Letter from a Region of My Mind," *New Yorker*, 17 November 1962, 59.

6. H. V. Savitch, "Black Cities/White Suburbs: Domestic Colonialism as an Interpretive Idea," *Annals of the American Academy of Political and Social Science* 439 (September 1978): 118–134; Thomas Borstelmann, *The Cold War and the Color Line: American Race Relations in the Global Arena* (Cambridge, MA: Harvard University Press, 2003); Mary Dudziak, *Cold War Civil Rights: Race and the Image of American Democracy* (Princeton, NJ: Princeton University Press, 2001); Timothy B. Tyson, *Radio Free Dixie: Robert F. Williams and the Roots of Black Power* (Chapel Hill: University of North Carolina, 1999).

7. Baldwin, "Letter from a Region of My Mind," 144.

8. Norman Podhoretz, "My Negro Problem—And Ours," *Commentary*, February 1963, 93–101, quote on 98. Podhoretz soon joined the swelling ranks of neoconservatives; Michael Kimmage, *The Conservative Turn: Lionel Trilling, Whittaker Chambers, and the Lessons of Anti-Communism* (Cambridge, MA: Harvard University Press, 2009), 291–302.

9. Norman Podhoretz, " 'My Negro Problem and Ours' at 50," *Commentary*, May 2013, 11–17.

10. Lerone Bennett Jr., "Introduction: The White Problem in America," *Ebony*, August 1965, later published as a book of the same name (Chicago: Johnson Publishing, 1966). Johnson Publishing Company owned not only *Ebony* magazine but also *Hue*, *Jet*, and *Negro Digest*.

11. Bennett, "White Problem in America," 36.

12. Ferry's friends thought of him as a "curmudgeon with . . . a razor-sharp wit, and always, a cause on his tongue": James Arthur Ward, *Ferrytale: The Career of W. H. "Ping" Ferry* (Stanford, CA: Stanford University Press, 2001), xv.

13. Ardrey to Ferry, 27 March 1968, Box 1, Folder 21, Robert Ardrey Papers, 1955–1980, Special Collections and University Archives, Rutgers University Libraries, New Brunswick, NJ (hereafter cited as Ardrey Papers). The only direct accusation of racism I have found came from William Burroughs (as related by Lionel Tiger in a letter to Ardrey), who called *African Genesis* "a racist book" because so much of his evidence "came from South African racists": Tiger to Ardrey, 1 November 1968, Box 1, Folder 23, Ardrey Papers. Burroughs claimed he had used Ardrey's book in crafting his account of the 1968 Democratic National Convention held in Chicago in which Homer Mandrill, a "purple-assed baboon" and conservative direct descendant of Dart's illustrious line of aggressive man-apes from the South African veldt, runs for president:

William Burroughs, "The Purple Better One" (read by Burroughs, audio CD, disc 3 of *The Best of William Burroughs*, Giorno Poetry Systems, 1998, 4 min., 11 sec.), in which he reflects on "The Coming of the Purple Better One," *Esquire*, November 1968, 89–91. On the justifiable association of social scientific research with social reform, see John P. Jackson Jr., *Social Scientists for Social Justice: Making the Case against Segregation* (New York: New York University Press, 2001).

14. Ardrey to Ferry, 5 February 1968, Box 1, Folder 21, Ardrey Papers. Ardrey was referring to Ferry's "standard lecture"—W. H. Ferry, "Farewell to Integration," Century 21 Lecture at Stanford University, 8 November 1967, published in *Center Magazine*, March 1968, 34–40 and *Liberator*, January 1968, 4–11.

15. On the complex politics of Ferry's proposal and other schemes, see Ward, *Ferrytale*, quote on 103.

16. As early as 1962, Baldwin repeated unsubstantiated rumors that members of the John Birch Society had contributed to the cause of black nationalism in an effort to keep the races separate: Baldwin, "Letter from a Region of My Mind."

17. Ardrey to Ferry, 5 February 1968, Box 1, Folder 21, Ardrey Papers.

18. For a nuanced analysis of scientific ideas about racial hybridization, see Paul Farber, *Mixing Races: From Scientific Racism to Modern Evolutionary Ideas* (Baltimore: Johns Hopkins University Press, 2010).

19. Ardrey's conviction that African Americans would eventually control urban centers and whites would be relegated to the suburbs echoed contemporary fictional and social scientific visions of the future; e.g., Sol Yurick's *The Warriors* (1965), later turned into a cult-classic film of the same name with a whiter cast of characters: Walter Hill, dir. (Paramount Pictures, 1979), 92 min. Afrofuturist science fiction offered a series of more hopeful cultural messages; e.g., Alondra Nelson, ed., "Afrofuturism," special issue, *Social Text* 20, no. 2 (2002): 1–146.

20. Based at the Center for the Study of Democratic Institutions in Santa Barbara, California, the *Center Magazine* published its first issue in the fall of 1967.

21. Ardrey to Ferry, 27 March 1968, Box 1, Folder 21, Ardrey Papers, all emphases in the original. I have found no evidence of these opinions in print, in *Center Magazine* or elsewhere.

22. Ferry to Ardrey, 5 April 1968, Box 1, Folder 21, Ardrey Papers. See Kwame Ture (formerly known as Stokely Carmichael) and Charles V. Hamilton, *Black Power: The Politics of Liberation* (1967; New York: Vintage, 1992). In 1968 Carmichael moved to Guinea, where he worked with Kwame Nukruma (now exiled from Ghana), the subject of Lionel Tiger's dissertation and the origins of his fascination with charisma.

23. Richard Hofstadter, "The Future of American Violence," *Harper's Magazine*, 1 April 1970, 47–53.

24. Towles and Turnbull, "White Problem in America," 7.

25. Towles and Turnbull, "White Problem in America," 8.

26. For example, Frantz Fanon, *The Wretched of the Earth*, preface by Jean-Paul Sartre, trans. Constance Farrington (1961; New York: Grove Weidenfeld, 1991); Julian E. Zelizer, introduction to *Kerner Report*, xiii–xxxvi.

27. "21 Beastly Days in Africa" (British Overseas Airways Corporation Advertisement), *Natural History*, June–July 1968, 9; "The Civilized Safari" (Braniff International's Amazon Safari), *Natural History*, June–July 1968, 15.

28. Towles and Turnbull, "White Problem in America," 14.

29. Towles and Turnbull, "White Problem in America," 16, 18.

30. Colin Turnbull, *The Mountain People* (New York: Simon and Schuster, 1972).

31. Turnbull's carefully crafted intimacy was just that. Towles appeared only briefly in his preface and in one photograph, despite living with Turnbull for much of the time he spent with the Ik; Turnbull, *Mountain People*, 12. In his biography of Turnbull, the anthropologist Robert Grinker calls it "a book about Colin as much as it was about the Ik"; Grinker, *In the Arms of Africa*, 214.

32. Turnbull, *Mountain People*, 157.

33. Turnbull, *Mountain People*, 261.

34. Turnbull, *Mountain People*, 286.

35. Colin Turnbull to Margaret Mead, 2 January 1966, Box C78, Folder 1966, Margaret Mead Papers and South Pacific Ethnographic Archives, 1838–1996, MSS32441, Library of Congress, Washington, DC.

36. Turnbull, *Mountain People*, 293. He reiterated this point in a brief essay, "Dying Laughing," in the *New York Times*, 29 November 1972, 45.

37. In a new dedication written for the UK edition, he wrote "For the Ik, whom I learned not to hate, and for Joe, who helped me to learn." Colin Turnbull, *The Mountain People* (London: Jonathan Cape, 1973).

38. Colin Turnbull, "Human Nature and Primal Man," *Social Research* 40, no. 3 (1973): 511–530.

39. David Hapgood for the *Washington Post* and *Times Herald*, 29 October 1972, BW1.

40. Mary Daniels, "The Ik; A Study of the Loveless People," *Chicago Tribune*, 5 November 1972, M3.

41. William Golding, *Lord of the Flies* (New York: Coward-McCann, 1954). See Grinker's careful account of *The Ik*'s construction, performance, and reception in *In the Arms of Africa*, 209–222.

42. Clyde Kuhn, "Technological Man and the Transformation of 'Primitives,'" *Insurgent Sociologist* 4, no. 3 (Spring 1974): 85–87.

43. Fredrik Barth, "On Responsibility and Humanity: Calling a Colleague to Account," *Current Anthropology* 15, no. 1 (1974): 99–102. Turnbull penned a brief reply in which he claimed to have sent a copy of the review along with an offer to resign to every institution with which he was professionally involved: "Reply," *Current Anthropology* 15, no. 1 (1974): 103.

44. Barth, "On Responsibility and Humanity," 102.

45. Peter J. Wilson, Grant McCall, W. R. Geddes, A. K. Mark, John E. Pfeiffer, James B. Boskey, and Colin Turnbull, "More Thoughts on the Ik and Anthropology [and Reply]," *Current Anthropology* 16, no. 3 (1975): 343–358. Curtis Abraham returned to the Ik and found smiling and happiness rather than despair. See also Grinker's thoughtful reflection on Turnbull's legacy in *In the Arms of Africa*: "The Mountain People Revisited," *New African*, February 2002, 34–36.

46. Ardrey to Michael Korda, Simon and Schuster, 21 May 1972, Box 2, Folder 8, Ardrey Papers.

47. Turnbull to Korda, 2 September 1972, as quoted in Grinker, *Into the Arms of Africa*, 179.

48. John B. Calhoun, "Plight of the Ik and Kaiadilt Is Seen as a Chilling Possible End for Man," *Smithsonian* , November 1972, 26–33.

49. Calhoun, "Plight of the Ik," 32.

50. Frederick Sartwell, "The Small Satanic Worlds of John B. Calhoun," *Smithsonian*, April 1970, 68–71, quote on 70. Writing on his own in *Smithsonian* a few months later, Calhoun discussed the article favorably and noted, "We, if we are to be called human, produce ideas as well as ourselves. Whenever we fail to produce new ideas and utilize them we commit suicide" (12), in "The Lemmings' Periodic Journeys Are Not Unique," *Smithsonian*, January 1971, 6–13.

51. Medical explanations of violence in individual cases remain important today but do not address the question of aggression and violence as properties of populations as conceived by Calhoun, Ardrey, Lorenz, or Morris.

52. Edmund Ramsden and Jon Adams, "Escaping the Laboratory: The Rodent Experiments of John B. Calhoun and Their Cultural Influence," *Journal of Social History* 42 (2009), 761–792; Edmund Ramsden, "From Rodent Utopia to Urban Hell: Population, Pathology, and the Crowded Rats of NIMH," *Isis* 102, no. 4 (2011): 659–688.

53. Ramsden and Adams provide a detailed discussion of publications discussing Calhoun's work—including the novel by Ron M. Linton, *Terracide* (1970); John Hersey, *My Petition for More Space* (1974); Harry Harrison, *Make Room! Make Room!* (1966), which was later made into the 1973 film *Soylent Green* starring Charlton Heston; and William F. Nolan and George Clayton Johnson, *Logan's Run* (1967), also made into a film of the same name in 1976.

54. John B. Calhoun, "Population Density and Social Pathology," *Scientific American*, February 1962, 139–148. The architect Oscar Newman found Ardrey's conception of defensible territories intriguing and used it as the basis of his building designs: Joy Knoblauch, "The Economy of Fear: Oscar Newman Launches Crime Prevention through Urban Design (1969–197x)," *Architectural Theory Review* 19, no. 3 (2015): 336–354. Knoblauch suggests, too, that Newman's designs paved the way for Broken Windows policing of the 1980s; "Defensible Space and the Open Society," *Aggregate Architectural History Collaborative*, vol. 2, March 2015, accessed 28 January 2018, http://www.we-aggregate.org/piece/defensible-space-and-the-open-society.

55. Paul R. Ehrlich [and Anne Ehrlich], *The Population Bomb* (Ballantine Books, 1968); Garret Hardin, "The Tragedy of the Commons," *Science* 162, no. 3859 (1968), 1243–1248, later expanded to book length in *Exploring New Ethics for Survival: The Voyage of the Spaceship Beagle* (New York: Viking Press, 1972).

56. "A Clash of Gloomy Prophets," *Time*, 11 January 1971, 56–57.

57. As quoted in "Clash of Gloomy Prophets," 56–57.

58. Paul Ehrlich and Anne H. Ehrlich, "The Population Bomb Revisited," *People* 6, no. 2 (1979): 21–24.

59. Barry Commoner, *The Closing Circle* (New York: Bantam Books, 1971).

60. Commoner, *Closing Circle*, quotes on 33, 39, 41, 45, and 280. On Hardin's pessimistic determinism about human nature, see Jason Oakes, "Garrett Hardin's Tragic Sense of Life," *Endeavour* 40, no. 4 (2016): 238–247; Michael Egan, "The Social Significance of the Environmental Crisis: Barry Commoner's *The Closing Circle*," *Organization Environment* 15, no. 4 (2002): 443–457; and Egan, *Barry Commoner and the Science of Survival: The Remaking of American Environmentalism* (Cambridge, MA: MIT Press, 2007).

61. James Walls, "Ecodoom," *Family Planning Perspectives* 5, no. 1 (1973): 64.

62. Paul Ehrlich and John Holdren, "The Closing Circle," *Bulletin of the Atomic Scientists* (May 1972): 16, 18–27; Barry Commoner, "On 'The Closing Circle,'" *Bulletin of the Atomic Sci-*

entists (May 1972): 17, 42–56; John Holdren and Paul Ehrlich, "One-Dimensional Ecology Revisited: A Rejoinder," *Bulletin of the Atomic Scientists* (June 1972): 42–45.

63. Constance Holden, "Ehrlich versus Commoner: An Environmental Fallout," *Science* 177, no. 4045 (1972): 245–247, quote on 247.

64. Hardin tried unsuccessfully to intervene by moving the debate to solidly scientific grounds: Garrett Hardin, "Population Skeletons in the Environmental Closet," *Bulletin of the Atomic Scientists* (June 1972): 37–41.

65. A. J. Miller, "Doomsday Politics: Prospects for International Co-operation," *International Journal* 28, no. 1 (Winter 1972–1973): 121–133, quote on 124.

66. J. D. Carthy and F. J. Ebling, eds., *The Natural History of Aggression* (New York: Academic Press, 1964). Far more contentious, from the perspectives of the participants in the conference, was whether nation-states could be considered a kind of individual for theorizing strategies of cooperation or conflict.

67. Paul Ehrlich and Jonathan Freedman, "Population, Crowding and Human Behavior," in *Aggression and Evolution*, ed. C. Otten (Lexington, MA: Xerox Publishing, 1973), 274–282; Jonathan Freedman, Alan Levy, Roberta Buchanan, and Judy Price, "Crowding and Human Aggressiveness," *Journal of Experimental Social Psychology* 8 (1972): 528–548; Jonathan Freedman, Simon Klevansky, and Paul Ehrlich, "The Effect of Crowding on Human Task Performance," *Journal of Applied Social Psychology* 1 (1971): 7–25. For more recent analysis, see Matthew Connelly, *Fatal Misconception: The Struggle to Control World Population* (Cambridge, MA: Harvard University Press, 2008).

68. The prisoner's dilemma is a classic game-theory scenario in which, for example, two friends are arrested and accused of a crime. The police prevent them from communicating and each is given the choice of staying silent or confessing. If both remain silent, they each suffer a minor penalty, but if one testifies against the other, the "defector" goes free and the silent partner goes to jail for a long time. If both defect, then both go to jail for slightly less time. Cooperating requires trust that your partner will not sell you out. Biologists transformed this logic into evolutionary theory by calculating the cumulative successes and failures of particular strategies for iterative game play; see Paul Erickson, *The World the Game Theorists Made* (Chicago: University of Chicago Press, 2015).

69. Desmond Morris, *The Human Zoo: A Zoologist's Study of the Urban Animal* (New York: McGraw-Hill, 1969).

70. As quoted in John R. Wilmoth and Patrick Ball, "The Population Debate in American Popular Magazines, 1946–90," *Population and Development Review* 18, no. 4 (1992): 631–668, quote on 649; "A Scientist Looks at 'The Human Zoo,'" *U.S. News & World Report*, 2 March 1970, 38.

71. C. Henry Kempe, Frederic N. Silverman, Brandt F. Steele, William Droegemueller, and Henry K. Silver, "The Battered-Child Syndrome," *Journal of the American Medical Association* 181, no. 1 (1962): 17–24; "Parents Who Beat Children: A Tragic Increase in Cases of Child Abuse Is Prompting a Hunt for Ways to Detect Sick Adults Who Commit Such Crimes," *Saturday Evening Post*, 6 October 1962, 30–35; Ray E. Helfer and C. Henry Kempe, eds., *The Battered Child* (Chicago: University of Chicago Press, 1968); Ian Hacking, "The Making and Molding of Child Abuse," *Critical Inquiry* 17, no. 2 (1991): 253–288.

72. Sir Julian Huxley, "The Age of Overbreed," *Playboy*, January 1965, 103, quote on 179.

73. Huxley, "Age of Overbreed," 181.

74. Richard Evans, "A Conversation with Konrad Lorenz about Aggression, Homosexuality, Pornography, and the Need for a New Ethic," *Psychology Today* 8 (November 1974): 82–87, quote on 85.

75. Konrad Lorenz, "Analogy as a Source of Knowledge," *Science* 185, no. 4147 (1974): 229–234, quote on 233.

76. Peter Matthiessen, "The Physical Environment," *Playboy*, January 1969, 90. On pro-choice sentiment among conservative Christians before the federal legalization of abortion in 1973, see Neil J. Young, *We Gather Together: The Religious Right and the Problem of Interfaith Politics* (New York: Oxford University Press, 2016).

77. D. J. Mulvaney, "The Prehistory of the Australian Aborigine," *Scientific American*, March 1966, 84–93.

78. See, for example, Edward S. Deevey Jr., "The Human Population," *Scientific American*, September 1960, 195–204; *Scientific American* reprinted the article at least twice (in 1965 and 1969).

79. Mulvaney, "Prehistory of the Australian Aborigine," 93.

80. Ehrlich and Holdren, "Closing Circle," 16–18.

81. Ehrlich and Holdren, "Closing Circle," 18.

82. Mark E. Borrello, *Evolutionary Restraints: The Contentious History of Group Selection* (Chicago: University of Chicago Press, 2010).

83. Morris argued for a similar transition, citing the shift from small bands of humans to larger settlements: Desmond Morris, "Status and Superstatus in the Human Zoo," *Playboy*, September 1969, 122, quote on 123; cf. Paul S. Martin, "Pleistocene Overkill," *Natural History*, December 1967, 32–38.

84. V. C. Wynne-Edwards, "Population Control in Animals," *Scientific American*, August 1964, 68–74, quote on 68.

85. By the end of the 1960s, Wynne-Edwards had run afoul of evolutionary theorists who attacked his theories as inherently flawed because of their reliance on group selection—a story to which we will return in Chapter 13. Even so, Ardrey was captivated by his grand vision. Ardrey used Wynne-Edwards's point about humanity's inability to regulate our population size together with the aggressive and territorial nature of humans as cornerstones of both *The Territorial Imperative* and *The Social Contract*.

86. Towles and Turnbull, "White Problem in America," 7.

Chapter Eleven. A Dangerous Medium

1. Pauline Kael, "The Current Cinema: Sam Peckinpah's Obsession," *New Yorker*, 29 January 1972, 80–85; see 85 for the source of *Playboy*'s paraphrase.

2. William Murray, "Playboy Interview: Sam Peckinpah," *Playboy*, August 1972, 66.

3. Konrad Lorenz similarly argued that by "releasing" aggression through playing sports, an individual's aggressive drive would be temporarily depleted, and he would become less likely to commit a violent crime. This conception of violence fit well with his "hydraulic" model of behavior, in which the urge to act would build within an individual, providing a powerful (if

unconscious) motivation to exhibit certain kinds of behavior. Richard W. Burkhardt Jr., *Patterns of Behavior: Konrad Lorenz, Niko Tinbergen, and the Founding of Ethology* (Chicago: University of Chicago Press), 2005, 311–315.

4. Murray, "Sam Peckinpah," 68.

5. Robert Ardrey, in "Dear Playboy," *Playboy*, November 1972, 18.

6. Arthur Knight, "Playboy Interview: Clint Eastwood," *Playboy*, February 1974, 57, quotes on 72.

7. The article Eastwood had in mind appeared the previous year—William Drummond, "San Quentin: An Inside View of Its Turmoil," *Los Angeles Times*, 18 March 1971, 1.

8. Knight, "Clint Eastwood," 72.

9. Leonard Berkowitz, "The Effects of Observing Violence," *Scientific American*, February 1964, 35–41. On similar postwar debates, see James Gilbert, *A Cycle of Outrage: America's Reaction to the Juvenile Delinquent in the 1950s* (New York: Oxford University Press, 1986), esp. 143–195.

10. The debates continued through the next decades. See a series of papers and letters to the editor that appeared in the pages of the *New England Journal of Medicine* in the mid-1970s: Anne R. Somers, "Violence, Television and the Health of American Youth," *NEJM* 294 (1976): 811–817; F. J. Ingelfinger, "Violence on TV: An Unchecked Environmental Hazard," *NEJM* 294 (1976): 837–838; G. Timothy Johnson and Murray Feingold, "TV Violence: A Call for Protest," *NEJM* 294 (1976): 1007; various letters to the editor in response to Somers, *NEJM* 295 (1976): 288–291; Murray Feingold and G. Timothy Johnson, "Television Violence—Reactions from Physicians, Advertisers and the Networks," *NEJM* 296 (1977): 424–427; various letters to the editor responding to Feingold and Johnson, *NEJM* 296 (1977): 1481–1482. Similar articles appeared in other medical journals: M. B. Rothenberg, "Effect of Television Violence on Children and Youth," *Journal of the American Medical Association* 234 (1975): 1043–1046. The blurring of reality and fantasy in the minds of children still worry media critics even if video games have replaced film and television as the locus of concern.

11. "Mom-Pop Film Code: G-M-R-X," *Variety*, 9 October 1968, 3. In 1970 the MPAA changed the M rating to a GP, which they thought sounded less dire: "Latest Survey Shows X-R-GP-G Better Understood by Public," *Variety*, 4 November 1970, 3. In 1971 they flipped the letters of that rating to PG; "GP Now PG (Parental Guidance)," *Variety*, 22 September 1971, 5.

12. Stephen Prince, *Classical Film Violence: Designing and Regulating Brutality in Hollywood Cinema, 1930–1968* (New Brunswick, NJ: Rutgers University Press, 2003).

13. R. K. Baker and S. J. Ball, *Mass Media and Violence*, Task Force Report of the National Commission on the Causes and Prevention of Violence (Washington, DC: Government Printing Office, 1969), 269.

14. Baker and Ball, *Mass Media and Violence*, 270. See, for example, David Culbert, "Television's Visual Impact on Decision-Making in the USA, 1968: The Tet Offensive and Chicago's Democratic National Convention," *Journal of Contemporary History* 33, no. 3 (1998): 419–449.

15. Baker and Ball, *Mass Media and Violence*, 242.

16. Baker and Ball, *Mass Media and Violence*, 323.

17. Baker and Ball, *Mass Media and Violence*, 326–327.

18. John Lindsay, "Race Relations," *Playboy*, January 1969, 90.

19. Bruno Bettelheim, "Children Should Learn about Violence," *Saturday Evening Post,* 11 March 1967, 10–12, quote on 10: "We delight in watching violence at the movies or on TV, and reading about it in the spy stories, and talking about it so that even the plight of the Negro is often a thin disguise for telling violent stories."

20. Baker and Ball, *Mass Media and Violence,* 244; Berkowitz, "Effects of Observing Violence."

21. Baker and Ball, *Mass Media and Violence,* 245.

22. Michael Couzens, "Reflections on the Study of Violence," *Law & Society Review* 5, no. 4 (1971): 583–604, on 585.

23. Couzens, "Reflections on the Study of Violence," 592–593.

24. Couzens, "Reflections on the Study of Violence," 590.

25. Judith Shatnoff, "A Gorilla to Remember," *Film Quarterly* 22, no. 1 (Autumn 1968): 56–62, quote on 56.

26. Pauline Kael, "Apes Must Be Remembered, Charlie," *New Yorker,* 24 February 1968, 37.

27. Pierre Boule, *La Planète des Singes* (Paris: Juilliard, 1963), *Monkey Planet,* trans. Xan Fielding (London: Secker & Warburg, 1964). Boulle wrote *La Planète des Singes* as a critique of France's crumbling colonialism in which civilization is saved by the apes it once rejected as primitive.

28. Susan Bridget McHugh notes that chimpanzees, orangutans, and gorillas were assigned to different racial and socioeconomic stereotypes according to the color of their skin. Blond, light-skinned orangutans were politicians, dark-haired but light-skinned chimpanzees formed a scientific class, and dark gorillas served as laborers and military; see "Horses in Blackface: Visualizing Race as Species Difference in 'Planet of the Apes,'" *South Atlantic Review* 65, no. 2 (2000): 40–72.

29. Shatnoff, "Gorilla to Remember," 57.

30. On the Americanization of race in the movie, see Eric Greene, *Planet of the Apes as American Myth: Race and Politics in the Films and Television Series* (Jefferson, NC: McFarland, 1996). To save on costs, filmmakers decided to endow the ape society with rustic nineteenth-century technology rather than attempting to duplicate the technological wonders of Boulle's novel; McHugh, "Horses in Blackface," 58.

31. Kael, "Apes Must Be Remembered, Charlie," 108.

32. Shatnoff, "Gorilla to Remember," 57.

33. Fred Myers, "Day of the Hunter," *Fact* 4, no. 2 (March–April 1967): 46–51, quote on 55. Sentiment against animal experimentation and factory farming was also gathering steam, finding a powerful manifesto in Peter Singer's *Animal Liberation: A New Ethics for Our Treatment of Animals* (New York: Random House, 1975).

34. On the relationship between the theology of Pierre Teilhard de Chardin and Arthur C. Clarke, see Peder Anker, "The Ecological Colonization of Space," *Environmental History* 10, no. 2 (2005): 239–268, esp. 262n18; Arthur C. Clarke, *2001: A Space Odyssey* (New York: Signet, 1968); Pierre Teilhard de Chardin, *The Future of Man,* trans. Norman Denny (London: Collins, 1964).

35. E. G. Marshall and Nicolas Noxon, *The Man Hunters,* 16 mm (MGM, 1970), 52 min.; Robert Poole, "*2001: A Space Odyssey* and the Dawn of Man," in *Stanley Kubrick: New Perspectives* (London: Black Dog, 2015), 174–197.

36. Shatnoff, "Gorilla to Remember," 58.

37. Vincent Canby, " 'Clockwork Orange' Dazzles the Senses and Mind," *New York Times*, 20 December 1971, 44.

38. James Kendrick, *Hollywood Bloodshed: Violence in 1980s American Cinema* (Carbondale: Southern Illinois University Press, 2009); J. David Slocum, ed., *Violence and American Cinema* (New York: Routledge, 2000); Stephen Prince, *Savage Cinema: Sam Peckinpah and the Rise of Ultraviolent Movies* (Austin: University of Texas Press, 1998).

39. By 1980, film critics were instead worried that young-adult viewers preferentially sought movies with a PG or R rating (the percentage of G movies declined precipitously in the first decade of the rating system's existence, from 32 percent to 12 percent of movies produced); Austin Bruce, "Rating the Movies," *Journal of Popular Film and Television* 7, no. 4 (1980): 384–399; "X for 'Clockwork,'" *Variety*, 15 December 1971, 4; "Mysterious New Film Ratings," *America*, 15 January 1972, 32; "MPAA Film Ratings, 1968–79," *Variety*, 7 November 1979, 24.

40. "Facesaver: Dr. Stern or Dr. Kubrick?" *Variety*, 30 August 1972, 4.

41. Some of Calhoun's facts would be incorporated into the discourse on man as an animal, especially his contention that rats (and people) could only interact meaningfully with a small group of individuals before becoming stressed—twelve. This number "twelve" was central to theories of hunting groups, too, especially in Antony Jay's *Corporation Man* (New York: Random House, 1971). John B. Calhoun, "The Social Use of Space," in *Physiological Mammalogy*, ed. W. Mayer and R. van Gelder (New York: Academic Press, 1963), 17–187; and Calhoun, "The Role of Space in Animal Sociology," *Journal of Social Issues* 22 (1966): 46–58.

42. Robert K. Baker and Sandra J. Ball, *Mass Media and Violence: A Report to the National Commission on the Causes and Prevention of Violence* (Washington, DC: US Government Printing Office: November 1969), 237.

43. For example, Michael Shedlin, "Police Oscar: *The French Connection*," *Film Quarterly* 25, no. 4 (Summer 1972): 2–9. Of course, this does not preclude reading films in other ways. See, e.g., Siegfried Kracauer, *Theory of Film: The Redemption of Physical Reality* (New York: Oxford University Press, 1960); and John Berger, *About Looking* (New York: Pantheon, 1980).

44. Sheldin, "Police Oscar," 3.

45. Pauline Kael, "Movies in Movies," *New Yorker*, 9 October 1971, 145–157, quote on 154.

46. Pauline Kael, "El Poto—Head Comics," *New Yorker*, 20 November 1971, 212–220, quotes on 219.

47. Pauline Kael, "Urban Gothic," *New Yorker*, 30 October 1971, 113–116, quote on 113.

48. Kael, "Urban Gothic," 116.

49. Kael, "Scavenging with Computers," *New Yorker*, 21 March 1970, 161–167, quote on 161.

50. Kael, "Scavenging with Computers," 161. Similar concerns were raised over the intent and reaction to Katsuhiro Ōtomo's *Akira* in 1988.

51. Kael, "Stanley Strangelove," *New Yorker*, 1 January 1972, 50–53, quote on 53.

52. Sheila Schwartz, "Science Fiction: Bridge between Two Cultures," *English Journal* 60, no. 8 (1971): 1043–1051, quote on 1044.

53. Stephen Farber, "The Old Ultra-Violence," *Hudson Review* (1972): 287–294, quote on 290. Fred M. Hechinger similarly condemned Kubrick's new effort as "the voice of fascism": see "A Liberal Fights Back," *New York Times*, 13 February 1972, D1.

54. Farber, "Old Ultra-Violence," 291.

55. Kael, "Stanley Strangelove," quote on 50.

56. Stanley Kubrick, "Now Kubrick Fights Back," *New York Times*. 27 February 1972, D1—a direct response to Hechinger's review two weeks earlier. Malcolm McDowell, the actor who played Alex, protested too, although his letter appeared ten pages deeper in the newspaper: *New York Times*, 27 February 1972, D11.

57. Robert Ardrey to Louis S. B. Leakey, 29 November 1971, Box 2, Folder 2 (I–L 1971), Robert Ardrey Papers, 1955–1980, Special Collections and University Archives, Rutgers University Libraries, New Brunswick, NJ (hereafter cited as Ardrey Papers). A signed contract with Wolper Productions (dated 1 October 1971) promising Ardrey $85,000 for his completed screenplay can be found in Box 2, Folder 3.

58. Ardrey to Mike [Hyde, Collins Publishers], 9 December 1974, Box 2, Folder 24 (C-F 1974), Ardrey Papers.

59. Green had begun his career as a writer and director for Time-Life and National Geographic Specials and worked with Peckinpah on *The Wild Bunch*, launching an impressive career behind the scenes at Hollywood. After arbitration through the Screen Writers Guild, sole writing credit belonged to Ardrey, but Green insisted he deserved credit for this, too.

60. "Pictures: New York Sound Track," *Variety*, 19 April 1972, 26.

61. "Pictures: Hollywood Production Pulse," *Variety*, 12 July 1972, 22; "Larry Turman as Wolper Pic Prez," *Variety*, 1 November 1972, 3. However, a follow-up article listed "The Ardrey Papers" in the future tense: it "will be produced and directed by Walon . . . Green"; see "Columbia Wolperizes," *Variety*, 21 February 1973, 23.

62. "The Ardrey Papers Pends with Wolper," *Variety*, 20 October 1971, 4, and "Set 'Ardrey Papers' as Wolper $7–800,000 Pic; Mel Stuart Will Direct," 17 November 1971, 27.

63. "'Ardrey Papers' Pic Now 'Up from Apes' for David Wolper," *Variety*, 10 April 1974, 28.

64. Robert Ardrey, *The Hunting Hypothesis: A Personal Conclusion Concerning the Evolutionary Nature of Man* (New York: Atheneum, 1976), published without the movie stills.

65. Hobe Morrison, "Ardrey Has Film Docu, Book Due; To Take Non-Writing Vacation," *Variety*, 16 April 1975, 89.

66. Ardrey to Edward H. Levi, president of the University of Chicago, 9 December 1974, Box 2, Folder 26 (I–M 1974), Ardrey Papers. In typical Ardrey fashion, he wrote to Levi using the film's less dramatic title, "Up from the Apes" on the same day he bragged about "The Animal Within" to Mike Hyde.

67. "Wolper's Heavier Entertainment for '75–'76 in Swing from Docus," *Variety*, 12 March 1975, 57.

68. "The Fourth Annual U.S.A. Film Festival in Dallas," *Variety*, 5 June 1974, 35.

69. When Ardrey died in 1980, his obituary in *Variety* noted, "There has been no recent announcement of the status of the project"; Hobe Morrison, "Robert Ardrey, Playwright Who Wrote Social Studies, Dies at 72," *Variety*, 23 January 1980, 96. Billed as "Animal Within," the film was also screened at Iran's fourth film festival in 1975: "Main & Auxiliary Screenings for Iran's Fourth Int'l Film Festival," *Variety*, 26 November 1975, 6. I have been unable to locate an extant copy of the film.

70. Kael, "Saint Cop," *New Yorker*, 15 January 1972, 78–82, quote on 81.

71. Ayn Rand, *The Fountainhead* (New York: Bobbs-Merrill, 1943); see also King Victor, *The Fountainhead* (Warner Bros. Pictures, 1949), 114 min., starring Gary Cooper and Patricia Neal.

72. Kael, "Saint Cop," quotes on 78.

73. These Supreme Court cases—Gideon v. Wainright, 372 U.S. 335 (1963), Escobedo v. Illinois, 378 U.S. 478 (1964), Miranda v. Arizona, 384 U.S. 436 (1966)— raised eyebrows among conservatives who argued that protecting criminals would slow the pace of justice; John Denton Carter, *The Warren Court and the Constitution: A Critical View of Judicial Activism* (Nashville, TN: Pelican Publishing, 1973).

74. Judy Fayard, "Who Can Stand 32,500 Seconds of Clint Eastwood? Just about Everybody," *Life*, 23 July 1971, 44–48, quote on 46.

75. Kael, "Saint Cop," quote on 78.

76. Kael, "Saint Cop," quote on 78. Kael was equally disdainful of Eastwood's sequel, Ted Post, dir., *Magnum Force* (1973); Pauline Kael, "Killing Time," *New Yorker*, 14 January 1974, 84–89.

77. Verdon Cummings, "Is 'Dirty Harry' a Right-Wing Melodrama?," *Human Events* 32, no. 14 (1972): 12.

78. *Sweet Sweetback's Baadasssss Song* (1971). "Blaxploitation" movies including Gordon Parks, dir., *Shaft* (Metro-Goldwyn-Mayer and Shaft Productions, 1971), 100 min.; Gordon Parks Jr., dir., *Super Fly* (Warner Brothers, 1972), 93 min.; William Crain, dir., *Blacula* (American International Pictures, 1972), 93 min.; Jack Hill, dir., *Coffy* (American International Pictures, 1973), 91 min. Van Peebles refused to submit his film to the MPAA for rating.

79. Donn C. Worgs, "'Beware the Frustrated . . .': The Fantasy and Reality of African American Violent Revolt," *Journal of Black Studies* 37, no. 1 (2006): 20–45, esp. 34. On the dearth of roles for African Americans in the movies prior to the 1970s, see Donald Bogle, *Toms, Coons, Mulattoes, Mammies, and Bucks: An Interpretive History of Blacks in American Films* (New York: Viking Press, 1973); Thomas Cripps, *Slow Fade to Black: The Negro in American Film* (New York: Oxford University Press, 1977); and more recently, Manthia Diawara, ed., *Black American Cinema: The New Realism* (New York: Routledge, 1993).

80. Harry M. Benshoff, "Blaxploitation Horror Films: Generic Reappropriation or Reinscription?," *Cinema Journal* 39, no. 2 (2000): 31–50, esp. 34; Ed Guerrero, *Framing Blackness: The African American Image in Film* (Philadelphia: Temple University Press, 1993), esp. "The Rise and Fall of Blaxploitation," 69–112.

81. Brandon Wander, "Black Dreams: The Fantasy and Ritual of Black Films," *Film Quarterly* 29, no. 1 (1975): 2–11, 5–6.

82. Pauline Kael, "Notes on Black Movies," *New Yorker*, 2 December 1972, 159–165, quote on 162. This critique is developed much more fully in Robert E. Wems Jr., *Desegregating the Dollar: African American Consumerism in the Twentieth Century* (New York: New York University Press, 1998).

83. Pauline Kael, "Notes on Black Movies," quote on 159.

84. Huey Newton, "He Won't Bleed Me: A Revolutionary Analysis of 'Sweet Sweetback's Baadasssss Song,'" *Black Panther*, no. 6 (19 June 1971); a copy can be found in the papers of the Dr. Huey P. Newton Foundation, Collection M0864, Box 50, Folder 5, Department of Special Collections, Stanford University Libraries.

85. Lerone Bennett Jr., "The Emancipation Orgasm: Sweetback in Wonderland," *Ebony*, September 1971, 106–118.

86. John Hartmann, "The Trope of Blaxploitation in Critical Responses to 'Sweetback,'" *Film History* 6, no. 3 (1994): 382–404. On Attica, see Heather Ann Thompson, *Blood in the Water: The Attica Prison Uprising of 1971 and Its Legacy* (New York: Pantheon Books, 2016).

87. Guy Hamilton, dir., *Live and Let Die* (United Artists, 1973), 121 min.

88. Wander, "Black Dreams."

89. David Walker, introduction to *Reflections on Blaxploitation: Actors and Directors Speak*, by David Walker, Andrew J. Rausch, and Chris Watson (Lanham, MD: Scarecrow Press, 2009): vii–x; Benshoff, "Blaxploitation Horror Films," 33; Alan Poussaint, "Stimulus/Response: Blaxpoitation Movies—Cheap Movies that Degrade Blacks," *Psychology Today*, February 1974, 22.

90. Eldredge Cleaver, *Soul on Ice* (New York: Dell, 1968); Frantz Fanon, *The Wretched of the Earth*, preface by Jean-Paul Sartre, trans. Constance Farrington (1963; New York: Grove Press, 1965).

91. Louis H. Masotti and Jerome R. Corsi, *Shoot-Out in Cleveland: Black Militants and the Police*, a report to the National Commission on the Causes and Prevention of Violence (May 1969): 93; Alondra Nelson, *Body and Soul: The Black Panther Party and the Fight against Medical Discrimination* (Minneapolis: University of Minnesota Press, 2011).

92. Fayard, "Who Can Stand?," 46.

93. Isaac Hayes, "Theme from *Shaft*," in *Shaft*, soundtrack album (Memphis, TN: Stax Recording Studios, 1971), 4:39 min.

94. Kael, "Current Cinema," 83. On Peckinpah's anti-intellectualism, see Fedor Hagenauer, "'Straw Dogs': Aggression and Violence in Modern Film," *American Imago* 30, no. 3 (Fall 1973): 221–248, 247. On Peckinpah's difficulty with interviews, see P. F. Kluge, "Master of Violence," *Life*, 11 August 1972, 47–53.

95. Kael, "Sam Peckinpah's Obsession," 85.

96. Kael, "Sam Peckinpah's Obsession," 83.

97. Hagenauer, "'Straw Dogs,'" citing Kael, "Sam Peckinpah's Obsession," and Paul Zimmerman, "Rites of Manhood," *Newsweek*, 20 December 1971, 80–87. See also Dan Yergin, who describes interviewing Peckinpah as an exercise in telling fact from fantasy, in "Peckinpah's Progress: From Blood to Killing in the Old West to Siege and Rape in Rural Cornwall," *New York Times Magazine*, 31 October 1971, 16.

98. Russell Kirk, "Movies Are Becoming More Depraved," *Human Events*, 18 July 1964, 11; James J. Kilpatrick, "Violent Crime Must Be Annihilated," *Los Angeles Times*, 15 December 1967, A7; Louise Sweeny, "'Ultraviolence' Assaults Movies: Freedom of Choice Is All-Important," *Christian Science Monitor*, 26 February 1972, 9; "Obscenity and Violence in the Mails, Movies & TV," *Phyllis Schlafly Report*, March 1977, 1–3.

Chapter Twelve. Moral Lessons

1. A few examples will suffice: the rise of Billy Graham and Jerry Falwell as televangelist stars; Donald Thompson's apocalyptic *A Thief in the Night* (Mark IV Pictures, 1973), 69 min., which featured a track from Larry Norman's Christian rock album *Upon this Rock* (Capitol Records, 1969); the popularity of Jack T. Chick's evangelical tracts and comic books (see Figure 23); Tim and Beverly LaHaye's explicit *Act of Marriage: The Beauty of Sexual Love* (Grand Rapids, MI: Zondervan, 1976); William Morris's *Scientific Creationism* (San Diego: Creation-Life Publishers, 1974); the founding of Oral Roberts University in 1963 and Jerry Falwell's Lynchburg Baptist College in 1971, now Liberty University. On conservative protests over school curricula, especially sex education and broader questions of multiculturalism, see William Mar-

tin, *With God on Our Side: The Rise of the Religious Right in America* (New York: Broadway Books, 1996).

2. John Rudolph, *Scientists in the Classroom: The Cold War Reconstruction of American Science Education* (New York: Palgrave Macmillan, 2002).

3. "Commercial Curriculum Development and Implementation in the United States," prepared by BCMA Associates for the National Science Foundation (1 May 1975), appendix 6 of the *Pre-college Science Curriculum Activities of the National Science Foundation: Report of Science Curriculum Review Team*, Vol. 2—(MWD-76-26), report prepared for the Committee on Science and Technology, US House of Representatives, 94th Cong., 1st Sess., Serial Q, 61-579 O (Washington, DC: US Government Printing Office, 1975), 277.

4. Richard Bumstead, "Man, A Course of Study," *Education* (September 1970): 20–29, quote on 22.

5. Box 36, Folder 1: "Lake City, Florida, Peter B. Dow—Man: A Course of Study Records," Monroe C. Gutman Library, Graduate School of Education, Harvard University, Cambridge, MA (hereafter cited as MACOS Records). Resistance to MACOS followed patterns established by protests over sex education and the creation of the Sex Information and Education Council of the United States (SIECUS) in 1964; see Martin, *With God on Our Side*, 100–116; Jeffrey P. Moran, *Teaching Sex: The Shaping of Adolescence in the United States* (Cambridge, MA: Harvard University Press, 2000).

6. Dorothy Nelkin, "The Proper Study of Mankind . . . : The MACOS Debate," in Nelkin, *Science Textbook Controversies and the Politics of Equal Time* (Cambridge, MA: MIT Press, 1977), 81–103.

7. Dick Hagwood, "Course in Social Studies Creates Furor in Columbia," *Florida Times-Union, Jacksonville*, 5 November 1970.

8. Dick Hagwood, "Minister Cites Objections to Social Studies Course," *Florida Times-Union, Jacksonville*, n.d., Box 36, Folder 1, MACOS Records.

9. This narrative is reconstructed from a workbook/reader created during and after the debates at Lake City called "Community Issues and Man: A Course of Study," Box 36, Folder 1, MACOS Records. See also Peter B. Dow's detailed account of the program's development and the resistance they faced in the 1970s: *Schoolhouse Politics: Lessons from the Sputnik Era* (Cambridge, MA: Harvard University Press, 1991).

10. "Of 360 Total Only 45 Students Drop 'Man' Course," *Lake City Reporter*, 4 December 1970.

11. In 1967, Education Services, Inc. (the nonprofit founded by Jerrold Zacharias to develop new science curricula) merged with the Institute for Educational Innovation, creating the Education Development Center; US Department of Health, Education, and Welfare, Office of Education, Bureau of Research, *Regional Educational Laboratories* (Washington, DC: US Government Printing Office, 1968), 8.

12. Dow expressed these thoughts after the next round of difficulties in Phoenix, but they reflect his commitment to open discussion even in 1970. Peter Dow to SSCP Executive Committee, memorandum, 21 June 1971, Box 39, Folder 2, MACOS Records.

13. Peter Dow to Don Koeller, memorandum, 10 December 1970, Box 36, Folder 1, MACOS Records.

14. Max Rafferty, *Guidelines for Moral Instruction in California Schools* (Sacramento, CA:

California State Department of Education, 1969). On debates over secular humanism in education, see also Martin, *With God on Our Side.*

15. Despite his conservative stance on sex education, the moral decline of American schools, and so much else, he also supported bilingual education in California schools and remained a staunch integrationist his entire career: Natalia Mehlman Petrzela, "Revisiting the Rightward Turn: Max Rafferty, Education, and Modern American Politics," *The Sixties: A Journal of History, Politics, and Culture* 6, no. 2 (2013): 143–171.

16. As quoted in Les Ledbetter, "Max L. Rafferty, 65, Conservative Who Ran California Schools, Dies," *New York Times,* 14 June 1982, D11.

17. "The Supreme Being—Man," *Time,* 17 August 1962; *Brooklyn Tablet,* 8 July 1965; and "A Humanist Manifesto," *New Humanist* 6, no. 3 (1933), as cited and excerpted in Rafferty, *Guidelines for Moral Instruction,* 36–37.

18. Rafferty, *Guidelines for Moral Instruction,* 37.

19. "Innovation's Perils," *Phoenix Gazette,* 18 October 1971, 6.

20. David Fitzpatrick, "Content of MACOS Class at Center of Controversy," *Arizona Republic,* 25 October 1971.

21. Both quotes from "Who Do We Eat?," a leaflet distributed before the October 28 School Board meeting in Phoenix, Box 36, Folder 1, MACOS Records.

22. On earlier instances of the fraught relationship between local school boards and evolution, see Edward Larson, *Summer for the Gods: The Scopes Trial and America's Continuing Debate over Science and Religion* (New York: Basic Books, 1997). On the rise of Christian conservatism, see Kevin Kruse, *One Nation under God: How Corporate America Invented Christian America* (New York: Basic Books, 2015).

23. Jerome Bruner: Phone Interview for Phoenix, 27 October 1971, Box 36, Folder 3, MACOS Records.

24. See Box 36, Folder 5: "Innovations and Perils Film Planning 1971" and Folder 6: "Innovations and Perils Correspondence," MACOS Records.

25. Edward C. Martin, "Talking Paper: EDC and Innovation," 24 April 1972, Box 36, Folder 6, MACOS Records.

26. Frances R. Link (CDA) to Janet Whitla (EDC) and Rita Holt (EDC), memorandum, "Innovation's Perils," 19 November 1974, Box 36, Folder 6, MACOS Records. Dow used the same phrase in the title of his dissertation, which was based on his experiences with MACOS: Peter Burton Dow, "Innovation's Perils: An Account of the Origins, Development, Implementation, and Public Reaction to 'Man: A Course of Study'" (EdD diss., Harvard University, 1979).

27. Dow, "Innovation's Perils," 482.

28. Dr. Onalee S. McGraw, "If More Parents Only Knew What Educators Are Doing with Your Federal Tax Dollars," *Human Events* 16 (14 August 1971): 624–625. Timothy Asch believed McGraw had instigated protests over MACOS in Phoenix in 1971: "Man, A Course of Study: Situating Tim Asch's Pedagogy and Ethnographic Films," in *Timothy Asch and Ethnographic Film,* ed. Ed Lewis (New York: Routledge, 2004), 57–74, on 69. McGraw earned her PhD in education from Georgetown University in 1970. In 1963, as "a 23-year-old Californian whose Presbyterian faith had eroded away," she met and later that year later married a "conservative Catholic from the Midwest who was 21 years my senior." Onalee McGraw, "My Life with Bill: The Mystery of Faithful Love," accessed 24 February 2016, http://www.forevermissed.com /william-francis-mcgraw/#lifestory.

29. Dow to Mr. Art Ware, Social Studies Coordinator, Bellevue Public Schools, 1 May 1972, Box 37, Folder 1, MACOS Records.

30. Peter Woolfson, "The Fight over MACOS—An Ideological Conflict in Vermont," *Council on Anthropology and Educational Quarterly* 5, no. 3 (August 1974): 27–30.

31. "She's a School Textbook Critic," *Atlanta Constitution*, 22 August 1975, 3C. On Gabler's influence in debates over education, see Joan DelFattore, *What Johnny Shouldn't Read: Book Censorship in America* (New Haven, CT: Yale University Press, 1992); and William Martin, "The Guardians Who Slumbereth Not: Meet Mel and Norma Gabler of Longview," *Texas Monthly*, November 1982, 145.

32. John A. Steinbacher, *The Child Seducers* (Fullerton, CA: Educator Publications, 1971); Steinbacher, *The Conspirators: Men against God* (Whittier, CA: Orange Tree Press, 1972).

33. Peter Dow to SSCP Executive Committee, memorandum, 21 June 1971, Box 39, Folder 2, MACOS Records.

34. L. Sprague de Camp, "End of the Monkey War," *Scientific American*, February 1969, 15–21. On the history of legal battles over evolution in the classroom, see Edward Larson, *Trial and Error: The American Controversy over Creation and Evolution* (New York: Oxford University Press, 1985).

35. Institute for Creation Research, *Acts and Facts* (June 1975), as quoted in Nelkin, *Science Textbook Controversies*, 54.

36. Onalee McGraw [misprinted as "McCraw"], director of curriculum, Citizens United for Responsible Education (CURE), Chevy Chase, "Letter to the Editor: 'Man: A Course of Study,'" *Washington Post*, 3 April 1973.

37. Torcaso v. Watkins, 367 U.S. 495 (1961).

38. US v. Seeger, 380 U.S. 193 (1965).

39. Abington School District v. Schempp, 374 U.S. 203 (1963). See also Christopher Toumey, "Evolution and Secular Humanism," *Journal of the American Academy of Religion* 61, no. 2 (1993), 275–301; and Toumey, *God's Own Scientists: Creationists in a Secular World* (New Brunswick, NJ: Rutgers University Press, 1994).

40. Henry M. Morris, *Scientific Creationism* (San Diego: Creation-Life Publishers, 1974), iii. This foreword does not appear in the "public school" edition. See also, Ronald Numbers, *The Creationists: From Scientific Creationism to Intelligent Design*, expanded ed. (Cambridge, MA: Harvard University Press, 2006).

41. Larson, *Trial and Error*, esp. "Legislating Equal Time, 1970–1981," 125–155.

42. *In the Spirit of '76: The Citizen's Guide to Politics* (Washington, DC: Third Century Publishers, 1975). On Conlan's association with Bright, see John G. Turner, *Bill Bright and Campus Crusade for Christ: The Renewal of Evangelicalism in Postwar America* (Chapel Hill: University of North Carolina Press, 2009), 160–166, who suggests that Bright began to distance himself from Conlan in 1976, after Conlan lost the bruising Republican primary for a seat in the US Senate to Sam Steiger.

43. This was the first photograph of the window allowed in a publication. The window was installed in the mid-1950s, around the same time that God was inscribed in the Pledge of Allegiance and on American paper money. On that history, see Daniel K. Williams, *God's Own Party: The Making of the Christian Right* (New York: Oxford University Press, 2010), 121.

44. Steven P. Miller, *Billy Graham and the Rise of the Republican South* (Philadelphia: University of Pennsylvania Press, 2011), 98.

45. John B. Conlan, "Frogs in the Budget," *Chicago Tribune*, 3 June 1974; Christine Russell, "Scientists Defend Studies of Foreign Frogs and Lizards," *Bioscience* 24, no. 10 (1974): 607.

46. Scientists began to pay attention to critiques of evolution as "secular humanist" in the early 1980s; Toumey, *God's Own Scientists*; Myrna Perez Sheldon, "Stephen Jay Gould, An Evolutionary Heretic" (unpublished manuscript, 2017).

47. Karen B. Wiley, "NSF Science Education Controversy: The Issues," *Social Science Education Consortium*, no. 26 (July 1976): 1–7.

48. Cong. Rec. H9496 (9 April 1975) (statement of Rep. John B. Conlan). See also *Hearings before the Special Subcommittee on the National Science Foundation of the Comm. on Labor and Public Welfare, United States Senate, Ninety-Fourth Cong., First Sess. S.1539 and S.1478 to Authorize Appropriations to the NSF, and for Other Purposes, H.R. 4723 Appropriations for Fiscal Year 1976* (14 March and 21 April 1975).

49. B. F. Skinner, *Science and Behavior* (New York: Macmillan, 1953), 402–403, as quoted by MACOS, *Seminar for Teachers* (Washington, DC: Education Development Center, 1970), 7.

50. *Hearings before the Special Subcommittee to the NSF of the Committee on Labor and Public Welfare* (14 March and 21 April 1975), 984.

51. E. L. Stanford, *The History of Calvary Baptist Church, Jackson, Mississippi* (Jackson, MS: Hederman Brothers, 1980), 257–258. See also https://www.youtube.com/watch?v=4W0sZ PDFcGU, accessed 11 March 2016; 281,000 people attended over the eight days of the crusade.

52. It is unlikely that Conlan was objecting to the consumption of raw meat, but rather to the highly visible ways in which that meat was procured and shared. In other words, his shock at the MACOS movies seems to have been as much about class as it was about religious commitment.

53. Joseph Cassidy, "Angry Congressman Charges . . . Shocking School Lessons Teach Children to Glorify Incest, Murder & Cannibalism," *National Enquirer*, 24 June 1975; see also Charles Laird, dir., *Through These Eyes* (Watertown, MA: Documentary Educational Resources, 2003), 55 min.

54. James J. Kilpatrick, "MACOS Described by One Who Will Teach It No More," *Baltimore Sun*, 24 April 1975, A21.

55. To combat this fear, one of Dow's key strategies, honed through iterative local skirmishes, involved sharing material with concerned parents at the beginning of the year.

56. Kilpatrick's columns appeared in newspapers across the country. On Kilpatrick's resistance to desegregation and position as an emerging conservative elite, see William P. Huswitt, *James J. Kilpatrick: Salesman for Segregation* (Chapel Hill: University of North Carolina Press, 2013). According to Huswitt, Kilpatrick shared with Podhoretz his belief that the social sciences should not be allowed to dictate civil rights policy and an underlying conviction that blacks and whites could never be equal (161–166).

57. James J. Kilpatrick, "Is Eskimo Sex Life a School Subject?," *Boston Globe*, 2 April 1975, 21. See also James J. Kilpatrick, "Bay State Fifth-Grade Teacher Calls New Study Lethal Brainwash," *Boston Globe*, 25 April 1975; "War Flares over Scientific Funds," *Boston Globe*, 20 January 1976.

58. James J. Kilpatrick, "National Science Foundation under Fire," *Boston Globe*, 27 March 1975, 31.

59. *Administration of the Science Education Project "Man: A Course of Study" (MACOS)*, GAO Report MWD-76-26, Report to the House Committee on Science and Technology by the Comptroller General of the United States. See also Philip M. Boffey, "Controversial Curriculum's Developers Face Tax Probe," *Chronicle of Higher Education*, 19 May 1975, 3.

60. Bruner later joked, "I guess I'm the only professor in the long history of Oxford University that ever sailed his own sailboat across the Atlantic Ocean to come occupy his chair." Bruner, interview by author, 4 October 2011.

61. Peter Dow to Jerome Bruner, 7 April 1975, Box 38, Folder 1, MACOS Records. Original in Unprocessed Accession 10823, Box 9B, Papers of Jerome Seymour Bruner, Harvard University Archives, Cambridge, MA (hereafter cited as Bruner Papers).

62. On Bruner's development of cognitive science and antipathy to behaviorism, see Jamie Cohen-Cole, *The Open Mind: Cold War Politics and the Sciences of Human Nature* (Chicago: University of Chicago Press, 2014).

63. Bruner to Dow, 8 April 1975, Box 38, Folder 1, MACOS Records.

64. Judith Kogan, "Congressmen Attack Program That DeVore Helped Develop," *Harvard Crimson*, 8 April 1975.

65. Irven DeVore, interview by author, 6 August 2011; cf. Irven DeVore, "Chimpanzee Behavior Revealing Clue to That of Humans," *Arizona Republic*, 17 December 1971, 57.

66. Bruner to Dow, 12 April 1975, Unprocessed Accession 10823, Box 9B, Bruner Papers. Bruner had found compelling Richard Hofstadter's *Anti-Intellectualism in American Life* (New York: Knopf, 1963). See also Dow, *Schoolhouse Politics*, 199–249; and John B. Conlan and Peter B. Dow, "Pro/Con Forum: The MACOS Controversy," *Social Education* 39, no. 6 (October 1975): 388–396.

67. John Walsh, "NSF: Congress Takes Hard Look at Behavioral Science Course," *Science* 188, no. 4187 (1975): 426–428. John Walsh covered news of the MACOS controversy for *Science* throughout this period, and his many articles reveal how keenly scientists worried about the events unfolding in Congress: e.g., "NSF and Its Critics in Congress: New Pressure on Peer Review," *Science* 188, no. 4192 (1975): 999–1001; "NSF: How Much Responsibility for Course Content, Implementation?" *Science* 190, no. 4215 (1975): 644–646; "NSF: Science Education Is Still in the Spotlight," *Science* 191, no. 4233 (1976): 1246–1249.

68. "House Orders Monthly Review of NSF," *Science News* 107, no. 16 (April 1975), 253; an amendment to H.R. 4723, *Hearings before the Special Subcommittee to the NSF of the Committee on Labor and Public Welfare* (14 March and 21 April 1975), 668; Deborah Shapley, "Proxmire Hits NSF Research Priorities, Funding Flexibility," *Science* 183, no. 4124 (1974): 498; Constance Holden, "Social Science at NSF Needs Pruning, Says Proxmire," *Science* 185, no. 4151 (1974): 597; "Proxmire vs. NSF: Economizing or Baiting?," *Science News* 107, no. 11 (March 1975), 165; Robert Chiovetti Jr. and Anna Marie Mulvihill, "Proxmire, NSF and Basic Research," *Science News* 107, no. 14 (April 1975): 230.

69. Some of these letters may have been sent in response to Dow's request that faculty associated with MACOS consider writing to their congressmen: Peter Dow, "An Open Letter to Friends of Man: A Course of Study," 4 April 1975, Box 39, Folder 3, MACOS Records.

70. An amendment to H.R. 4723, *Hearings before the Special Subcommittee to the NSF of the Committee on Labor and Public Welfare* (14 March and 21 April 1975), 702.

71. Haith to Kennedy (12 May 1975), p. 747–748, and Paul J. Davis to Kennedy (12 May 1975),

p. 745, in *Hearings before the Special Subcommittee to the NSF of the Committee on Labor and Public Welfare*.

72. "Political Winds Still Buffeting NSF," *Science News* 107, no. 26 (June 1975): 412.

73. Barbara B. Herzstein to <u>Man: A Course of Study</u> Development and Revision Staff, re: Plan for Revision, 28 April 1975, Box 21, Folder 3, MACOS Records.

74. Nelkin, *Science Textbook Controversies*, 108. At the time, Dow worried that CDA might go bankrupt as a result of the controversy, but MACOS was only one program among many and the educational development company weathered the storm intact. Dow to Bruner, 19 June 1975, Box 38, Folder 1, MACOS Records.

75. Appendix 6 of the *Pre-college Science Curriculum Activities of the National Science Foundation: Report of Science Curriculum Review Team*, Vol. 2—(MWD-76-26), report prepared for the Committee on Science and Technology, US House of Representatives, 94th Cong., 1st Sess., Serial Q, 61-579 O (Washington, DC: US Government Printing Office, 1975), 203.

76. NCSS, "The MACOS Question. Views of 'Man: A Course of Study' and the Roles of the National Science Foundation and the Federal Government in Curriculum Development and Implementation. A Statement by the National Council for the Social Studies, June 20, 1975," *Social Education* 39 (November–December 1975): 445–450.

77. John Walsh, "National Science Foundation: Criticism from Conlan, GAO," *Science* 191, no. 4229 (1976): 830–832.

78. Karen B. Wiley, "NSF Science Education Controversy: The Issues," *Social Science Education Consortium*, no. 26 (July 1976): 1–7.

79. Wayne W. Welch, "Twenty Years of Science Curriculum Development: A Look Back," *Review of Research in Education* 7 (1979): 282–306.

80. Marc Rothenberg, "Making Judgments about Grant Proposals: A Brief History of the Merit Review Criteria at the National Science Foundation," *Technology and Innovation* 12 (2010): 189–195; *National Science Foundation Peer Review*, report prepared for the Subcommittee on Science, Research, and Technology of the Committee on Science and Technology, US House of Representatives, 94th Cong., 2nd Sess., 63-458 (Washington, DC: Government Printing Office, January 1976).

81. Guy Stever served as director of the NSF throughout these proceedings. Read his account of the "unbelievable misjudgments" (230) of the MACOS Program in *In War and Peace: My Life in Science and Technology* (Washington, DC: Joseph Henry Press, 2002): 228–234.

82. Welch, "Twenty Years," 303.

83. Onalee McGraw, *Secular Humanism and the Schools: The Issue Whose Time Has Come* (Washington, DC: The Heritage Foundation, 1976).

84. McGraw, *Secular Humanism and the Schools*, 4. In establishing "humanism" as a religious doctrine, she cited not only the original "Humanist Manifesto" published in the *New Humanist* 6 no. 3 (1933) but also the more recent "Humanist Manifesto II," *Humanist* 33, no. 5 (1973). Paul Kurtz, editor of the *Humanist*, wrote that Conlan's position (and McGraw's) amounted to an objection to "modernism itself": John Dart, "Charges of Secular Humanism in Public Schools Spread and Provoke Dissension," *Los Angeles Times*, 11 September 1976, 31.

85. McGraw, *Secular Humanism and the Schools*, 7–9. The support provided by Conlan's authority as a US Representative appears to have been more important to McGraw than priority.

86. Marguerite Michaels, "Public School Book Censors Try It Again," *Boston Globe*, 25 November 1979, AB4–AB5.

87. Dorothy Nelkin, "The Science-Textbook Controversies," *Scientific American*, April 1976, 33–39.

88. Nelkin, *Science Textbook Controversies*, 150.

89. Nelkin, *Science Textbook Controversies*, 151.

90. Mark Turin, "Interview with Asen Balikci," 12 January 2003, video file, published 11 November 2004, University of Cambridge DSpace, accessed 2 January 2017, http://www.dspace .cam.ac.uk/handle/1810/1467.

91. Asen Balikci, interview by Peter Dow, 17–18 October 1974, Box 42, Folder 6, MACOS Records.

92. Balikci, interview by Peter Dow.

93. Turin, "Interview with Asen Balikci."

94. Irven DeVore, interview by Peter Dow, 20 October 1974, Box 42, Folder 12, MACOS Records.

95. Both the filmic ethnographer Timothy Asch and the evolutionary biologist Robert Trivers have attested to the program's influence on their later work: Nancy C. Lutkehaus, "Man: A Course of Study: Situating Tim Asch's Pedagogy and Ethnographic Films," in *Timothy Asch and Ethnographic Film*, ed. E. D. Lewis (New York: Routledge, 2003), 57–73; Robert Trivers, *Natural Selection and Social Theory: Selected Papers of Robert Trivers* (New York: Oxford University Press, 2002), 56–58.

96. "International Distribution," Box 34, Folder 1, MACOS Records; "Man a Course of Study in Australia," Unprocessed Accession 11380, Box 35 (old Box 21A), Bruner Papers; R. A. Smith and J. Knight, "MACOS in Queensland: The Politics of Educational Knowledge," *Australian Journal of Education* 22, no. 3 (October 1978): 225–248; A. J. Hepworth, "Values Education—Some New South Wales Experiences," *Journal of Moral Education* 48, no. 3 (1979): 193–201. *Desert People* and *People of the Australian Western Desert* were directed by Ian Dunlop, anthropologically advised by Robert Tonkinson, and photographed by Richard Tucker in 1965 (Commonwealth Film Unit, Australian Institute of Aboriginal Studies, 1967). By the 1980s, only smaller segments of the films were available for distribution—*Cooking Kangaroo* (17 min.), *Fire Making* (7 min.), *Gum Preparation, Stone Flaking, Djagamara Leaves Badjar* (19 min.), *Old Camp Sites at Tike Tika* (12 min.), *Sacred Boards and an Ancestral Site* (8 min.), *Seed Cake Making and General Camp Activity* (21 min.), *Spear Making, Boys' Spear Fight* (10 min.), *Spear Thrower Making, Including Stone Flaking and Gum Preparation* (34 min.), *Spinning Hair String, Getting Water from Well, Binding Girl's Hair* (13 min.)—see James C. Pierson, *American Anthropologist*, n.s., 88, no. 1 (March 1986): 269–271. In the end, however, MACOS was discontinued in the late 1970s under pressure from conservatives. See "Sex Education on Queensland Deserves Framework," *Courier-Mail* (Brisbane), 8 April 2009. For attempts to introduce MACOS in England, see David Jenkins, "MAN: A Course of Study," *Design Issues E203 14-15 of the Educational Studies Course* (Open University Press, 1976): 59–100; Jean Rudduck, "Dissemination as the Encounter of Cultures," *Research Intelligence* 3, no. 1 (1977): 3–5; David Jenkins, "Man: A Course of Study," in *Curriculum Research and Development in Action*, ed. L. Stenhouse (London, Heinemann: 1980): 215–224.

97. "Was ist der Mensch?" (1980), Box 34, Folder 3, MACOS Records; "Was ist der

Mensch?" Unprocessed Accession 11380, Box 35 (old Box 21A), Bruner Papers. West German educators may have initially encountered the program at American military schools. See Clara Hicks, interview by Peter Dow, 5 November 1974, Box 42, Folder 18, MACOS Records.

98. See Janet Whitla, interview by Peter Dow, 23 March 1976, Box 43, Folder 12, MACOS Records.

99. Richard Morin, "Mud Slinging, Arizona-Style," *Washington Post*, 7 September 1976, A1. Goldwater and Reagan, although initially friendly, would later fall out over political differences.

100. David Anable, "Four Churchmen Condemn Religious Zeal in Politics," *Christian Science Monitor*, 26 October 1976, 4.

101. Conlan continued to raise funds for expressly Christian candidates, including for Ronald Reagan: John Dart, "50 Christians Agree to Raise Reagan Funds," *Los Angeles Times*, 3 December 1979.

102. John W. Whitehead and John Conlan, "The Establishment of the Religion of Secular Humanism and Its First Amendment Implications," *Texas Tech Law Review* 10 (1979): 1, 61.

103. Constance Holden, "Republican Candidate Picks Fight with Darwin," *Science* 209, no. 4462 (1980): 1214.

104. John Walsh, "Science Education Redivivus," *Science* 219, no. 4589 (1983): 1198–1199, quote on 1198.

105. Conlan briefly appears in Sarah Barringer Gordon, *The Spirit of the Law: Religious Voices and the Constitution in Modern America* (Cambridge, MA: Belknap Press of Harvard University Press, 2010): 152–156; Sara Diamond, *Spiritual Warfare: The Politics of the Christian Right* (New York: Black Rose Books, 1990), 49–50; Garry Wills, *Under God: Religion and American Politics* (New York: Simon & Schuster, 1990), 18–19; Susan Harding, *The Book of Jerry Falwell: Fundamentalist Language and Politics* (Princeton, NJ: Princeton University Press, 2000), 164–165.

106. Tim LaHaye, *Battle for the Mind: A Subtle Warfare* (Old Tappan, NJ: Fleming H. Revell, 1980); Tim LaHaye and Jerry B. Jenkins, *Left Behind: A Novel of the Earth's Last Days* (Wheaton, IL: Tyndale House, 1995), followed by fifteen additional books in the series.

107. Jerry Falwell, *Listen, America!* (New York: Doubleday and Company, 1980); Francis A. Schaeffer, *A Christian Manifesto* (1981; Wheaton, IL: Crossway Books, 2005); James Hitchcock, *What Is Secular Humanism? How Humanism Became Secular, and How It Is Changing Our World* (Ann Arbor, MI: Servant Books, 1982); Phyllis Schlafly, ed., *Child Abuse in the Classroom* (Alton, IL: Pere Marquette Press, 1984).

Part Five. Death of the Killer Ape

William D. Hamilton, "Selection of Selfish and Altruistic Behavior in Some Extreme Models," in *Man and Beast: Comparative Social Behavior*, ed. J. Eisenberg (Washington, DC: Smithsonian Institution Press, 1971), 57–91, quote on 83.

1. This group moved into one of the two acre-and-a-half quadrant enclosures. With more money, Hamburg and Goodall planned to build an additional two enclosures, rounding out the planned six-acre arena. Even in 1973, it seemed likely that once completed, at least one of these new quadrants would house baboons. Joan Hinman, "Primate Center Will Permit Sci-

entific Observation," *Stanford Daily*, 1 November 1973. Today, the planned quadrants remain unbuilt.

2. Sandy White, "Chimps' Big Move to Freedom," *Stanford Daily*, 26 April 1974.

3. Terrie McDonald, "Primate Specialist Discusses Work," *Stanford Daily*, 1 December 1971.

4. On the long-standing difficulty of negotiating this balance, see Robert Kohler, *Landscapes and Labscapes: Exploring the Lab-Field Border in Biology* (Chicago: University of Chicago Press, 2002).

5. Donna Haraway, *Primate Visions: Gender, Race, and Nature in the World of Modern Science* (New York: Routledge, 1989); Shirley C. Strum and Linda M. Fedigan, *Primate Encounters: Models of Science, Gender, and Society* (Chicago: University of Chicago Press, 2000); Susan Sperling, "Baboons with Briefcases: Feminism, Functionalism, and Sociobiology in the Evolution of Primate Gender," *Signs* 17, no. 1 (1991): 1–27.

6. Cable, originally listed as "Limited official use," US Embassy Dar es Salaam to Department of State, 1975 May 22, 07:30 (Thursday): 1975DARES01609. This cable and the others I cite later have since been declassified and are available online through the United States National Archives, Record Group 59: General Records of the Department of State, Central Foreign Policy Files, 1973– , https://aad.archives.gov/aad/index.jsp.

7. David A. Hamburg and Jane van Lawick-Goodall to Wenner-Gren Foundation, 4 April 1973, BANC 98/132C Box 1, Folder 19: "Hamburg, David—Correspondence, 1966–1995," Sherwood L. Washburn Papers, Bancroft Library, University of California, Berkeley (hereafter cited as Washburn Papers).

8. Derek Bryceson to Hazel Eickworth, "Gombe Stream Centre Financing and Accounting," 22 November 1974, Box 439, Folder 27, David Hamburg Papers, 1949–2003, CA#0005, Carnegie Collection, Rare Book and Manuscript Library, Columbia University, New York (hereafter cited as Hamburg Papers).

9. Most prominently, research at Gombe was funded by the Grant Foundation, administered through Stanford University. Additional funds came from the Wenner-Gren Foundation, the Leakey Foundation, and the Harry Frank Guggenheim Foundation. This money supported one graduate student, Anne Pusey, officially advised by David Hamburg, and also Richard Wrangham, who earned his PhD at Cambridge while funded by Hamburg at Stanford. See Anne Pusey, "Age-Changes in the Mother-Offspring Association of Wild Chimpanzees," in *Recent Advances in Primatology*, vol. 1, ed. D. J. Chivers and J. Herbert (London: Academic Press, 1978), 119–123; and Pusey, "Inter-community Transfer of Chimpanzees in Gombe National Park," in *The Great Apes*, ed. D. A. Hamburg and E. R. McCown (Menlo Park, CA: Benjamin/Cummings, 1979), 464–479. Foundations were reluctant to give the money to an individual directly, preferring to run the finances through an associated university, at first Cambridge, but by the mid-1970s, Stanford.

10. Richard Wrangham, speaking at "An Oral History of Primatology, Jane Goodall, at Cambridge," University of Cambridge, Personal Histories Project, 28 April 2011, accessed 8 February 2018, https://sms.cam.ac.uk/media/1332629.

11. Louise Brodie and Robert Hinde, interview transcript, recorded at St. John's College, University of Cambridge, 9 February 2010, in *National Life Stories: An Oral History of British Science* (London: British Library) 97–100, accessed 31 January 2018, http://sounds.bl.uk/related-content/TRANSCRIPTS/021T-C1379X0008XX-0000A0.pdf.

12. William McGrew, speaking at "An Oral History of Primatology, Jane Goodall, at Cambridge."

13. Georgina Montgomery, *Primates in the Real World: Escaping Primate Folklore and Creating Primate Science* (Richmond: University of Virginia Press, 2015).

14. David Hamburg, "Ancient Man in the Twentieth Century," in *The Quest for Man*, ed. Vanne Goodall (New York: Praeger Publishers, 1975), 26–53, quotes on 27. Vanne Morris-Goodall, Jane's mother and another frequent visitor at Gombe, arrived in Tanzania with her daughter in 1960. British Tanzanian authorities would not grant Jane permission to be at Gombe by herself—she needed an older companion. Over the years, Vanne developed her own interest in human evolution, producing the edited volume *The Quest for Man*, with contributions by (additionally) Theodosius Dobzhansky, Irenaüs Eibl-Eibesfeldt, Jane Goodall, and two chapters by John Napier (filling in for Louis Leakey after his death). She also penned a coming-of-age adventure story of a young British woman in Africa, *Beyond the Rain Forest* (London: Collins, 1967).

15. David Hamburg, "An Evolutionary Perspective on Human Aggressiveness," in *The Field of Psychiatry: Essays in Honor of Roy R. Grinker, Sr.*, ed. D. Offer and D. Freedman (New York: Basic Books, 1972), 30–43, quote on 30. See also, David Hamburg, *A Model of Prevention: Life Lessons* (New York: Routledge, 2016).

16. Hamburg, "Ancient Man," 46.

17. Lisa Stampnitzky, *Disciplining Terror: How Experts Invented "Terrorism"* (New York: Cambridge University Press, 2013): 2–21. As a concept, terrorism has a long history, but Stampnitzky argues that these political concerns in the early 1970s gave rise to terrorism "experts"; see, e.g., Bruce Howard, "Living with Terrorism," *Washington Post*, 18 July 1976, 25.

18. David Fromkin, "The Strategy of Terrorism," *Foreign Affairs* 53, no. 4 (1975): 683–698, on 683.

19. W. Tapley Bennet Jr., "U.S. Initiatives in the United Nations to Combat International Terrorism," *International Lawyer* 7, no. 4 (1973): 752–760; Stampnitzky, *Disciplining Terror*.

20. David Hamburg, "Aggressive Behavior of Chimpanzees and Baboons in Natural Habitats," *Journal of Psychiatric Research* 8 (1971): 385–398; Hamburg, "Evolutionary Perspective on Human Aggressiveness," 40–41.

21. David Hamburg and Elizabeth McCown, eds., *The Great Apes*. Vol. 5 of *Perspectives on Human Evolution* (Menlo Park, CA: Benjamin/Cummings, 1979).

22. Joseph L. Popp and Irven DeVore, "Aggressive Competition and Social Dominance Theory: Synopsis," in *Great Apes*, ed. Hamburg and McGown, 317–340.

23. Sherwood Washburn, "Evolution of a Teacher," *Annual Review of Anthropology* 12 (1983): 1–24. For more on the connections between ethology and psychiatry, see Anthony Storr, *Human Aggression* (New York: Atheneum, 1968); he dedicated the book, "To Konrad Lorenz with admiration and affection." As an analyst and writer, rather than an academic, Storr circulated in different professional circles and was never funded by the Harry Frank Guggenheim Foundation; Robert J. Trotter, "Human Behavior: Do Animals Have the Answer?," *Science News* 105, no. 17 April 1974): 274; and Marga Vicedo, "The Social Nature of the Mother's Tie to Her Child: John Bowlby's Theory of Attachment in Post-War America," *British Journal for the History of Science* 44, no. 2 (2011): 401–426. Bowlby also spent the 1957–58 academic year at Stanford's Center for Advanced Study in the Behavioral Sciences, where he met both Hamburg and Washburn.

24. Later published as Sherwood Washburn, ed., *Social Life of Early Man* (Chicago: Aldine, 1961). The Wenner-Gren Foundation also provided symposium support for Richard B. Lee and Irven DeVore, eds., *Man the Hunter* (Chicago: Aldine, 1968), although they met instead at the University of Chicago's Center for Continuing Education. See also David Hamburg, "Emotions in the Perspective of Human Evolution," in *Expression of the Emotions in Man*, ed. Peter H. Knapp (New York: International Universities Press, 1963), 300–317.

25. Irven DeVore, ed., *Primate Behavior: Field Studies of Monkeys and Apes* (New York: Holt, Rinehart and Winston, 1965).

26. David Hamburg and Elizabeth McCown, "Introduction," in *Great Apes*, ed., Hamburg and McCown, 9.

27. Clifford J. Jolly, "The Seed-Eaters: A New Model of Hominid Differentiation Based on a Baboon Analogy," *Man* 5, no. 1 (1970): 15–26. Discussed in Linda Marie Fedigan, "Theories on the Evolution of Human Social Life," in Fedigan, *Primate Paradigms: Sex Roles and Social Bonds* (1982; Chicago: University of Chicago Press, 1992), 307–323.

28. Cited by John Pfeiffer, *The Emergence of Man* (New York: Harper & Row, 1972); Fedigan, *Primate Paradigms* and others.

29. Elizabeth Marshall Thomas, *The Harmless People* (New York: Knopf, 1959); Colin Turnbull, *The Forest People* (New York: Simon and Schuster, 1962).

30. Colin M. Turnbull, *The Mountain People* (New York: Simon and Schuster, 1972); Robert Gardner and Karl G. Heider, *Gardens of War* (New York: Random House, 1968).

31. Napoleon A. Chagnon, *Yąnomamö: The Fierce People* (New York: Rinehart and Winston, 1968); Timothy Asch and Napoleon A. Chagnon, *Yąnomamö Series* (Watertown, MA: Documentary Education Resources, 1968–1987).

32. Compare: R. Brian Ferguson, *Yanomami Warfare: A Political History* (Ann Arbor, MI: School of American Research Press, 1995); Patrick Tierney, *Darkness in El Dorado: How Scientists and Journalists Devastated the Amazon* (New York: Norton, 2001); Napoleon A. Chagnon, *Noble Savages: My Life among Two Dangerous Tribes—the Yąnomamö and the Anthropologists* (New York: Simon and Schuster, 2013).

33. Richard E. Leakey and Roger Lewin, *Origins: What New Discoveries Reveal about the Emergence of Our Species and Its Possible Future* (New York: Dutton, 1977) and *People of the Lake: Mankind and Its Beginnings* (Garden City, NY: Anchor Press, 1978); Richard E. Leakey, *The Making of Mankind* (New York: Dutton, 1981). The most dramatic of Richard Leakey's efforts resulted in a five-part television documentary series called *The Making of Mankind* (BBC in association with Time-Life Incorporated, 1983): "1. In the Beginning"; "2. A Human Way of Life"; "3. A New Era"; "4. Settling Down"; "5. Survival of the Species."

34. Peter Stoler, "Books: Animal Paragon," *Time*, 14 August 1978, 86.

35. Richard Leakey, "Discarding the Concept of Man as 'Killer Ape,'" *U.S. News and World Report*, 15 February 1982, 62.

36. C. Loring Brace, "Review," *American Anthropologist* 81, no. 3 (1979): 702–704, quote on 704.

37. Glynn L. Isaac and Elizabeth R. McCown, eds., *Human Origins: Louis Leakey and the East African Evidence* (Menlo Park, CA: W. A. Benjamin, 1976).

38. Philip Morrison, reviews of *Human Origins: Louis Leakey and the East African Evidence*, by Glynn L. Isaac and Elizabeth R. McCown; and *Leakey's Luck: The Life of Louis Seymour Bazett Leakey, 1903–1972*, by Sonia Cole, *Scientific American*, September 1976, 212–218, quote on 217.

39. E. O. Wilson, *Sociobiology: A New Synthesis* (Cambridge, MA: Belknap Press of Harvard University Press, 1975). Because ants exhibit highly social, if stereotyped, behavior, myrmecologists have often used their knowledge of the social insects as an evidentiary basis for normative claims about human behavior. See Charlotte Sleigh, *Six Legs Better: A Cultural History of Myrmecology* (Baltimore: Johns Hopkins University Press, 2007); and Abigail Lustig, "Ants and the Nature of Nature in Auguste Forel, Erich Wasmann, and William Morton Wheeler," in *The Moral Authority of Nature*, ed. Lorraine Daston and Fernando Vidal (Chicago: University of Chicago Press, 2003), 282–307.

40. On Wilson and the reception of *Sociobiology*, see Ullica Segerstråle, *Defenders of the Truth: The Battle for Science in the Sociobiology Debate and Beyond* (New York: Oxford University Press, 2000).

41. On the gendered politics of research on X and Y chromosomes, see Sarah Richardson, *Sex Itself: The Search for Male and Female in the Human Genome* (Chicago: University of Chicago Press, 2013). On the legacy of eugenics, see Dorothy Nelkin and Susan Lindee, *The DNA Mystique: The Gene as a Cultural Icon* (Ann Arbor: University of Michigan Press, 2004) and Nathaniel Comfort, *The Science of Human Perfection: How Genes Became the Heart of American Medicine* (New Haven, CT: Yale University Press, 2012).

Chapter Thirteen. The New Synthesis

1. Nina McCain, "Sociobiology: New Theory on Man's Motivation," *Boston Globe*, 13 July 1975, A1; Edward O. Wilson, *Sociobiology: The New Synthesis* (Cambridge, MA: Belknap Press of Harvard University Press, 1975).

2. Richard Wrangham, interview by author, 7 May 2012, Harvard University.

3. Wrangham, author interview.

4. See especially Richard Dawkins's *The Selfish Gene* (New York: Oxford University Press, 1976), published the year after *Sociobiology*. On the power of genetic metaphors in the life sciences, see Lily E. Kay, *Who Wrote the Book of Life? A History of the Genetic Code* (Palo Alto, CA: Stanford University Press, 2000).

5. William D. Hamilton, "Selection of Selfish and Altruistic Behavior in Some Extreme Models," in *Man and Beast: Comparative Social Behavior*, ed. J. Eisenberg (Washington, DC: Smithsonian Institution Press, 1971), 57–91; Sarah A. Swenson, " 'Morals Can Not Be Drawn From Facts but Guidance May Be': The Early Life of W. D. Hamilton's Theory of Inclusive Fitness," *British Journal for the History of Science* 48, no. 4 (2015): 543–563.

6. Evolutionary explanations of altruism, especially when connected to ideas of group selection and addressed with the tools provided by game theory, has recently attracted the attention of many historians of science: e.g., Paul Erickson, *The World the Game Theorists Made* (Chicago: University of Chicago Press, 2015); Paul Erickson, Judy L. Klein, Lorraine Daston, Rebecca Lemov, Thomas Sturm, and Michael Gordin, *How Reason Almost Lost Its Mind: The Strange Career of Cold War Rationality* (Chicago: University of Chicago Press, 2013); Ullica Segerstråle, *Nature's Oracle: The Life and Work of W. D. Hamilton* (New York: Oxford University Press, 2013); Mark Borello, *Evolutionary Restraints: The Contentious History of Group Selection* (Chicago: University of Chicago Press, 2010); Oren Harman, *The Price of Altruism: George Price and the Search for the Origins of Kindness* (New York: W. W. Norton, 2010); and, on similar questions a century

earlier, Piers J. Hale, *Political Descent: Malthus, Mutualism, and the Politics of Evolution in Victorian England* (Chicago: University of Chicago Press, 2014).

7. Hamilton, "Selection of Selfish and Altruistic Behavior," 60; Sarah A. Swenson, " 'From Man to Bacteria': W. D. Hamilton, the Theory of Inclusive Fitness, and the Post-War Social Order," *Studies in History and Philosophy of Biological and Biomedical Sciences* 49 (2015): 45–54.

8. Hamilton, "Selection of Selfish and Altruistic Behavior," 80.

9. William D. Hamilton, "The Evolution of Altruistic Behaviour," *American Naturalist* 97 (1963): 354–356; and Hamilton, "The Genetical Evolution of Social Behavior," parts 1 and 2, *Journal of Theoretical Biology* 7 (1964): 1–52. The calculations change if the species is haplo-diploid, like bees, ants, or wasps; and in the case of inbreeding, or clonal groups, and other interesting cases.

10. Hamilton, "Selection of Selfish and Altruistic Behavior," 83.

11. Robert Trivers, "Self-Deception in Service of Deceit," in *Natural Selection and Social Theory: Selected Papers of Robert Trivers* (Oxford: Oxford University Press, 2002), 257.

12. E. O. Wilson cites all four men in *Sociobiology*, but Hamilton and Williams feature far more prominently than Maynard Smith and Price. On Trivers's growing interest in the research of Hamilton, Maynard Smith, Price, and Williams, see Paul Erickson, "Rationality without Reason," in *World the Game Theorists Made*, 204–239. In recalling his visit to Harvard in the early 1970s, Maynard Smith remembered entering DeVore's office and discovering a promising game of poker: Ullica Segerstråle, *Defenders of the Truth: The Battle for Science in the Sociobiology Debate and Beyond* (New York: Oxford University Press, 2000), 81. On the social and intellectual complexities of British game theory, see Harman, *Price of Altruism*.

13. George C. Williams, *Adaptation and Natural Selection: A Critique of Some Current Evolutionary Thought* (Princeton, NJ: Princeton University Press, 1966).

14. See also John Maynard Smith, "The Evolution of Alarm Calls," *American Naturalist* 99, no. 904 (1965): 59–63.

15. See Borrello's careful explication of these events and intellectual positions in *Evolutionary Restraints*.

16. Vero Copner Wynne-Edwards, *Animal Dispersion in Relation to Social Behavior* (New York: Hafner, 1962).

17. Robert Ardrey, *The Social Contract: A Personal Inquiry into the Evolutionary Sources of Order and Disorder* (New York: Atheneum, 1970). Not everyone immediately jumped on the anti-group-selection bandwagon, e.g., Richard Lewontin, "Adaptation and Natural Selection," *Science* 152, no. 3720 (1966): 338–339. This rhetoric has changed dramatically only in the last few years. The doyen of neo-Darwinism, E. O. Wilson, now suggests group selection likely does play an important role in the evolution of social behavior: Natalie Angier, "Edward O. Wilson's New Take on Human Nature," Smithsonian, April 2012, accessed 5 February 2018, http://www.smithsonianmag.com/science-nature/Edward-O-Wilsons-New-Take-on-Human-Nature.html.

18. Robert Trivers, "The Evolution of Reciprocal Altruism," *Quarterly Review of Biology* 46 (1971): 35–57; quotation from Robert Trivers, *Natural Selection and Social Theory: Selected Papers of Robert Trivers* (New York: Oxford University Press, 2002), 5.

19. For example, Marcel Mauss, *The Gift*, trans. W. D. Halls (London: Routledge, 1990). A few years later Donna Haraway suggested that sociobiology was written in the language of

"classical capitalist political economy": see "Animal Sociology and a Natural Economy of the Body Politic, Part 2: The Past is the Contested Zone: Human Nature and Theories of Production and Reproduction in Primate Behavior Studies," *Signs* 4, no. 1 (1978): 37–60, quote on 57.

20. Robert Trivers, "Parental Investment and Sexual Selection," in *Sexual Selection and the Descent of Man, 1871–1971*, ed. Bernard Campbell (Chicago: Aldine, 1972), 136–179.

21. Zoologists hailed Trivers's paper as having single-handedly resurrected Darwin's concept of sexual selection that had lain fallow for almost a century (even if such claims overlooked the complexity of the intellectual landscape of the intervening years); Erika Lorraine Milam, "Selective History: Writing Female Choice into Organismal Biology," in *Looking for a Few Good Males: Female Choice in Evolutionary Biology* (Baltimore, MD: Johns Hopkins University Press, 2010), 135–159.

22. Peter Ellison, "A Conversation with Irven DeVore," *Annual Reviews Conversations* (14 November 2012): quotes on 9, 10, accessed 5 February 2018, http://www.annualreviews.org /userimages/contenteditor/1357748046867/annurev-conversations-010913-100006.pdf.

23. Sarah Blaffer Hrdy, "Myths, Monkeys, and Motherhood: A Compromising Life," in *Leaders in Animal Behavior: The Second Generation*, ed. Lee Drickamer and Donald Dewsbury (Cambridge: Cambridge University Press, 2010), 343–374, quotes on 352; her comment echoes Thomas Henry Huxley's moniker as "Darwin's bulldog."

24. Margo Miller, "First the Ants, Then a Pulitzer," *Boston Globe*, 17 April 1979, 25.

25. Trivers, *Natural Selection and Social Theory*, 125.

26. Richard Wrangham, interview by author, 7 May 2012, Harvard University; Richard Wrangham, speaking at "An Oral History of Primatology, Jane Goodall, at Cambridge," University of Cambridge, Personal Histories Project, 28 April 2011, accessed 8 February 2018, https:// sms.cam.ac.uk/media/1332629.

27. Patricia Adair Gowaty, "Watcher: The Development of an Evolutionary Biologist," in *Leaders in Animal Behavior*, ed. Drickamer and Dewsbury, 309–342, quote on 318.

28. Robert L. Trivers, "Parent-Offspring Conflict," *American Zoologist* 14, no. 1 (1974): 249–264.

29. Trivers, "Parental Investment and Sexual Selection."

30. Trivers, *Natural Selection and Social Theory*, 126.

31. Trivers's two other publications of the early 1970s were Robert Trivers and Daniel Willard, "Natural Selection of Parental Ability to Vary the Sex Ratio of Offspring," *Science* 179, no. 4068 (1973): 90–92; and Trivers, "Parent-Offspring Conflict," 249–262.

32. McCain, "Sociobiology," A4.

33. Wilson, *Sociobiology*, 28–29.

34. Wilson, *Sociobiology*, 155.

35. On sociobiologists' use of genetic language, see Henry Howe and John Lyne, "Gene Talk in Sociobiology," *Social Epistemology* 6, no. 2 (1992): 109–163.

36. Trivers, foreword to *Selfish Gene*. John Pfeiffer's review of Dawkins' book similarly called it "first popular introduction to the new evolution," in "The New Evolution," *New York Times*, 27 February 1977, 222.

37. Dawkins, *Selfish Gene*, 1.

38. Dawkins, *Selfish Gene*, 2.

39. Alfred, Lord Tennyson, *In Memorium A. H. H.* (1849).

40. Dawkins, *Selfish Gene*, 7, 9, 147.

41. Richard Dawkins, "Growing Up in Ethology," in *Leaders in Animal Behavior*, ed. Drickamer and Dewsbury, 189–218.

42. Peter Medawar "Pro Bono Public," *Spectator*, 15 January 1977, 353; Steven Pinker, *Blank Slate: The Modern Denial of Human Nature* (New York: Penguin Paperbacks, 2003), 124; Dawkins, "Growing Up in Ethology," 207.

43. "Change in Wenner-Gren Foundation Conference Program," *Current Anthropology* 19, no. 3 (1978): 654.

44. Erika Lorraine Milam, "The Equally Wonderful Field: Ernst Mayr and Organismic Biology," *Historical Studies in the Natural Sciences* 40, no. 3 (2010): 279–317. The Committee on Research on Problems of Sex stopped funding research in the mid-1960s, too: Michael Pettit, Darya Serykh, and Christopher D. Green, "Multispecies Networks: Visualizing the Psychological Research of the CRPS," *Isis* 106, no. 1 (2015): 121–149.

45. In the first three years of their tenure, Fox and Tiger provided funds for research on (their categories) basic social processes, animal behavior, children, cross-cultural studies, and prehistoric human behavior. Over two-thirds fell into the first two categories. The board of the HFGF independently donated $100,000 to the Leakey Foundation and $250,000 to the American Museum of Natural History for the Harry Frank Guggenheim Hall of Minerals and Earth Minerals, which remains open today.

46. For example, Nicolas Blurton-Jones, Napoleon Chagnon, Irven DeVore, Robin Fox, Jane Goodall, Desmond Morris, Donald Symons, Robert Trivers, and Sherwood Washburn. (They passed on funding Adrienne Zihlman; Zihlman, interview by author, 24 October 2011.)

47. Robin Fox, "Overview of Research-Grant Activity of the Harry Frank Guggenheim Foundation," attached to Donald R. Griffin to Peter O. Lawson-Johnson, 9 November 1978; Chapter 8 of the *History of the HFGF* (Harry Frank Guggenheim Foundation, 25 West 53rd Street, New York, NY 10019), 6.

48. Fox, "Overview of Research-Grant Activity," 6; Lionel Tiger asserted much the same sentiment in my interview with him on 10 November 2011.

49. Numerically, the grants broke down as follows: 1972–1976: basic social processes (11 grants), animal behavior (6), children (3), cross-cultural (2), and prehistoric human behavior (2); 1973–1978: human and animal ethology (27), sociobiology (13), cognition and expression (9), brain and behavior (9), sociological and historical studies (7), and hormones and behavior (6).

50. For example, John Buettner-Janusch, ed., *Origins of Man: Physical Anthropology* (New York: Wiley, 1966); Phyllis C. Jay, ed., *Primates: Studies in Adaptation and Variability* (New York: Holt, Rinehart and Winston, 1968); Russel Tuttle, ed., *Functional and Evolutionary Biology of Primates* (Chicago: Aldine, 1972); Francisco Salzano, ed., *The Role of Natural Selection in Human Evolution* (Amsterdam: North-Holland Publishing Company, 1975); David Hamburg and Elizabeth McCown, eds., *The Great Apes* (Menlo Park, CA: Benjamin/Cummings, 1979).

51. Susan Lindee and Joanna Radin, "Patrons of the Human Experience: A History of the Wenner-Gren Foundation for Anthropological Research," *Current Anthropology* 57, no. S14 (2016): S218–S301; "The Story and People of Wenner-Gren," Wenner-Gren Foundation, accessed 22 August 2017, http://www.wennergren.org/history/story-and-people-wenner-gren.

52. For a thorough history of the NSF's funding of biology in this period, see Toby Appel,

Shaping Biology: The National Science Foundation and American Biological Research, 1945–1975 (Baltimore: Johns Hopkins University Press, 2002), esp. "Allocating Resources to a Divided Science," 207–234.

53. Mark Solovey, "Senator Fred Harris's National Social Science Foundation Proposal: Reconsidering Federal Science Policy, Natural Science—Social Science Relations, and American Liberalism during the 1960s," *Isis* 103, no. 1 (2012): 54–82. See also Daniel Lee Kleinman and Mark Solovey, "Hot Science/Cold War: The National Science Foundation after World War II," *Radical History Review* 63 (1995): 110–139. The Social Science Research Council would not have looked kindly on this research either: Mark Solovey, "Riding Natural Scientists' Coattails onto the Endless Frontier: The SSRC and the Quest for Scientific Literacy," *Journal of the History of the Behavioral Sciences* 40, no. 4 (2004): 393–422.

54. Box 433: "Rationale, 1959–1975," David Hamburg Papers, 1949–2003, CA#0005, Carnegie Collection, Rare Book and Manuscript Library, New York.

55. For a full exposition of the differences between British and American sociobiology- inclined scientists, see Segerstråle, *Defenders of the Truth*, and for the intellectual variation among contributors, see Drickamer and Dewsbury, eds., *Leaders in Animal Behavior*.

56. The list of invitees included Robin Fox and Irven DeVore, as organizers, and Richard Alexander, Anthony Ambrose, Frederick Barth, Bernardo Bernardi, Mireille Bertrand, Norbert Bischof, Napoleon Chagnon, John Crook, Richard Dawkins, Mildred Dickeman, Meyer Fortes, Jack Goody, François Héritier, William Irons, Albert Jaquard, Jeffrey Kurland, Roger Masters, John Maynard Smith, Andrew Strathern, Lionel Tiger, Robert Trivers, Mario von Cranach, and Richard Wrangham. On his disappointment with the conference, see Robin Fox, *Encounter with Anthropology*, 2nd ed. (New Brunswick, NJ: Transaction Publishers, 1991), 309.

57. Wilson, *Sociobiology*, 547.

58. Wilson, *Sociobiology*, 575.

59. Washburn's talk was recorded and later broadcast by radio: "Science Story: A Condemnation of Sociobiology," Laurie Garrett, producer, Science Story for KPFA Berkeley, 11 July 1977, 29 min., Pacifica Radio Archives #AZ0027.07. Ironically, Wilson saw himself (and other organismal biologists) relegated to second-class status among Harvard biologists because he was not a molecular biologist working in a laboratory: see Edward O. Wilson, "The Molecular Wars," in *The Naturalist* (Washington, DC: Shearwater Press, 1995), 218–237.

Chapter Fourteen. The Old Determinism

1. Raymond A. Sokolov, "Talk with Stephen Jay Gould," *New York Times*, 20 November 1977, BR4; Myrna Perez Sheldon, "Stephen Jay Gould, An Evolutionary Heretic" (unpublished manuscript, 2017).

2. Stephen Jay Gould, "The Nonscience of Human Nature," *Natural History*, April 1974, 21–24.

3. Gould, "The Nonscience of Human Nature."

4. Ullica Segerstråle, *Defenders of the Truth: The Battle for Science in the Sociobiology Debate and Beyond* (New York: Oxford University Press, 2000).

5. See Marshall Sahlins, *The Use and Abuse of Biology: An Anthropological Critique of Sociobi-*

ology (Ann Arbor: University of Michigan Press, 1976); Ashley Montagu, ed., *Race and IQ* (New York: Oxford University Press, 1975); Ashley Montagu, *The Nature of Human Aggression* (New York: Oxford University Press, 1976); Ethel Tobach and Betty Rosoff, eds., *Genes and Gender* (New York: Gordian Press, 1978) and *Genes and Gender II: Pitfalls on Research in Sex and Gender* (New York: Gordian Press, 1979).

6. Sarah Bridger, *Scientists at War: The Ethics of Cold War Weapons Research* (Cambridge, MA: Harvard University Press, 2015).

7. On the deeply held and conflicting investments of moral value in the debates over socio-biology, see Segerstråle's nuanced analysis in *Defenders of the Truth*.

8. Edward O. Wilson, "Dialogue, the Response: Academic Vigilantism and the Political Significance of Sociobiology," *BioScience* 26, no. 3 (1976): 183–190.

9. Wilson, "Dialogue," 189, 190.

10. "Against 'Sociobiology,'" *New York Review of Books*, 13 November 1975, 43.

11. Sociobiology Study Group of Science for the People, "Dialogue, the Critique: Sociobiol-ogy—Another Biological Determinism," *BioScience* 26, no. 3 (1976), 182–186; Ann Arbor Sci-ence for the People Editorial Collective, *Biology as a Social Weapon* (Minneapolis: Burgess Publishing Company, 1977).

12. Kelly Moore, *Disrupting Science: Social Movements, American Scientists, and the Politics of the Military, 1945–1975* (Princeton, NJ: Princeton University Press, 2008).

13. Genetic Engineering Group, "Book Review: Sociobiology—*The Skewed Synthesis*," *Sci-ence for the People* 7, no. 6 (1975): 28–30.

14. Arthur R. Jensen, "How Much Can We Boost I.Q. and Scholastic Achievement?," *Har-vard Educational Review* 39, no. 1 (1969): 1–123.

15. Raymond Fancher, *The Intelligence Men: Makers of the IQ Controversy* (New York: Nor-ton, 1985): 185–201; Leila Zenderland, "*The Bell Curve* and the Shape of History," *Journal of the History of the Behavioral Sciences* 33, no. 2 (1997): 135–139.

16. Richard Herrnstein, "IQ," *Atlantic Monthly*, September 1971; and Herrnstein, *IQ and the Meritocracy* (Boston: Little, Brown, 1973); "About This Issue," *Science for the People* 6, no. 2 (1974): 3.

17. Richard J. Herrnstein and Charles Murray, *The Bell Curve: Intelligence and Class Structure in American Life* (New York: Free Press, 1994).

18. "XYY: Fact or Fiction," *Science for the People* 6, no. 5 (September 1974): 22–24; "Actions on XYY Research," *Science for the People* 7, no. 1 (1975): 4.

19. "The XYY Controversy Continued," *Science for the People* 7, no. 4 (1975): 28–32, quote on 28. See also "Genetic Engineering Pamphlet" distributed by *Science for the People*.

20. For example, Jon Beckwith and J. King, "The XYY Syndrome: A Dangerous Myth," *New Scientist*, 14 November 1974, 474.

21. Genetic Engineering Group, "Book Review: Sociobiology," 28.

22. Genetic Engineering Group, "Book Review: Sociobiology," 30; they cited Ashley Mon-tagu's edited collection *Man and Aggression* (New York: Oxford University Press, 1968; 2nd ed., 1973) as especially useful in countering misunderstandings of human evolutionary history, as well as Alexander Alland Jr., *The Human Imperative* (New York: Columbia University Press, 1972).

23. Science for the People continued to publish articles decrying sociobiology as sexist, ideological, and just plain bad science: "Biological Determinism Attacked at Ann Arbor Conference: Does Biology Explain Violence? Sex Roles? Competition?" *Science for the People* 8, no. 1 (1976): 14–15; "Sociobiology: Tool for Oppression," *Science for the People* 8, no. 2 (1976): 7–9; Barbara Chasin, "Sociobiology: A Sexist Synthesis," *Science for the People* 9, no. 3 (1977): 27–31; Freda Salzman, "Are Sex Roles Biologically Determined?" *Science for the People* 9, no. 4 (1977): 27–32; Richard Lewontin, "Biological Determinism as an Ideological Weapon," *Science for the People* 9, no. 6 (1977): 36–38; Tedd Judd, "Naturalizing What We Do: A Review of the Film *Sociobiology: Doing What Comes Naturally*," *Science for the People* 10, no. 1 (1978): 16–19; Jon Beckwith and Bob Lange, "AAAS: Sociobiology on the Run," *Science for the People* 10, no. 2 (1978): 38–39; Dan Atkins, Jack Dougherty, Walda Katz Fishman, and Frank Rosenthal, "Towards a Renewed and Expanded SftP Role among Science Teachers," *Science for the People* 10, no. 4 (1978): 34–35; Freda Salzman, "Sociobiology: The Controversy Continues," *Science for the People* 11, no. 2 (1979): 20–27.

24. *Sociobiology: Doing What Comes Naturally* (1976; New York: Document Associates, 1978), 21 min. The film appears to have aired on television in Hong Kong several years earlier: "Television Guide," *Sunday Post-Herald* (*South China Morning Post*), 13 January 1974, 21.

25. Richard Wrangham, author interview, 7 May 2012, Harvard University.

26. Irven DeVore, "DeVore Explains Sociobiology Film Interviews," *Anthropology Newsletter* 18, no. 8 (1977): 2, 14.

27. DeVore, "DeVore Explains Sociobiology Film Interviews," 2.

28. Doug Boucher, Fred Gifford, Sue Porter, Scott Schneider, and John Vandermeer (Sociobiology Study Group, Ann Arbor Science for the People), "Sociobiology Critics Speak Out," *Anthropology Newsletter* 18, no. 10 (1977): 19–20.

29. *Nova*, "Transcript: The Human Animal," Season 4, Episode 10 (23 March 1977). Thank you to Keith Luf at the WGBH Media Library and Archives (Boston, MA) for providing me with the transcript.

30. See, for example, their willingness to sign a group letter condemning sociobiology in the pages of the *New York Review of Books*—Anthony Leeds, Barbara Beckwith, Chuck Madansky, David Culver, Elizabeth Allen, Herb Schreier, Hiroshi Inouye, Jon Beckwith, Larry Miller, Margaret Duncan, Miriam Rosenthal, Reed Pyeritz, Richard C. Lewontin, Ruth Hubbard, Steven Chorover, and Stephen Jay Gould, "Against 'Sociobiology,'" *NYRB* 22, no. 18 (1975), in reaction to Conrad Hal Waddington's original review of the book in "Mindless Societies," *NYRB* 22, no. 13 (1975), and Edward O. Wilson's response, "For 'Sociobiology,'" *NYRB* 22, no. 20 (1975). On the complex professional relationships at Harvard during this time, see Segerstråle, *Defenders of the Truth*.

31. On the long history of heredity in the life sciences, see Staffan Müller-Wille and Hans-Jörg Rheinberger, *A Cultural History of Heredity* (Chicago: University of Chicago Press, 2012). On the different and contradictory meanings of "gene" in twentieth-century biology, see Peter J. Beurton, Raphael Falk, and Hans-Jörg Rheinberger, eds., *The Concept of the Gene in Development and Evolution: Historical and Epistemological Perspectives* (New York: Cambridge University Press, 2000). On the controversies surrounding behavioral genetics and claims about the heredity of human behavior, see Aaron Panofsky, *Misbehaving Science: Controversy and the Development of Behavior Genetics* (Chicago: University of Chicago Press, 2014).

32. On debates over "mother love," see Marga Vicedo, *The Nature and Nurture of Love: From Imprinting to Attachment in Cold War America* (Chicago: University of Chicago Press, 2013).

33. He based his comments for *Nova* on his dissertation research: Lionel Tiger and Joseph Shepher, *Women in the Kibbutz* (New York: Harcourt Brace Jovanovich, 1975).

34. "Council Refuses to Condemn Sociobiology," *Anthropology Newsletter* 18, no. 1 (1977): 1.

35. Richard C. Lewontin, "The Principle of Historicity in Evolution," in *Mathematical Challenges to the Neo-Darwinian Interpretation of Evolution*, ed. Paul S. Moorhead and Martin M. Kaplan (Philadelphia: Wistar Institute, 1967), 81–94; Stephen Jay Gould, *Wonderful Life: The Burgess Shale and the Nature of History* (New York: Norton, 1989), 51; Stephen Jay Gould and Richard C. Lewontin, "The Spandrels of San Marco and the Panglossian Paradigm: A Critique of the Adaptationist Programme," *Proceedings of the Royal Society of London B* 205 (1979): 581–598. See, too, Theodosius Dobzhansky, "Darwinian Evolution and the Problem of Extraterrestrial Life," *Perspectives in Biology and Medicine* 15, no. 2 (1972): 157–175; and George Gaylord Simpson, "The Nonprevalence of Humanoids," *Science* 143, no. 3608 (1964): 769–754. For a lucid discussion of these issues, see John Beatty, "Chance Variation and Evolutionary Contingency," in *The Oxford Handbook of Philosophy of Biology*, ed. Michael Ruse (Oxford: Oxford University Press, 2008), 189–211.

36. Myrna Perez Sheldon, "Stephen Jay Gould, An Evolutionary Heretic"; Michael Shermer, "Stephen Jay Gould as Historian of Science and Scientific Historian, Popular Scientist and Scientific Popularizer," *Social Studies of Science* 32, no. 4 (2002): 489–524.

37. Stephen Jay Gould, "Play It Again, Life," *Natural History* 95, no. 2 (1986): 18–26. On the vitality of this metaphor, see David Sepkoski, " 'Replaying Life's Tape': Simulations, Metaphors, and Historicity in Stephen Jay Gould's View of Life," *Studies in History and Philosophy of Science Part C: Studies in History and Philosophy of Biological and Biomedical Sciences* 58 (2016): 73–81.

38. Jeffrey Kurland, "Film Review—Sociobiology: 'The Human Animal,' " *American Journal of Physical Anthropology* 61 (1983): 507–508, quote on 507.

39. Kurland, "Film Review—Sociobiology: 'The Human Animal,' " 508. In the review Kurland says "histological" but I cannot help but wonder if he meant "historical."

40. Jeffrey Kurland, "Film Review—*Sociobiology: Doing What Comes Naturally*," *American Journal of Physical Anthropology* 61 (1983): 267–268.

41. Kurland, "Film Review—*Sociobiology: Doing What Comes Naturally*," 268.

42. Ellipses in original. On the Piltdown fraud, see Roger Lewin's running commentary in *Bones of Contention: Controversies in the Search for Human Origins* (Chicago: University of Chicago Press, 1987).

43. Eleanor Leacock, "Society and Gender," in *Genes and Gender*, ed. Tobach and Rosoff, 75–85, from advertising material for *Sociobiology: Doing What Comes Naturally*, quoted on 81.

44. On the Nobel Prize and von Frisch, see Tania Munz, *The Dancing Bees: Karl von Frisch and the Discovery of the Honeybee Language* (Chicago: University of Chicago Press, 2016). On Lorenz and Tinbergen, see Richard W. Burkhardt, *Patterns of Behavior: Konrad Lorenz, Niko Tinbergen, and the Founding of Ethology* (Chicago: University of Chicago Press, 2005).

45. Rosoff and Tobach, "Prologue," in *Genes and Gender*, 8.

46. Rosoff and Tobach, "Prologue," in *Genes and Gender*, 7.

47. Helen Block Lewis, "Psychology and Gender," in *Genes and Gender*, ed. Tobach and Rosoff, 63–73.

48. Rosoff and Tobach, "Epilogue," in *Genes and Gender*, 89.

49. Rosoff and Tobach, "Prologue," in *Genes and Gender II*, 7.

50. Ruth Bleier, "Social and Political Bias in Science: An Examination of Animal Studies and Their Generalizations to Human Behavior and Evolution," in *Genes and Gender II*, ed. Tobach and Rosoff, 49–70.

51. Jane Beckman Lancaster, *Primate Behavior and the Emergence of Human Culture* (New York: Holt, Rinehart and Winston, 1975); Thelma Rowell, *The Social Behaviour of Monkeys* (Baltimore: Penguin, 1972), and "The Concept of Dominance," *Behavioral Biology* 11, no. 2 (1974): 131–154; Vinciane Despret, "Culture and Gender Do Not Dissolve into How Scientists 'Read' Nature: Thelma Rowell's Heterodoxy," in *Rebels, Mavericks, and Heretics in Biology*, ed. Oren Harman and Michael Deitrich (New Haven, CT: Yale University Press, 2008), 338–355.

52. Sarah Blaffer Hrdy, "Myths, Monkeys, and Motherhood: A Compromising Life," in *Leaders in Animal Behavior: The Second Generation*, ed. Lee Drickamer and Donald Dewsbury (Cambridge: Cambridge University Press, 2010), 343–374, quote on 364.

53. Hrdy, "Myths, Monkeys, and Motherhood," 357.

54. See especially Patricia Adair Gowaty, ed., *Feminism and Evolutionary Biology Boundaries, Intersections and Frontiers* (Boston: Springer, 1997), based on a 1994 symposium of the same name.

55. Donna Haraway, *Primate Visions: Gender, Race, and Nature in the World of Modern Science* (New York: Routledge, 1989); Shirley C. Strum and Linda M. Fedigan, eds., *Primate Encounters: Models of Science, Gender, and Society* (Chicago: University of Chicago Press, 2000).

56. Sarah Blaffer Hrdy, interview by author, 17 October 2011, Citrona Farms, Winters, CA; Robin Fox, *Participant Observer* (New York: Transaction Publishers, 2004), 561.

57. For DeVore's account, see Segerstråle, *Defenders of the Truth*, esp. 81–84.

58. Segerstråle, *Defenders of the Truth*, 79–83, Robert Trivers, "Interview," *Omni*, July 1985, 77–111, esp. 78.

59. For a detailed first-hand account of the events, including the quotes below, see Segerstråle, *Defenders of the Truth*, 22–26. The proceedings of the symposium were later published: George W. Barlow and James Silverberg, eds., *Sociobiology, Beyond Nature/Nurture? Reports, Definitions, and Debate* (Boulder, CO: Westview Press, 1980).

60. For the nineteenth-century parallel, see Robert Nye, "Medicine and Science as Masculine 'Fields of Honor,'" in *Women, Gender, and Science: New Directions*, ed. Sally Gregory Kohlstedt and Helen Longino, *Osiris* 12 (Chicago: University of Chicago Press, 1997), 60–79. On the legacy of these debates, see Aaron Panofsky, *Misbehaving Science: Controversy and the Development of Behavior Genetics* (Chicago: University of Chicago Press, 2014).

61. Wilson's next book, *On Human Nature* (Cambridge, MA: Harvard University Press, 1978), won the 1979 Pulitzer Prize for General Non-Fiction, further cementing his reputation as an excellent writer for nonscientific audiences.

62. Raymond A. Sokolov, "Talk with Stephen Jay Gould," *New York Times*, 20 November 1977, BR4.

63. Carl Sagan had just published *The Dragons of Eden: Speculations on the Evolution of Human Intelligence* (New York: Random House, 1977) and was already known as a great popularizer of science.

Chapter Fifteen. Human Nature

1. Sherwood Washburn and Phyllis Jay, ed., *Perspectives on Human Evolution*, 5 vols. (New York: Holt, Rinehart and Winston, 1968–1975): Vol. 1, *Perspectives on Human Evolution*, ed. Washburn and Jay; Vol. 2, *Perspectives on Human Evolution*, ed. Washburn and [Jay] Dolhinow; Vol. 3, *Human Origins: Louis Leakey and East African Evidence*, ed. Glynn Isaac and Elizabeth McCown; Vol. 4, *Human Evolution: Biosocial Perspectives*, ed. Washburn and McCown; Vol. 5, *The Great Apes*, ed. David A. Hamburg and McCown.

For a historical perspective on shifting primate models after the Second World War, see Susan Sperling, "Baboons with Briefcases vs. Langurs in Lipstick: Feminism and Functionalism in Primate Studies," in *Gender at the Crossroads of Knowledge: Feminist Anthropology in the Postmodern Era*, ed. Micaela di Leonardo (Berkeley: University of California Press, 1992), 204–234; Sperling, "The Troop Trope: Baboon Behavior as a Model System in the Postwar World," in *Science Without Laws: Model Systems, Cases, Exemplary Narratives*, ed. Angela N. H. Creager, Elizabeth Lunbeck, and M. Norton Wise (Durham, NC: Duke University Press, 2007), 73–89; and Donna Haraway, *Primate Visions: Gender, Race, and Nature in the World of Modern Science* (New York: Routledge, 1989); Shirley Strum and Linda Fedigan, eds. *Primate Encounters: Models of Science, Gender, and Society* (Chicago: University of Chicago Press, 2000).

2. Linda Marie Fedigan, "Theories on the Evolution of Human Social Life," in *Primate Paradigms: Sex Roles and Social Bonds* (Chicago: University of Chicago Press, 1982), 307–323.

3. Ann Pusey, interview by author, 23 May 2012.

4. In addition to Fedigan's research, see Jeanne Altmann, *Baboon Mothers and Infants* (Cambridge, MA: Harvard University Press, 1980); and Barbara Smuts, *Sex and Friendship in Baboons* (New York: Aldine, 1985).

5. See Jane Goodall's description of these events in "Life and Death at Gombe," *National Geographic*, May 1979; and *The Chimpanzees of Gombe: Patterns of Behavior* (Cambridge, MA: Belknap Press of Harvard University Press, 1986), 206.

6. All quotes this paragraph, Hamburg to Faculty, Staff, and Friends of the Department of Psychiatry, 11 August 1972, BANC 98/132C Box 1, Folder 19: "Hamburg, David—Correspondence, 1966–1995," Sherwood L. Washburn Papers, Bancroft Library, University of California, Berkeley (hereafter cited as Washburn Papers).

7. Richard Wrangham, "Artificial Feeding of Chimpanzees and Baboons in Their Natural Habitat," *Animal Behaviour* 22 (1974): 83–93; Wrangham, "Feeding Behavior of Chimpanzees in Gombe National Park Tanzania," in *Primate Ecology*, ed. T. H. Clutton-Brock (London: Academic Press, 1977), 504–538.

8. David A. Hamburg and Jane van Lawick-Goodall to Wenner-Gren Foundation, 4 April 1973, BANC 98/132C Box 1, Folder 19: "Hamburg, David—Correspondence, 1966–1995," Washburn Papers.

9. Hamburg and Lawick-Goodall to Wenner-Gren Foundation, 4 April 1973.

10. Louise Roche, *The Program in Human Biology at Stanford University: The First 30 Years* (Stanford, CA: Program in Human Biology, 2001), https://humbio.stanford.edu/sites/default/files/alumni_/humbiohistory.pdf.

11. Washburn to Hamburg, 27 February 1969, BANC 98/132C Box 1, Folder 19: "Hamburg, David—Correspondence, 1966–1995," Washburn Papers.

12. Of the $1,936,000 given by the Ford grant, $1.6 million was allocated to the creation of

four endowed professorships (matched by the university). The remaining money went to general operating funds, substantially supplemented by the university who covered expenses for all office staff and facilities.

13. *Campus Report*, 18 March 1970, as quoted in Roche, *Program in Human Biology*, 9.

14. Hamburg to Faculty, Staff, and Friends of the Department of Psychiatry.

15. Word of the kidnapping reached the US Embassy through an American Missionary in Kigoma; Unclassified Cable, US Embassy Dar es Salaam to Department of State, 1975 May 20, 13:55 (Tuesday): 1975DARES01577. Goodall has written about these events many times but always in brief: e.g., Jane Goodall, with Phillip Berman, *Reason for Hope: A Spiritual Journey* (New York: Grand Central Publishing, 1999), 113–119. The most thorough account of the kidnapping and its aftermath can be found in Dale Peterson, "Domesticity and Disaster," in *Jane Goodall: The Woman Who Redefined Man* (New York: Mariner Books, 2006), 541–561.

Other Westerners who escaped capture included US students Larry Goldman, Craig Packer, Helen Neely, Susan Loeb, Emily Polis, Jim Baugh, Ann Pierce, Phyllis Lee, Michelle Trudeau, and Joan Silk, as well as UK citizens Tony Collins, Juliette Oliver, and Ann Pusey: Unclassified cable US Embassy Dar es Salaam to Department of State, 1975 May 21, 08:00 (Wednesday): 1975DARES01586. Packer and Pusey were in Kenya for medical treatment and reported to the US Embassy in Nairobi on May 22 when they heard the news; Ben Wieder, "At Duke, an Evolutionary Anthropologist Plumbs Jane Goodall's Research Trove," *Chronicle of Higher Education*, 26 June 2011; Limited Official Use cable, US Embassy Nairobi to US Embassy Dar es Salaam, 1975 May 22 07:32 (Thursday): 1975NAIROB04126.

Jane Goodall, Emily Polis, Jim Baugh, and Jonathan Wainwright (an additional British student not mentioned in the above cable) had been at Gombe and in their official statements after the fact "praised courage of Tanzanian guards who badly beaten when refused to disclose location of Goodal [sic] and 'other white people at the camp.' " They speculated that they had time to find cover because their houses were on higher ground (closer to the edge of the lake) than those who were kidnapped. Stanford chartered a plane to bring the nine people associated with the university from Kigoma to Nairobi: Unclassified cable US Embassy Nairobi to US Embassy Dar es Salaam, 1975 May 23 07:45 (Friday): 1975NAIROB04178.

16. Jane Goodall, *Through a Window: My Thirty Years with the Chimpanzees of Gombe* (Boston: Houghton Mifflin Company, 1990), 65. In accounts written at the time and after, the Westerners in hiding were awed by the fortitude of those camp members who at great peril to their own safety protected other people from being kidnapped (76–78).

17. Limited Official Use cable US Embassy Dar es Salaam to Department of State, 1975 May 25, 16:50 (Sunday): 1975DARES01693. The PRP was especially interested in two men: Gabriel Yumbu (captured in 1973 and presumed dead by the US Embassy) and Brigadier General Saleh Kilenga (captured on 2 March 1975); Confidential cable US Embassy Dar es Salaam, 1975 June 21 12:09 (Saturday): 1975DARES02204.

18. Limited Official Use cable US Embassy Dar es Salaam to Department of State, 1975 May 25, 16:50 (Sunday): 1975DARES01693.

19. "Government Rejects Responsibility for Kidnaped Students," Foreign Broadcast Information Service (FBIS) Daily Reports, Paris AFP-1975-05-28, as published in *Daily Report, Sub-Saharan Africa*, FBIS-SAF-75-104, on 1975-05-29, p. B4. When contacted by Carter, the Tanzanian government immediately indicated they would consider prisoner release but could not be

seen to negotiate directly with the PRP—that would have to be handled by the Americans. Classified cable US Department of State to the Netherlands, 1975 May 29, 14:27 (Thursday): 1975STATE124769.

20. On behind-the-scenes negotiating with terrorists, see Guy Olivier Faure, "Negotiating with Terrorists: A Discrete Form of Diplomacy," *Hague Journal of Diplomacy* 3 (2008): 179–200; and David Tucker, "Responding to Terrorism," *Washington Quarterly* 21, no. 1 (1998): 103–117. Refusing to make concessions to terrorists was first articulated as US policy only in 1971. See also Richard Clutterbuck, *Kidnap and Ransom: The Response* (London: Faber & Faber, 1978). According to Lisa Stampnitzky, policy experts initially conceptualized kidnappings like this one as a tool of political counterinsurgents or guerillas. By the 1980s, they would be redefined as acts of irrational terrorism: *Disciplining Terror: How Experts Invented 'Terrorism'* (Cambridge: Cambridge University Press, 2013).

21. Smith's letter was reproduced in Bill Grueskin, "Kidnap Date Etched in Student's Memory," *Palo Alto Times*, 27 May 1975.

22. [Emily Polis] Briarcroft, "May 19, 1975," 15 February 2010, accessed 6 February 2018, http://briarcroft.wordpress.com/2010/02/15/may-19-1975/.

23. [Emily Polis] Briarcroft, "Forty Years Ago Today," 19 May 2015, accessed 6 February 2018, https://briarcroft.wordpress.com/2015/05/19/forty-years-ago-today/.

24. "Interview with Ambassador W. Beverly Carter Jr.," Association for Diplomatic Studies and Training Foreign Affairs Oral History Project, Ralph J. Bunch Legacy: Minority Officers, interviewed by Celestine Tutt, 30 April 1981, accessed 26 August 2017, https://www.loc.gov/item/mfdipbib001527/.

25. Joshua Lederberg to Thomas Schelling, 28 June 1975; Shelling to Lederberg, 2 July 1975. Lederberg later wrote and asked for Schelling to return the original letter he had sent: see Schelling to Lederberg, 15 August 1975, Box 25, Folder 31: "Schelling, Thomas C.," Joshua Lederberg Papers, Profiles in Science, US National Library of Medicine, Bethesda, MD, accessed 16 March 2018, https://profiles.nlm.nih.gov/ps/retrieve/Collection/CID/BB.

26. In anticipation of the students' release, Henry Kissinger strongly felt everyone concerned should "decline to comment" on the negotiations that led to their release, especially that ransom money had been paid and on "the role played by the USG in effecting the release." Kissinger advised the official spokesperson to say that "the U.S. government does not pay ransom and does not condone payment of ransom by other parties." Confidential cable, 1975 June 25, 21:31 (Wednesday): 1975STATE149719.

27. The US government originally suspected that Kilenga preferred to remain in Tanzania (confidential cable, US Embassy Dar es Salaam to Department of State, 1975 July 12, 08:32 [Saturday]: 1975DARES02475), but the Tanzanian government released him in mid-July and he returned to Zaire a week later (confidential cable, US Embassy Dar es Salaam to Department of State, 1975 July 25, 10:30 [Friday]: 1975DARES02674).

28. Secret cable, US Embassy Dar es Salaam to Department of State, 1975 June 28, 05:21 (Saturday): 1975DARES02313. By mid-June, the PRP was also keen to improve the way radio news broadcasts described the PRP, including calling the party "large" and "well-organized," as well as indicating it was "controlling a large area." Confidential cable US Embassy Dar es Salaam to Department of State, 1975 July 11, 10:32 (Friday): 1975DARES02458.

29. Smith returned to Kigoma the night of July 25–26, his transit across the lake possibly

delayed by up to ten days due to inclement weather that turned the crossing into a "tremendous barrier." Limited official use cable, US Embassy Dar es Salaam to Department of State, 1975 July 25, 22:20 (Friday): 1975DARES02686.

30. "Interview with Ambassador William Beverly Carter Jr.," Tutt, 30 April 1981.

31. "Beverly Carter, 61; Held High Positions as a U.S. Diplomat," *New York Times*, 11 May 1982, B14.

32. Tutt and Carter, "Interview."

33. Dale Peterson, *Jane Goodall: The Woman Who Redefined Man* (New York: Houghton Mifflin, 2006), 550.

34. Peterson, *Jane Goodall*, 554–555.

35. Peterson, *Jane Goodall*. In my interview with Anne Pusey (23 May 2012), she also mentioned Goodall and Hamburg's falling out as a result of the events at Gombe.

36. Brian C. Aronstam, "Out of Africa," *Stanford Magazine*, July–August 1998.

37. Although all Westerners left Gombe after the kidnapping, publications from these students based on the data they had already gathered took several years to publish. See, for example, Joan Silk, "Patterns of Food Sharing among Mother and Infant Chimpanzees at Gombe," *Folia Primatologica* 29 (1978): 129–141; Silk, "Feeding, Foraging, and Food Sharing of Immature Chimpanzees," *Folia Primatologica* 31 (1979): 123–142; Silk, "Kidnapping and Female Competition in Captive Bonnet Macaques," *Primates* (1980): 100–110; Richard Wrangham, "Sex Differences in Chimpanzee Dispersion," in *The Great Apes*, ed. D.A. Hamburg and E. R. McCown (Menlo Park: Benjamin/Cummings, 1979), 481–489; Wrangham, "On the Evolution of Ape Social Systems," *Social Science Information* 18, no. 3 (1979): 335–368; Wrangham, "Sociobiology: Modification with Dissent," *Biological Journal of the Linnean Society* 13 (1979): 171–177; Richard Wrangham and B. B. Smuts, "Sex Differences in the Behavioural Ecology of Chimpanzees in Gombe National Park, Tanzania," *Journal of Reproduction and Fertility,* suppl., 28 (1980): 13–31. For a more extensive discussion of these students' future careers and continued collaborations, see Donna Haraway, *Primate Visions*, 167–170, who notes that the deep ties between students connecting Cambridge to Harvard to Stanford to Gombe resembled those bonds of alliance the researchers sought to uncover in observations of their animal subjects. Pusey regretted how little time Hamburg had for advising after he left Stanford in 1975. She earned her PhD in 1978.

38. William McGrew, speaking at "An Oral History of Primatology, Jane Goodall, at Cambridge," University of Cambridge, Personal Histories Project, 28 April 2011, accessed 8 February 2018, https://sms.cam.ac.uk/media/1332629.

39. Richard Wrangham, speaking at "An Oral History of Primatology, Jane Goodall, at Cambridge"; Wrangham, interview by author, , 7 May 2012, Harvard University.

40. Barbara Boardman Smuts, "Special Relationships between Adult Male and Female Olive Baboons (*Papio anubis*)" (PhD diss., Stanford University, 1982). David Hamburg served as her doctoral advisor. Trivers is the first person she mentions in her acknowledgments: "I owe a great intellectual debt to Bob Trivers who, by introducing me to evolutionary theory, provided a framework to guide my thinking about animal behavior" (iii). She received funding from the HFGF, the Grant Foundation, the L.S.B. Leakey Foundation, and the Wenner-Gren Foundation for Anthropological Research.

41. Donald Kennedy to Jane Goodall, 8 December 1975, Folder 27, Box 439, David Hamburg

Papers, 1949–2003, CA#0005, Carnegie Collection, Rare Book and Manuscript Library, Columbia University, New York.

42. David Hamburg, interview by author, 13 January 2017; David Hamburg, *A Model of Prevention: Life Lessons* (Boulder: Paradigm Publishers, 2015), 56–64.

43. Jane Goodall, "Infant Killing and Cannibalism in Free-Living Chimpanzees," *Folia Primatologica* 28, no. 4 (1977): 259–282.

44. Sarah Blaffer Hrdy, "Male-Male Competition and Infanticide among the Lanurs (*Presbytis entellus*) of Abu, Rajasthan," *Folia Primatologica* 22, no. 1 (1974): 19–58. These arguments found wider readership in Sarah Blaffer Hrdy, "Infanticide as a Primate Reproductive Strategy," *American Scientist* 65, no. 1 (1977): 40–49.

45. Sarah Blaffer Hrdy, interview by author, 17 October 2011, Citrona Farms, Winters, CA.

46. Ann Pusey, interview by author, 23 May 2012.

47. Phyllis Dolhinow, "Normal Monkeys?" *American Scientist* 65, no. 3 (1977): 266.

48. Dolhinow to Washburn, 30 October 1963, BANC 98/132c, Series 1, Folder 24, Washburn Papers. She wrote that she had recently given a short talk on "survival of the gentlest" at the Wenner-Gren. "This was one of the Peace and Anthropology series going on all over. My contribution was a sketch of non-human primate behavior, in direct contrast to the 'Killer-ape' origin of human behavior etc."

49. In the wild, female chimpanzees typically give birth every five or six years.

50. Goodall, "Infant Killing."

51. Amanda Rees offers an in-depth analysis of this ongoing debate and of the subsequent fallout for field research in *The Infanticide Controversy: Primatology and the Art of Field Science* (Chicago: University of Chicago Press, 2009).

52. Anne Pusey, Jennifer Williams, and Jane Goodall, "The Influence of Dominance Rank on the Reproductive Success of Female Chimpanzees," *Science* 277, no. 5327(1997): 828–831, quote on 830.

53. Craig Packer, *Into Africa* (Chicago: University of Chicago Press, 1994), 170.

54. Jane Goodall, *Chimpanzees of Gombe: Patterns of Behavior* (Belknap Press of Harvard University Press, 1986), 338.

55. Jane Goodall, A Bandora, Emilie Bergman, C Busse, H Matama, et al. "Intercommunity Interactions in the Chimpanzee Population of the Gombe National Park," in *The Great Apes*, ed. David Hamburg and E. R. McCown (Menlo Park, CA: Benjamin/Cummings, 1979), 13–53.

56. Goodall, *Chimpanzees of Gombe*, 514.

57. Margaret Power, *The Egalitarians—Human and Chimpanzee: An Anthropological View* (Cambridge: Cambridge University Press, 1991); Power, "The Authoritative Gombe Chimpanzee Studies," in *Race and Other Misadventures: Essays in Honor of Ashley Montagu in His Ninetieth Year*, ed. Larry T. Reynolds and Leonard Lieberman (New York: General Hall, 1996), 260–280. The debate has continued: e.g., Richard Wrangham and Dale Peterson, *Demonic Males: Apes and the Origins of Human Violence* (Boston: Houghton Mifflin, 1996) vs. Donna Hart and Robert W. Sussman, *Man the Hunted: Primates, Predators, and Human Evolution* (Cambridge, MA: Westview Press, 2005), 207–211; Hart and Sussman, however, were primarily concerned with demonstrating that early humans were far more likely to have been prey than predator.

58. Ashley Montagu, foreword to Power, *Egalitarians*, xv.

59. Richard Wrangham, "Artificial Feeding of Chimpanzees and Baboons in their Natural Habitat," *Animal Behaviour* 22 (1974): 83–93; Wrangham, "Feeding Behavior of Chimpanzees in Gombe National Park Tanzania," in *Primate Ecology*, ed. T. H. Clutton-Brock (London: Academic Press, 1977), 504–538.

60. It is tempting to read Power's arguments in the context of 1980s debates over welfare in the United States. In that light, Montagu's endorsement of her views, in accord with a political perspective defined by antiracism in the 1950s, fits awkwardly with the changing political context of the Reagan years. See Marisa Chappell, *The War on Welfare: Family, Poverty, and Politics in Modern America* (Philadelphia: University of Pennsylvania Press, 2010).

61. Ashley Montagu, ed., *Learning Non-aggression: The Experience of Non-literate Societies* (New York: Oxford University Press, 1978).

62. Goodall, *Chimpanzees of Gombe.*

63. Goodall, *Through a Window*, 126.

64. Pusey, author interview.

65. Hamburg to "Friends and Colleagues," 27 October 1975, BANC 98/132C Box 1, Folder 19 "Hamburg, David—Correspondence, 1966–1995," Washburn Papers.

66. On the latter point, see "People and Places," *BioScience* 27/1 (1977): 69.

67. "David Hamburg: President-Elect of the AAAS," *Science* 221, no. 4609 (1983): 431–432; Hamburg, author interview.

68. David Quammen, "Jane, Fifty Years at Gombe," *National Geographic*, October 2010, 110.

69. Jane Goodall, "Infant Killing," 281.

70. Jane Goodall, "Life and Death at Gombe," *National Geographic*, June 1979, 598; see also Goodall, *Through a Window*, acknowledgments, 294–295.

71. William McGrew, speaking at University of Cambridge, Personal Histories Project, 28 April 2011, Transcription, Part 1, 21 December 2011.

72. Georgina Montgomery, "Inclusion and Indigenous Researchers: the Africanization of the Amboseli Baboon Project," in *Primates in the Real World*, 106–122. For example, the Leakey Foundation established the Franklin Mosher Baldwin Memorial Fellowship Program in 1977 to help graduate students from Africa pursue graduate training abroad in paleoanthropology and primatology. The program soon expanded eligibility to include students from other developing countries.

73. "Roving bands of young males" from Robin Fox, interview by author, 9 November 2011.

74. Hamburg continued his conversations with primatologists, writing in 1986 that "human behavioral tendencies grow deep in the soil of primate evolution." David Hamburg, foreword to Barbara Smuts, Dorothy Cheney, Robert Seyfarth, Richard Wrangham, and Thomas Struhsaker, eds., *Primate Societies* (Chicago: University of Chicago Press, 1987), vii.

Coda

John Berger, "Why Look at Animals?," in Berger, *About Looking* (New York: Pantheon, 1980): 3–28, quote on 6.

1. Many of the papers presented at the conference were also published in full: *Essays on Violence*, ed. J. M. Ramirez, Robert A. Hinde, and J. Groebel (Seville, Spain: University of Seville Publications, 1987).

2. Robin Fox, "The Seville Declaration: Anthropology's Auto-da-Fé," *Academic Questions* (Fall 1988): 35–47.

3. Robin Fox, *Anthropology Newsletter* 28, no. 8 (1987), as quoted in Fox, "Seville Declaration," 38.

4. On Americans' perception of Lysenkoism in the United States, see William deJong-Lambert and Nikolai Krementsov, eds., "The Lysenko Controversy and the Cold War," special issue, *Journal of the History of Biology* 43, no. 3 (2012).

5. Fox, "Seville Declaration," 39.

6. Fox, "Seville Declaration," 43.

7. Jonathan Benthall, "Fox among the Lambs," *Anthropology Today* 5, no. 3 (1989): 1–2.

8. Ullica Segerstråle, *Defenders of the Truth: The Battle for Science in the Sociobiology Debate and Beyond* (New York: Oxford University Press, 2000).

9. Nasser Zakariya, *A Final Story: Science, Myth, and Beginnings* (Chicago: University of Chicago Press, 2017).

10. Patricia Adair Gowaty, ed., *Feminism and Evolutionary Biology* (New York: Chapman & Hall, 1997).

11. My research throughout builds on insights from Donna Haraway and other scholars to show how scientists constructed and negotiated the twinned evolutionary binaries of female and male, animal and human. Donna Haraway, *Primate Visions: Gender, Race, and Nature in the World of Modern Science* (New York: Routledge, 1989).

12. On "animal studies" as a scholarly focus among humanists, for example, see Lorraine Daston and Gregg Mitman, eds., *Thinking with Animals: New Perspectives on Anthropomorphism* (New York: Columbia University Press, 2005), as well as scholarly works by Tim Ingold, Linda Kalof, Eben Kirksey, Nigel Rothfels, Peter Singer, Cary Wolfe, and others.

13. Cynthia Moss, *Elephant Memories: Thirteen Years in the Life of an Elephant Family*, with a new afterword (Chicago: University of Chicago Press, 2000); Janet Mann, Richard C. Connor, Peter L. Tyack, and Hal Whitehead, eds., *Cetacean Societies: Field Studies of Dolphins and Whales* (Chicago: University of Chicago Press, 2000); Barbara King, *How Animals Grieve* (Chicago: University of Chicago Press, 2013). Less commonly, this perspective has extended to fishes and invertebrates: Victoria Braithwaite, *Do Fish Feel Pain?* (New York: Oxford University Press, 2010); Peter Godfrey Smith, *Other Minds: The Octopus, the Sea, and the Deep Origins of Consciousness* (New York: Farrar, Straus and Giroux, 2016).

14. Even Edward O. Wilson has changed his thinking; see *The Meaning of Human Existence* (New York: Liveright Publishing Corporation, 2015). See also Brian Skyrms, *Evolution of the Social Contract* (Cambridge: Cambridge University Press, 1996); Elliott Sober and David Sloan Wilson, *Unto Others: The Evolution and Psychology of Unselfish Behavior* (Cambridge, MA: Harvard University Press, 1998); Donna Hart and Robert Sussman, *Man the Hunted: Primates, Predators, and Human Evolution* (New York: Westview Press, 2005); Sarah Blaffer Hrdy, *Mothers and Others: The Evolutionary Origins of Mutual Understanding* (Cambridge, MA: Belknap Press of Harvard University Press, 2009); Frans De Waal, *The Age of Empathy: Lessons for a Kinder Society* (New York: Harmony Books, 2009); Martin A. Nowak with Roger Highfield, *SuperCooperators: Altruism, Evolution, and Why We Need Each Other to Succeed* (New York: Free Press, 2011); Philip Kitcher, *The Ethical Project* (Cambridge, MA: Harvard University Press, 2011); Christopher Boehm, *Moral Origins: The Evolution of Virtue, Altruism, and Shame* (New York:

Basic Books, 2012); John Horgan, *The End of War* (San Francisco, CA: McSweeny's Books, 2012); Stefan Klein, *Survival of the Nicest: How Altruism Made Us Human and Why It Pays to Get Along*, trans. David Dollenmayer (New York: Experiment, 2014); David Sloan Wilson, *Does Altruism Exist? Culture, Genes, and the Welfare of Others* (New Haven, CT: Yale University Press, 2015).

15. For example, Matt Ridley, *The Origins of Virtue: Human Instincts and the Evolution of Cooperation* (New York: Viking, 1997).

16. For example, Agustín Fuentes, *Evolution of Human Behavior* (New York: Oxford University Press, 2009), and Fuentes, *Race, Monogamy, and Other Lies They Told You: Busting Myths About Human Nature* (Berkeley: University of California Press, 2012); Jonathan Marks, *What It Means to be 98% Chimpanzee: Apes, People, and Their Genes* (Berkeley: University of California Press, 2002); and Marks, *Tales of the Ex-Apes: How We Think about Human Evolution* (Berkeley: University of California Press, 2015).

17. Sarah Blaffer Hrdy, *The Woman That Never Evolved*, with a new preface (1981; Cambridge, MA: Harvard University Press, 1999); Marga Vicedo, *The Nature and Nurture of Love: From Imprinting to Attachment in Cold War America* (Chicago: University of Chicago Press, 2013).

18. Linda Fedigan, *Primate Paradigms: Sex Roles and Social Bonds* (Montréal: Eden Press, 1982); Shirley Strum, *Almost Human: A Journey into the World of Baboons*, with a new introduction and epilogue (1987; Chicago: University of Chicago Press, 2001); Mary Ellen Morebeck, Alison Galloway, and Adrienne Zihlman, eds., *The Evolving Female* (Princeton, NJ: Princeton University Press, 1997); Maryanne L. Fisher, Justin R. Garcia, and Rosemarie Sokol Chang, *Evolution's Empress: Darwinian Perspectives on the Nature of Women* (New York: Oxford University Press, 2013).

19. These are not the only books on the market, of course. Others continue to emphasize the aggressive nature of chimpanzees and humans: Richard Wrangham and Dale Peterson, *Demonic Males: Apes and the Origins of Human Violence* (New York: Houghton Mifflin, 1996); Matt Ridley, *The Red Queen: Sex and the Evolution of Human Nature* (New York: HarperCollins, 1993). Wrangham and Peterson pithily noted that the Seville Declaration was "quite a statement. Its motives were clearly on the side of the angels. But liking the idea doesn't make it right" (176). A few years later, the cognitive psychologist Steven Pinker derided the Seville Statement as bromides and disinformation in his best-selling *How the Mind Works* (New York: Norton, 1997), 51.

20. For example, Michael Tomasello, Joan B. Silk, Carol S. Dweck, Brian Skyrms, and Elizabeth Spelke, *Why We Cooperate* (Cambridge, MA: MIT Press, 2009).

INDEX

Note: Illustrations are indicated with bold page numbers

A NOTE ON THE TYPE

This book has been composed in Arno, an Old-style serif typeface in the classic Venetian tradition, designed by Robert Slimbach at Adobe.